PRIX DE L'ARC DE TRIOMPHE

1965–1982

Also by Arthur FitzGerald

PRIX DE L'ARC DE TRIOMPHE 1920-1948
(with Michael Seth-Smith)
PRIX DE L'ARC DE TRIOMPHE 1949-1964

PRIX DE L'ARC DE TRIOMPHE

1965–1982

The Official History by

ARTHUR FITZGERALD

SIDGWICK & JACKSON
LONDON

Picture Acknowledgements

The Author and Publishers are grateful to the following for permission to reproduce their photographs: APRH/P. Bertrand et Fils, pages 8, 20, 34, 47, 61, 73, 80, 86, 98, 110, 122, 133, 145, 157, 168, 172, 178, 185, 192, 197, 205 and 217; The European Racehorse, pages 1 (APRH/P. Bertrand et Fils), 14 (APRH/P. Bertrand et Fils), 26 (APRH/P. Bertrand et Fils), 40 (Fiona Forbes), 53 (Fiona Forbes), 67, 92 (Agence Recoupé), 103 (R. Anscomb), 116 (APRH/P. Bertrand et Fils), 127, 150 (Alec Russell) and 162; E.G. Byrne, page 138; Gerry Cranham, page 210

First published in Great Britain in 1983
by Sidgwick and Jackson Limited

ISBN 0-283-99036-8

Typeset by Tellgate Ltd, Shaftesbury, Dorset
Printed in Great Britain by
Biddles Ltd, Guildford, Surrey
for Sidgwick and Jackson Limited
1 Tavistock Chambers, Bloomsbury Way
London WC1A 2SG

CONTENTS

Author's Note

In this book all the horses are referred to as they appeared in either the French, English or Irish stud books at birth, therefore, for example, Sea Bird is not named Sea Bird II nor is Nijinsky Nijinsky II.

Author's Acknowledgements

I cannot overestimate the assistance that has been given to me by Monsieur Jean Romanet, the Directeur-Général of the Société d'Encouragement, who read and corrected the manuscript and whose advice has been invaluable. Similarly his son, Louis Romanet, and other members of the staff of the Société d'Encouragement assisted me in every possible way.

My very sincere thanks also go to Alec Head, Criquette Head, the late François Mathet, Ian Balding, David McCall, the Aga Khan, Daniel Wildenstein, Charles St George and Robert Sangster, who very kindly read the chapters concerning the winners they trained, owned or managed, and corrected many of my errors.

I must also thank Jean-Michel de Choubersky and his wife Gabrielle for their hospitality when I was writing this book, and the Directors and Editor of *The European Racehorse* for allowing me to use their library where Biddy Shennan and Emma Kelway-Bamber were particularly helpful.

Caroline Sturdy and my sister, Geraldine FitzGerald, who respectively typed part of the manuscript and the results and pedigrees were both understanding and patient with my difficult handwriting. Claudia Rotticci did an excellent job in arranging to have the original manuscript photocopied.

Finally a special word of thanks must go to Richard Berens who read the original manuscript and without whose corrections and suggestions this volume would have contained many mistakes.

Arthur FitzGerald
9 June, 1983

Hotel Leon d'Oro,
Pollone, Italy

Sea Bird 1965

Sea Bird

The Field

The quality and class of the field for the 1965 Prix de l'Arc de Triomphe was truly outstanding. It included four Derby winners: Sea Bird (Epsom Derby), Reliance (Prix de Jockey Club), Meadow Court (Irish Sweeps Derby) and Anilin (Soviet Derby). In addition to this brilliant array of talent the best three-year-old in the United States, Tom Rolfe, the winner of the Preakness Stakes and the American Derby, had travelled across the Atlantic to take on the best in Europe while the classic fillies were represented by Blabla (Prix de Diane). Moreover, with possibly three exceptions, the other fourteen participants had proved themselves to be no sluggards. Nonetheless it looked as if the race was only going to concern two French three-year-old colts, Sea Bird and Reliance, who were both unbeaten in 1965.

Jean Ternynck's Sea Bird had been a late-maturing two-year-old. His trainer Etienne Pollet, therefore, did not introduce him to a racecourse until 2 September 1964, when, after a slow start, he won the Prix de Blaison over 1400 metres at Chantilly by a short head. He beat Blabla 2 lengths in the Critérium de Maisons-Laffitte a little over a fortnight later. Sea Bird joined his stable-companion, Grey Dawn, the winner of the Prix Morny and Prix de la Salamandre, in the field for the Grand Critérium. Grey Dawn inevitably started favourite and he duly won by 2 lengths from Sea Bird, but had the latter's jockey Larraun not given his mount so much to do, the result might well have been much closer. It was certainly clear that Jean Ternynck possessed a colt of undoubted classic potential.

Sea Bird made his seasonal début in the Prix Greffulhe at Longchamp on 4 April and contemptuously humbled the fitter Corduroy and Pasquin by 3 lengths and a length.

So impressive was his victory that the English bookmakers immediately installed him as ante-post favourite for the Derby, although no definite decision concerning his participation at Epsom had been made. In the Prix Lupin at Longchamp on 16 May, Sea Bird's opponents included Grey Dawn's conqueror in the Poule d'Essai des Poulains, Cambremont, and the winner of the Prix Noailles, Diatome. Three hundred metres out Sea Bird challenged Diatome and without his jockey Pat Glennon even touching him with his whip, Jean Ternynck's colt trounced him and Cambremont by 6 lengths and half a length. The praise for this budding new champion was unanimous. Diatome's trainer, Geoffrey Watson, considered Sea Bird to be 'an exceptional horse in the same class as Ribot and Exbury'. The very experienced racereader Claude F. Dreyfus was adamant that he had not seen 'a horse like that since Pharis', whilst Pollet himself felt that along with Right Royal and Hula Dancer he was 'the best horse I have trained in my career'. It is therefore hardly surprising that such unqualified praise caused Sea Bird to start a short-priced favourite for the Derby.

The French champion won England's premier classic as convincingly and comfortably as he had done the Prix Lupin. He coasted up to the leader, I Say, a furlong and a half from home and in a couple of strides the issue was decided. Glennon had merely to push him out with hands and heels until he eased him fifty yards from the winning post. The Irish colt, Meadow Court, ridden by Lester Piggott came from behind to take second place from I Say inside the distance. The official margin of victory was 2 lengths and 1½ lengths, but almost everyone felt these distances to be merely academic, so great was Sea Bird's superiority over his rivals. Pollet had in fact asked Glennon to ease Sea Bird before the post in order to avoid the road crossing half a furlong after the finish.

The universally acclaimed champion once more demonstrated his prowess in the Grand Prix de Saint-Cloud, in which he beat the outsiders Couroucou and Francilius 2½ lengths and 2 lengths, with Free Ride fifth and Demi-Deuil unplaced. Once again Glennon eased his mount well before the winning post and it was clear he could have totally spreadeagled the field had he been asked to do so.

The fact that Sea Bird did not start at long odds-on, but at 12 to 10, for the Prix de l'Arc de Triomphe was the result of the presence of François Dupré's Reliance in the field. This full brother to Match and half-brother to Relko had carried all before him in France in 1965. He had never raced as a juvenile and on his début on 19 April he easily won the Prix de Marronniers (2400 metres) at Longchamp. In the Prix Hocquart Reliance showed himself to be a colt of great potential as he had 5 lengths to spare over Carvin, who had won the Critérium de Saint-Cloud and had been runner-up to Grey Dawn in the Prix Morny the previous season. Reliance's next objective was the Prix du Jockey Club which had always eluded both François Dupré and his trainer François Mathet. Tantième had been a controversial runner-up to Scratch; Bella Paola had just failed behind Tamanar in 1958 and Match had been second to Right Royal in 1961. In addition Dupré's Tanerko, Midnight Sun and Bonjour had all taken third place at Chantilly. Yves Saint-Martin always had Reliance up with the leaders and he took the lead half way up the straight. He was going so easily that Saint-Martin did not have to resort to the use of his whip and he comfortably held the challenges of Diatome and Carvin by three-quarters of a length and 2 lengths. In spite of the ease of his victory, doubts were cast on its merit due to its apparent narrowness. These were quickly dispelled by Reliance's performance in the Grand Prix de Paris on 27 June.

Dupré's colt was once again challenged by Diatome in the final stages, but this time Saint-Martin gave him a couple of cracks of the whip inside the distance. Reliance lengthened his stride and beat his rival a comfortable length, with Vianen and Carvin 6 lengths and a neck away third and fourth. Reliance was then rested until the Prix Royal Oak on 12 September when he defeated the Irish trained Ragazzo and Carvin by an easy three-quarters of a length and 3 lengths. He thus arrived for his confrontation with Sea Bird unbeaten in five outings, which had included three classic victories. Strictly on a line through Diatome, he had several pounds to find on Jean Ternynck's champion, but he still represented the most formidable opponent Sea Bird would have to face in 1965.

The fact, however, that the field for the 1965 Prix de l'Arc de Triomphe contained no less than eight foreign challengers demonstrated that the two French cracks were not considered invincible abroad. The Irish trainer, Paddy Prendergast, had three runners: Meadow Court, his pacemaker Khalife, and Ragazzo. Prior to the Epsom Derby, Meadow Court had been runner-up to Western Wind in the Gladness Stakes at the Curragh and had filled the same position behind the subsequently disappointing Ballymarais in the Dante Stakes at York. He had been no match for Sea Bird at Epsom, but in the Irish Derby at the Curragh on 26 June he convincingly beat Convamore and Wedding Present 2 lengths and three-quarters of a length. Meadow Court then proceeded to establish himself as the best and most improved three-year-old in the British Isles when he won the King George VI and the Queen Elizabeth Stakes at Ascot by 2 lengths and 5 lengths from the Hardwicke Stakes winner, Soderini, and the Coronation Cup victor, Oncidium. Lester Piggott's comment after this race – 'If we meet Sea Bird again, he will not beat Meadow Court' – was enthusiastically reiterated by his trainer. Nevertheless Glennon, who had been at Ascot as a spectator, merely regretted that Jean Ternynck's colt had been an absentee. It was, therefore, one of the great shocks of the 1965 season when Provoke slammed Meadow Court, on whom odds of 11 to 4 had been laid, by 10 lengths in the St Leger at Doncaster. It is true that the going that afternoon was exceptionally heavy and that Meadow Court, owned in partnership by Bing Crosby, Frank McMahon and the Canadian Max Bell, was all at sea all the way up the straight. Afterwards Prendergast felt that he had made a mistake in running Meadow Court in preference to Ragazzo, who revelled in such conditions.

Rory More O'Ferrall's Ragazzo was very lightly raced. He won the Ballysax Maiden Stakes (10f) by 1½ lengths from Donato at the Curragh on 21 April, but did not appear again until the Great Voltigeur Stakes at York on 18 August, when he beat Ballymarais and Craighouse by 2 lengths and a head. His performance behind Reliance in the Prix Royal Oak was most meritorious for such a lightly-raced colt, but he was still considered some way inferior to Meadow Court except in very testing conditions.

Khalife, Meadow Court's pacemaker, won a minor ten-furlong event at the Curragh on 15 August but, while he had no chance at Longchamp on 3 October, he had still run third to Craighouse and Alcalde in the Irish St Leger, more than sufficient credentials for a pacemaker.

There were two challengers from England, the four-year-olds Soderini and Oncidium. The previous season Soderini had been third in the St Leger, a head and 4 lengths behind Indiana and Patti, with Oncidium only fifth. Soderini won the John Porter Stakes at Newbury by three-quarters of a length from Cassim on 10 April but was then beaten 1½ lengths by Oncidium in the Coronation Cup at Epsom with Homeward Bound, the 1964 Oaks heroine, 5 lengths away third. He defeated Earldom a neck in the Hardwicke Stakes at Royal Ascot with Ardaban last. In the King George VI and the Queen Elizabeth Stakes he gained his revenge on Oncidium, beating him 5 lengths into third place, but he was still 2 lengths adrift of Meadow Court on the line. In the Goodwood Cup (21f) Soderini was not disgraced when he failed to give a lot of weight to Apprentice. On 24 September at Ascot he ran the useful Super Sam to a neck. Super Sam, who was to run in the Washington D.C. International, was receiving 12 lbs that afternoon but Oncidium, who finished third, merely a neck behind Soderini after he had appeared to be slightly squeezed for room in the straight, may have been unlucky.

Lord Howard de Walden's Oncidium was a very frustrating horse for he had two ways of running. In 1964 he had won the Lingfield Derby Trial, but had then run below expectations behind Santa Claus in the Derby. He had again disappointed in the St Leger, but in October won the Jockey Club Stakes (16f) at Newmarket. His best performance in 1965 had been his victory in the Coronation Cup at Epsom. When he was third in the King George VI and the Queen Elizabeth Stakes, he ran his race in snatches and never reached a challenging position. He failed to concede Court Gift 21 lbs at Newbury on 14 August, but a fortnight later he put his best foot forward and beat the Eclipse Stakes winner, Craighouse, 15 lengths at Goodwood. Oncidium was unquestionably a horse of considerable ability when he consented to show it. Nevertheless it was difficult to envisage either him or Soderini lowering the colours of the likes of Sea Bird, Reliance or Meadow Court.

The sole Italian challenger was the three-year-old Marco Visconti. He won his first two starts in the spring and then was most unfortunate in the Derby Italiano. Through no fault of his own, he was left 25 lengths at the start and he was unable to catch the field up until the turn into the straight. His performance in finishing third, only a length and 4 lengths behind Varano and Ben Marshall was, therefore, all the more meritorious and all those present regarded him as the moral winner of the Derby Italiano. In the Gran Premio di Milano Marco Visconti was runner-up to Accrale, who had won the Gran Premio d'Italia, with Varano beaten into third place. Moreover he was beaten by the narrowest of short heads. The Italian colt's limitations, however, were exposed when he was unplaced to Diatome in the Prix du Prince d'Orange at Longchamp in September and it was very difficult to regard him in the same light as Ribot or Molvedo.

For the first time in its history, there was a Russian challenger for the Prix de l'Arc de Triomphe, the four-year-old champion Anilin. As a two-year-old he won three starts from five outings including the Kalinin Memorial, the Soviet equivalent of the Grand Critérium. As a three-year-old Anilin proved himself to be the most outstanding Russian colt for a very long time. He won all his four starts in Moscow at distances from 1600 metres to 2400 metres, including the Soviet Derby. He also won twice in East Berlin and the Robert Pferdmenges Rennen (1800 metres) in Cologne. On his final outing in 1964 in the Washington D.C. International Stakes at Laurel Park, the Russian champion was third, 4½ lengths and 9 lengths behind Kelso and Gun Bow, but trailing behind him that afternoon were Belle Sicambre, Veronese and Biscayne, all classic winners in their respective countries: France, Italy and Ireland. As a four-year-old before arriving in Paris, Anilin was unbeaten in three outings in Moscow over ten, twelve and sixteen furlongs. While unlikely to trouble the favourites, Anilin seemed sure to give a good account of himself at Longchamp. The fact that he ran at all was slightly fortuitous. The Voshod Stud thought that they did not have sufficient money to afford the expenses involved. It then remembered that the Soviet Embassy had approximately £18,500 which had been won by their trotters Apex Hanover and Granit earlier in the year at Vincennes, and so the challenger was dispatched.

The other superpower, the United States, was also represented by one colt, Raymond Guest's three-year-old Tom Rolfe, who was by Ribot. He won the Preakness Stakes by a neck from Dapper Dan and was only run out of the Belmont Stakes by Hail to All by the same margin in the final strides. Although he had been beaten 2 lengths and a neck by Lucky Debonair and Dapper Dan in the Kentucky Derby, many had considered him an unlucky loser. Tom Rolfe won all his four races prior to arriving in Paris, the Citation Handicap (8f), the Chicagoan Handicap (9f), the Arlington Classic and finally on 13 September the American Derby at Arlington Park. As a result of these victories, Tom Rolfe was considered the best three-year-old colt in the United States. It is related that his trainer, Frank Whiteley Junior, was not as enthusiastic about the trip to Longchamp as was his owner. After his victory in the American Derby, Whiteley is reported to have commented that Tom Rolfe was a small horse who needed a rest. The Prix de l'Arc de Triomphe was to be his thirteenth race of the season. Nevertheless it seemed certain that he would provide a very reliable guide to the relative merits of the 1962 European and American generations and Raymond Guest can only be applauded for his very sportsmanlike action of sending his champion across the Atlantic to challenge the cream of Europe's thoroughbreds.

The best of the remaining French runners for the 1965 Prix de l'Arc de Triomphe was unquestionably Baron Guy de Rothschild's Diatome. As a juvenile he had won his sole outing, the Prix Trespas (1800 metres), at Le Tremblay. On 25 April he easily took the Prix Noailles by 2½ lengths from Champière, but on his next three runs had to be content with the runner-up's position, first behind Sea Bird in the Prix Lupin and then behind Reliance in the Prix du Jockey Club and in the Grand Prix de Paris. He gained his just reward, however, in the Prix du Prince d'Orange at Longchamp on 19 September when he beat Acer, Corfino and Timmy Lad 2½ lengths, half a length and the same, with Marco Visconti unplaced. While it was difficult to envisage Diatome reversing the form with either Sea Bird or Reliance, it seemed certain that he would be a strong contender for minor honours on 3 October.

Baron Guy de Rothschild was also

represented by the four-year-old Free Ride, who had been disappointing after his victory in the 1964 Prix Hocquart. On his reappearance in the Prix Boïard at Saint-Cloud on 20 March, however, Free Ride conclusively defeated Sigebert by 4 lengths. He again put up a very game and sterling display when he beat Michilmackinac and Sigebert three-quarters of a length and a short head in the Prix Ganay on 11 April. His only other run in 1965 was when he was fifth, about 5¾ lengths behind Sea Bird, in the Grand Prix de Saint-Cloud on 4 July. Free Ride was nevertheless a horse who ran best when fresh, and so could not be totally discarded from the calculations for the minor placings, although his official role was that of pacemaker for Diatome.

The French three-year-old fillies were represented by Blabla who had run so well against Sea Bird on his racecourse début. She had won the Prix de Bagatelle at Longchamp on 9 May very easily. She then beat Cantilene 1½ lengths in the Prix de Diane at Chantilly. Blabla was rested until the Prix de la Nonette at Longchamp in early September, when lack of fitness was the probable cause for her failure to give 5 lbs and 11 lbs to Tinganeve and Cantilene. Blabla started favourite for the Prix Vermeille but unfortunately fell during the course of the race and brought down Cantilene, whose jockey, Yves Saint-Martin, had a very nasty fall. The race was won with the utmost authority by the Noel Murless-trained filly Aunt Edith, who had 8 lengths to spare over Dark Wave with the Oaks heroine, Look Look, three-quarters of a length away third and the Oaks d'Italia and Premio Regine Elena victress, Tadolina, half a length back fourth. Tadolina was an intended runner for the Prix de l'Arc de Triomphe, but after this comprehensive defeat she was withdrawn. In fact, none of the first three in the Prix Vermeille ran in the 1965 Prix de l'Arc de Triomphe. It was almost certain that the colts of the 1962 generation were superior to the fillies, and Blabla's fall could only have impaired her chance of putting up a reasonable performance against them on 3 October.

Jean Boutillier's Carvin had won the Critérium de Saint-Cloud as a juvenile and had finished runner-up to Grey Dawn in the Prix Morny. Reliance had, however, conclusively put him in his place on three occasions in 1965, in the Prix du Jockey Club, the Grand Prix de Paris and the Prix Royal Oak. Carvin's best performance before the Prix de l'Arc de Triomphe had been when he won the Grand Prix de Vichy by a facile 4 lengths, but it still seemed unlikely that he would seriously trouble the principals.

André Rueff's Demi-Deuil won four races in the spring and summer of 1965. Before running unplaced to Sea Bird in the Grand Prix de Saint-Cloud, he took the Prix Cappiello (12f) at Saint-Cloud in April and beat Sailor and Impact a short head and a neck in the Grand Prix du Printemps on the same course on 7 June with Francilius, Sigebert and Emerald unplaced. Although Demi-Deuil beat Prince Paladin 4 lengths in the Grosser Preis von Baden, it seemed very probable that he would encounter opposition that would be too strong for him in the Prix de l'Arc de Triomphe.

Marcel Boussac had two runners in the 1965 Prix de l'Arc de Triomphe, Emerald and Ardaban. Emerald, whose dam Esméralda had run so gamely behind Verso II in the race in 1943, had won the minor Prix de Chambly at Saint-Cloud on 10 May. He was unsighted behind Demi-Deuil in the Grand Prix du Printemps, but on 14 July he beat Royal Avenue and Sailor a length and a short neck in the Prix Maurice de Nieuil. As he was so lightly raced, should there be a complete suprise on 3 October, Emerald was possibly the most likely outsider to spring it. His stable-companion, Ardaban, was also impeccably bred; his dam Asmèna had won the 1950 Oaks. Ardaban, who had won the Grand Prix de Marseille in November 1964, had been most disappointing in his four outings in 1965. He was out of the frame in his first three outings, including finishing last to Soderini in the Hardwicke Stakes. He had managed third place, 2½ lengths and 2 lengths behind Sigebert and Corfino, in the Prix Henri Foy at Longchamp on 3 September with Francilius unplaced, but that performance was some way inferior to that required for a live contender for the Prix de l'Arc de Triomphe.

Sigebert's best display in 1965 had in fact been his victory in the Prix Henri Foy. He was held by both Free Ride and Emerald on his running in the Prix Boïard, Prix Ganay and Prix Maurice de Nieuil. In the Prix d'Harcourt his stable-companion Frontin had beaten him, and he had been a disappointing third in the Prix Gontaut-Biron at Deauville, so on balance it looked as if he would be out of his depth as well on 3 October.

Similarly it was difficult to give either Francilius or Timmy Lad much chance in the 1965 Prix de l'Arc de Triomphe. Francilius had won three races, including the Prix du Point du Jour and the Prix de la Fouilleuse, before his slightly fortuitous third in the Grand Prix de Saint-Cloud, but overall his form was not good enough to entitle him to have a serious chance on 3 October. Timmy Lad, who won the 1964 Prix du Conseil Municipal, was clearly held on his running with Diatome in the Prix du Prince d'Orange. He had run well when runner-up to Sailor in the Grand Prix de Deauville and he had won the Prix de la Chamière (10f) on the Normandy course a few day previously. The opposition he was to encounter in the Prix de l'Arc de Triomphe was, however, of quite a different calibre.

Shortly after Sea Bird's victory in the Epsom Derby, the American breeder John Galbreath came to France and watched Jean Ternynck's champion gallop at Chantilly with his owner and his trainer Etienne Pollet. The latter attended the gallop in spite of the death of his mother the previous night. So impressed was Galbreath that he immediately concluded an agreement to lease Sea Bird as a stallion for $300,000 a year. Sea Bird was, however, to continue to run in Jean Ternynck's colours until after the Prix de l'Arc de Triomphe, which was to be his last race. Although he did not have a preliminary race in September, during the week before his greatest test Pollet expressed himself well satisfied with Sea Bird.

The Final Preparations

Tom Rolfe arrived in Paris on Tuesday 28 September accompanied by his own blacksmith and a retired sheriff to guard him. Frank Whiteley also brought his own hay and water over from the United States. In the third week of September the following advertisement appeared in the French racing press: 'W. Shoemaker, 7 stone 12 lbs, is looking for a horse to ride on Sunday 26 September and Saturday 2 October in order to familiarise himself with the Longchamp course.' Though unquestionably the best jockey in the United States, it cannot have been to Tom Rolfe's advantage that Shoemaker was so unfamiliar with Longchamp, a fact emphasised by his refusal to ride Prince Royal II the previous year. In addition, Tom Rolfe was not permitted to wear the spiked American racing shoes to which he was accustomed. But as can be seen from the thoroughness of their preparations, his connections were endeavouring to give the Ribot colt the best possible chance.

The Russian challenger, Anilin, had a very difficult journey. He was delayed a total of twelve hours, first at the frontier bestween West Germany and Belgium and then at that between Belgium and France. Yves Saint-Martin, who had taken a holiday in Spain after his serious fall in the Prix Vermeille, was very confident in the week leading up to the Prix de l'Arc de Triomphe that Reliance would be able to defeat Sea Bird. In spite of his failure behind Diatome, Marco Visconti impressed many onlookers when he did his final gallop the Tuesday before the race, while it was also reported that Carvin had worked well.

Both Prince Royal II, who had won the 1964 Prix de l'Arc de Triomphe, and Le Fabuleux, who had started favourite, were absentees in 1965. Le Fabuleux had fractured a fetlock in his final gallop before the Prix Ganay in the spring, but fortunately the injury was not serious enough for it to be necessary for him to be put down. It nevertheless precluded him from lining up with the other contenders on Sunday 3 October. Prince Royal II also sustained such a serious tendon injury in training in April 1965 that he had to retire to stud in the United States immediately.

Sunday 3 October was a sunny and warm day, but as a result of the large quantity of rain the previous week the track was dead, which would certainly test the stamina of all the contestants to the utmost. There was an enormous contingent of visitors from the United States at Longchamp that afternoon and none was more fêted than the part owner of Meadow Court, Bing Crosby. One of the visitors remarked that 'there have probably never been more of us Americans here in France since 6 June 1944'. Few of the spectators, however, paid much attention to the small hatless man in a grey suit who was to see Sea Bird run for the first time, but he was none other than John Galbreath, who had arranged to lease the French champion for such an enormous sum by the standards of 1965.

The Race

In the parade Sea Bird appeared even more nervous than he had done at Epsom; sweat was pouring off him, almost drenching his white blaze and two white stockings. Nevertheless there was no mistaking the power in his strong muscular quarters. Reliance, who was also sweating, did not please all the critics. While all admitted he had a good head and powerful quarters, some considered his fetlocks too large and his pasterns too long. Anilin seemed to be a nice-looking tall colt in contrast to the diminutive Tom Rolfe who was also nervous in the paddock. The American challenger, whom Shoemaker regarded as the best three-year-old he had ridden since Gallant Man, was led to the start by his lad mounted on a horse alongside him, as was the fashion in the United States; it was a sight never before witnessed at Longchamp.

Oncidium was a little reluctant to enter the starting stalls, but after very little delay the runners got away to a perfect start with the exception of Meadow Court's pacemaker Khalife, who was last to leave the gate. His jockey Roe rushed him through the field but he never was able to fulfil his role. Marco Visconti took the lead immediately and after a furlong the Italian challenger held a definite advantage from Blabla, Ardaban, Anilin and Khalife, who was vainly trying to reach his rightful position at the head of affairs. This group was closely tracked by Emerald, Tom Rolfe, Sea Bird and Reliance. After two furlongs Marco Visconti was still setting a strong gallop to the leading group, which had been joined by Meadow Court who was lying about seventh, while Diatome could be seen improving to head the mid-division which included Soderini and Ragazzo. By the time the field reached the top of the hill both Blabla and Khalife had begun to drop out and Marco Visconti just headed Anilin, with Ardaban third. Then came Sea Bird a little in front of a group consisting of Tom Rolfe, Emerald, Meadow Court and Reliance, with Timmy Lad, Diatome and Free Ride tracking them.

As the field raced down the hill towards the straight, it was clear that the long-time leader Marco Visconti was coming under pressure and that Anilin, on his heels, was going much the better. A glance behind, however, showed Sea Bird, galloping as usual with his head held high, still on a tight rein, merely cantering in third place. He was followed by Tom Rolfe and Reliance together and the improving Diatome.

Jean Ternynck's champion sailed past Anilin and the struggling Marco Visconti effortlessly, tracked by the unbeaten Reliance. As the two French cracks quickly detached themselves from the remainder it appeared for a few strides that the battle of giants that so many spectators had expected was about to take place. But as soon as Glennon gave Sea Bird his head, the Derby winner literally left Reliance standing and the only question remaining was how far he would win. In spite of veering from the rails to the middle of the course in the final two hundred metres and, although Glennon began to pat his mount to acknowledge his victory some way from the finish, Sea Bird was an incredible 6 lengths clear of Reliance at the winning post. The fact that the hitherto unbeaten Reliance trounced his other rivals by a massive 5 lengths was only further proof of Sea Bird's astounding superiority. Diatome had run gamely and dourly to head the rest of the field a short neck in front of his fast-finishing stable-companion Free Ride, who had deprived Anilin of fourth place by half a length. Thus French horses had filled the first four places. Tom Rolfe, who had weakened in the closing stages, finished a highly creditable sixth, 5 lengths behind the Russian champion, for it was the first time he had run on turf. Demi-Deuil was another 6 lengths behind the American challenger with Carvin eighth, Meadow Court, whom Piggott afterwards said had not been suited by the soft going, ninth and Marco Visconti tenth.

All those who had been privileged to be present at Longchamp that afternoon and see Sea Bird's spreadeagling victory were unanimous in their praise. A delighted John Galbreath exclaimed, 'I would never have believed that a horse could win so overwhelmingly and with such ease against some of the best horses in the world. He may perhaps be even better than Ribot.' Pollet had instructed Glennon to make his effort as late as possible. When the Australian jockey began to pull away at the beginning of the straight, Pollet briefly felt that he might have been going too soon. Glennon, however, felt at that moment that Sea Bird was going so well that 'I thought it pointless to wait any longer'. He added that he had always been confident and since there was not a single horse to challenge Sea Bird, it was still impossible to know how good he really was. In his opinion Sea Bird was 'undoubtedly the greatest horse I've ever seen,

Sea Bird (19) outclassing Reliance (15) by 6 lengths with Diatome (10) and Free Ride (1) 5 lengths and a short neck behind third and fourth

let alone ridden'. Breasley, whose mount Oncidium was never going well, thought 'Sea Bird a fantastic horse' while Yves Saint-Martin called him 'un crack prodigieux' and Willie Shoemaker was equally laudatory. Superlatives abounded that afternoon at Longchamp: 'the best three-year-old I've ever seen' and 'the best horse in the world, not excepting Kelso' were two typical comments. The remark made by Reliance's trainer, François Mathet, 'L'autre est vraiment le meilleur' was possibly the greatest praise, for had his charge not been of the same generation as Sea Bird, he would almost certainly have won the Prix de l'Arc de Triomphe.

When Sea Bird and Glennon returned to the unsaddling enclosure they inevitably received a tremendous ovation. His owner, the textile magnate Jean Ternynck, was a son of the first cousin of Henry Ternynck whose Massine had won the same race forty-one years earlier. Although Jean Ternynck did not have a large number of broodmares at his stud, Nôtre-Dame de l'Isle dans l'Eure, he had been exceptionally successful as an owner and breeder. His filly Camarée, who had won the 1000 Guineas in 1950, and Sanctus, who took the 1963 Prix du Jockey Club and the Grand Prix de Paris, were the two most successful products of his stud prior to Sea Bird.

For Sea Bird's trainer, Etienne Pollet, who was a cousin of Jean Ternynck, it was a second victory in the Prix de l'Arc de Triomphe. In 1953 he had achieved the remarkable feat of saddling the first two, La Sorellina and Silnet. Pollet had an exceptional record as a trainer of two-year-olds; by 1965 he had saddled the winners of the Prix Morny and Prix Robert Papin three times, of the Grand Critérium six times, and of the Prix de la Salamandre on eleven occasions. In addition, he had won the Prix du Jockey Club, the Grand Prix de Paris and the 1000 Guineas twice and his runners had also captured a 2000 Guineas, a King George VI and the Queen Elizabeth Stakes, an Irish Sweeps Derby and an Ascot Gold Cup, prior to the arrival of Sea Bird. Glennon was fulsome in his praise of Etienne Pollet after his triumph in Europe's richest race. He related that before an important gallop, Pollet covered the whole of the ground on foot to ensure there were no holes or other hazards. For a man to take such meticulous care, it is no surprise that Pollet's record as a trainer was studded with so many successes.

Pat Glennon had arrived in France in August 1963 from Ireland, where he had been stable jockey to Vincent O'Brien. He had ridden seventy-four winners in France; besides Sea Bird he had had important successes on Le Mesnil, Neptunus and Grey Dawn. Unfortunately he left France to return to Australia for good on Sunday 11 October. He suffered from weight problems and in addition his wife apparently preferred living in their native land to France. His modesty was epitomised when he stated that his success in Europe had been 'beyond my wildest dreams'.

The Rest of the Season

A few days after Glennon's departure, Sea Bird left France to take up his duties at stud in the United States, while Reliance was also retired. The other participants in the 1965 Prix de l'Arc de Triomphe who did run again that autumn more than upheld its form. Anilin easily beat Kronzengue and Lagopède in the Grosser Preis von Europa at Cologne on 18 October. The Russians thought so highly of their champion that they froze his sperm and used it to artificially inseminate mares after his death. On 5 November Demi-Deuil, 26 lengths behind Sea Bird at Longchamp, had 6 lengths to spare over the very useful Atilla, the Gran Premio del Jockey Club winner, with the good Italian horse, Veronese, third in the Premio Roma.

Ardaban was far from disgraced when 2 lengths runner-up to Hamman, who was receiving 10 lbs, in the Prix du Conseil Municipal. It was the result of the Washington D.C. International at Laurel Park, however, on 11 November that really emphasised the outstanding merit of Sea Bird's victory at Longchamp. Diatome and Carvin took on the best American older horse, Roman Brother, and the Canadian champion, George Royal. A furlong out, Diatome's task looked hopeless as he was pocketed on the rails and Carvin was in the process of catching and passing Roman Brother. Miraculously Deforge extricated his mount and Diatome produced such an astounding burst of speed that he caught Carvin in the very last stride and beat him a short head, with Roman Brother 1¼ lengths back third and George Royal fourth. It was a just recompense for his consistency throughout the season, but also it was a huge compliment to Sea Bird, who had dismissed him by 11 lengths at Longchamp.

Pedigree and Stud Record

Sea Bird's sire, Dan Cupid, had been a somewhat unpredictable and disappointing colt on the racecourse. As a two-year-old he initially showed great promise and won his first three races, all over 1000 metres, the Prix de la Reine Blanche, the Prix de Bois and the Prix de Sabonville. He was then disappointingly beaten 4 lengths when fourth to Taboun in the Prix Robert Papin and could only manage third place, 2½ lengths and the same, behind Oceanic and Taboun in the Prix Morny. In the Middle Park Stakes, however, he was only a length behind the best English two-year-old, Masham; this performance made some consider Dan Cupid as a possible classic contender. Others, however, pointed to the fact that he did not really possess a staying pedigree. Dan Cupid's sire, Native Dancer, had been a ten-furlong horse, and in 1958 had not the reputation that he now possesses, while his dam, Vixenette, was a daughter of Sickle, who also did not stay beyond ten furlongs. As a three-year-old Dan Cupid won the Prix de Saint James over 1400 metres in the spring and also the Prix Pourtales (1000 metres). Nevertheless he only managed sixth place behind Taboun in the 2000 Guineas and also ran a disappointing tenth behind Parthia in the Epsom Derby, although his jockey Poincelet reported afterwards that he had been very unlucky in running. It was all the more surprising, therefore, and indeed a measure of his unpredictability, that when he was challenged by Herbager a furlong from the finish of the Prix du Jockey Club he ran on so gamely and fought so tenaciously that he only failed to retain his advantage by a short head. Behind him that afternoon were such useful horses as Midnight Sun and Le Loup Garou. Dan Cupid never again reproduced that form and returned to his errant ways when failing behind Memorandum and Sallymount in the Prix Eugene Adam, and when a very dismal fifth behind Herbager in the Prix du Prince d'Orange. In the autumn he reverted to sprinting but finished last but one in the Prix de la Forêt, and an ignominious last in the Premio Chiusura. When retired to stud, therefore, it seemed very unlikely that Dan Cupid would sire one of the greatest French colts this century.

Similarly Sea Bird's female line was not anything to inspire optimism in Jean Ternynck. The future champion was born on 8 March 1962 at the Haras de Victot in Calvados, where his dam, Sicalade, had been sent to be covered by the useful but by no means outstanding Beau Prince. The most promising aspect of her birth in 1956 was that she was by Sicambre who in that year was to achieve his first classic success as a sire; his filly Sicarelle won the Oaks at Epsom. Sicalade only ran twice, when she was a two-year-old. On her début she was a respectable fifth at Longchamp in October, and a week later she deadheated for second place in a field of twenty-six in the Prix de la Colline (1800 metres) at Maisons-Laffitte. She then developed leg trouble and never appeared on the racecourse again. Unfortunately Sicalade suffered from a circulatory problem. The birth in 1961 of her first foal, the unraced Status Quo by El Relicario, was far from easy and she suffered considerable pain and distress. Sea Bird's foaling also was difficult for Sicalade. When her third foal, Syncom by Beau Prince was born, she again underwent considerable pain and immediately afterwards her circulation trouble became more pronounced. In 1963 Sicalade was not in foal and in November Ternynck and Pollet decided the best course of action was to sell her to the butcher in Andelys. Thus the residents of Andelys found themselves eating the dam of Sea Bird when he was only nineteen months old. His half-brother, Syncom, certainly did not possess a tenth of the ability of Sea Bird, but managed to win seven small races under both Rules.

Sicalade's dam, Marmelade by Maurépas, ran only once when she was unplaced. In her first six years at stud she only produced three foals: her first was put down after a training accident when still unraced; her second was Sicalade; and her third was born dead. Her fourth foal, the filly Marmite by Norman, was placed on the flat and over obstacles, whilst her 1961 colt Magnae, by El Relicario, won three minor races at provincial tracks in 1964 and 1965. He was Marmelade's last foal as she was barren to Verrières in 1962 and 1963 before she was destroyed.

The first mare of any consequence in Sea Bird's female line wsa his third dam, Couleur by Biribi. She did not win on the flat, but in 1943 she won a hurdle race at Nice and was placed there over both hurdles and fences. Henri Ternynck acquired her in foal to Maurépas and with a filly foal at foot by the same sire, later named Camargue, in 1946. Camargue won the Prix de Malleret at Longchamp and became the dam of Dalaba, whose four victories included the Prix de Royallieu. Ternynck had certainly made a fortunate purchase as the foal whom Couleur had been carrying when he bought her was the filly Camarée. She won two races as a two-year-old and in 1950 she was victorious in both the 1000 Guineas at Newmarket and the Prix Vanteaux for Jean Ternynck. Couleur only got two other moderate winners, Conakry by Tourment, once successful on the flat, and Caddor by Tifinar, who won two steeplechases. Until Sea Bird, however, none of Couleur's descendants since Camarée had given any indications of classic potential.

Sea Bird's fourth dam, Colour Bar by Colorado, had also been a disappointment on the racecourse. Running in the colours of Major J. S. Courthauld, Colour Bar had run five times in selling plates and had only twice managed third place. She was sold to Major Dermot McCalmont for 750 guineas and her first filly foal by Cyclonic was a maiden. Colour Bar next produced a grey colt by Tetrametra called Quarteroon, whose three victories included the Woodcote Stakes at Epsom. In 1936 Colour Bar was exported to France in foal to Tetrametra. The purchaser was Simon Guthmann, the owner of the 1926 Prix de l'Arc de Triomphe winner, Biribi. The Tetrametra foal, Tetrabar, won three races as a juvenile including the Prix des Foals and the Grand Critérium d'Ostende. Colour Bar then got two fillies by Biribi, the first a maiden and the second Couleur. Her final foal, Le Bosc Giard by Plassy, won a total of nine races in France and England soon after the war. Colour Bar was by then dead, as she was one of the many broodmares who disappeared in the fighting in Normandy in 1944.

Colour Bar's dam, Lady Disdain, ran twice unplaced in the colours of Colonel Hornung as a two-year-old, once at the Epsom spring meeting. Unfortunately she then met with a training accident and retired to stud. After breeding a non-winner for Colonel Hornung, he sold her to Major Courthauld. In all, Lady Disdain produced eight foals of which three were winners, two in England and one in Belgium. The best was the useful filly, Crosspatch by Tetrametra, whose four victories included the Molecomb Stakes at Goodwood.

Lady Disdain's dam, Lady Phoebe by Orby, ran six times as a juvenile. Lady Phoebe won twice, including the Acorn Stakes at Epsom, was runner-up on three occasions, including once to Tetrametra in the National Breeders Produce Stakes at Sandown, and she also ran third to Tetrametra in the Lavant Stakes at Goodwood. Lady Phoebe bred five winners, the best of which was Apple Sammy, who won two races and ran third to Colorado and Coronach in the 1926 2000 Guineas.

Sea Bird eventually traced back to Donnetta who, in addition to winning thirteen races, bred two classic winners, Diadem (1000 Guineas) and Diophon (2000 Guineas). His female line was noted for horses whose main asset was speed. On the other hand, his immediate female ancestors would not have encouraged an owner to feel that Sicalade had given birth to one of the finest thoroughbreds this century.

Sea Bird remained at stud in the United States until 1972, when he returned to France for the 1973 covering season. Most unfortunately he died on 15 March 1973 of an intestinal blockage at the Haras du Petit Tellier. His record at stud and his reputation as a stallion, therefore, suffered from his premature death. On the other hand, it must be admitted that, during his short career as a stallion, many of Sea Bird's stock suffered from character problems. A fine example was the early career of Sea Pigeon. He won the Duke of Edinburgh Stakes as a two-year-old and was clearly a colt with classic potential. He was most reluctant to show his undoubted ability and was a disappointing sixth in the Epsom Derby behind Morston. When he was gelded however, Sea Pigeon was a totally reformed character. His weight-carrying achievements in high-class handicaps like the Chester Cup and the Northumberland Plate, and his remarkable achievements over hurdles, including his two victories in the Champion Hurdle, only go to underline what he might have achieved had he shown his true potential as a three-year-old.

Sea Bird's best produce unquestionably was Allez France, whose successes included the Prix de l'Arc de Triomphe, the Prix de Diane, the Prix Vermeille, the Poule d'Essai des Pouliches, the Critérium des Pouliches, to mention but the most important. His best colt was probably Gyr, the winner of the Grand Prix de Saint-Cloud, the Prix Hocquart and the Prix Daru; he was also runner-up to Nijinsky in the Derby, and fourth to Sassafras in the Prix de l'Arc de Triomphe. Another of his colts, Arctic Tern, won the Prix Ganay, the Prix Thomas Bryon and the Prix de Fontainebleau, was runner-up in the Prix Lupin, the Prix Dollar and Prix Maurice Nieuil, as well as finishing third in the Eclipse Stakes and Prix d'Harcourt. Amongst Sea Bird's other notable produce were Great Heron (Prix de la Porte Maillot, the FitzWilliam Stakes, 2nd Irish 2000 Guineas and 2nd Observer Gold Cup), Pretty Puffin (Galtres Stakes), Open Season (Railway Stakes), Sylve (Prix de Saint Firmin) and Mr Long (Prix Juigné). In the United States Sea Bird's best progeny was probably Dubassoff, the winner of the American Derby.

It bears repeating that Sea Bird's regrettable premature death at an early age of eleven certainly did not give him sufficient time to prove himself as a sire. Nevertheless his superlative career on the racecourse, when he earned 2,268,649 francs in France and £65,301 in England, stands as testament to his brilliance. For those who watched him trounce his highly distinguished opponents in the 1965 Prix de l'Arc de Triomphe, there remains the certainty that they had seen one of the finest thoroughbreds they were ever likely to be privileged to see.

PRIX DE L'ARC DE TRIOMPHE 2400 metres
at Longchamp on Sunday 3 October 1965. For three-year-olds and upwards. Going: Good.
1,000,000 francs and 78 per cent of the entry fees to the winner; 200,000 fr. and 12 per cent of the entry fees to the 2nd; 100,000 fr. and 6 per cent of the entry fees to the 3rd; 50,000 fr. and 4 per cent of the entry fees to the 4th. 100,000 fr. to the breeder of the winner; 20,000 fr. to that of the 2nd; 10,000 fr. to that of the 3rd. Value to the winner: 1,084,747 francs.

SEA BIRD	ch.c.	by Dan Cupid – Sicalade by Sicambre	3.55½k	Jean Ternynck	T.P. Glennon	1
RELIANCE	b.c.	by Tantième – Relance by Relic	3.55½k	François Dupré	Y. Saint-Martin	2
DIATOME	br.c.	by Sicambre – Dictaway by Honeyway	3.55½k	Baron Guy de Rothschild	J. Deforge	3
FREE RIDE	b.c.	by Free Man – Croix du Risque by Wild Risk	4.60k	Baron Guy de Rothschild	L. Heurteur	4
ANILIN	b.c.	by Element – Analogichnaya by Agregat	4.60k	Haras Voshod	N. Nasibov	5
TOM ROLFE	b.c.	by Ribot – Pocahontas by Roman	3.55½k	Raymond Guest	W. Shoemaker	6
DEMI-DEUIL	b.c.	by Mourne – Laetitia by Lacaduv	4.60k	André Rueff	H. Samani	7
CARVIN	b.c.	by Marino – Coraline by Fine Top	3.55½k	Jean Boutillier	J. Massard	8
MEADOW COURT	ch.c.	by Court Harwell – Meadow Music by Tom Fool	3.55½k	Max Bell	L. Piggott	9
MARCO VISCONTI	ch.c.	by Antonio Canale – Magonza by Traghetto	3.55½k	Alarico Palmieri	C. Marinelli	10
EMERALD		by Auriban – Esméralda by Tourbillon	4.60k	Marcel Boussac	R. Poincelet	
ARDABAN		by Auriban – Asmèna by Goya	4.60k	Marcel Boussac	H. Mathelin	
SODERINI		by Crepello – Matuta by Tantième	4.60k	L.L. Lawrence	G. Lewis	
SIGEBERT		by Alizier – Senones by Prince Bio	4.60k	Mme Jean Stern	F. Head	
FRANCILIUS		by Philius – La Française by Casterari	4.60k	Mme Léon Volterra	J.C. Desaint	
ONCIDIUM		by Alcide – Malcolmia by Sayani	4.60k	Lord Howard de Walden	A. Breasley	
TIMMY LAD		by Tim Tam – Providence by Easton	4.60k	Mme P.A.B. Widener	M. Larraun	
KHALIFE		by Montaval – Jill Scott by Jock Scot	3.55½k	Max Bell	J. Roe	
RAGAZZO		by Macherio – Mill Baby by Mazarin	3.55½k	Rory More O'Ferrall	J. Mercer	
BLABLA		by Lavandin – Barina by Balthazar	3.54k	Gabriel Brun	M. Depalmas	

20 runners
SEA BIRD trained by Etienne Pollet; bred by Jean Ternynck.
Distances: 6 lengths; 5 lengths; short neck; ½ length; 5 lengths; 6 lengths; 1½ lengths; ½ length. Time: 2 mins 35.52 seconds.
Pari-Mutuel: Win: 2fr. 20. Places: 1fr. 80; 1fr. 70; 2fr. 20.
S.P.: 12 to 10 SEA BIRD; 9 to 2 Reliance; 15 to 2 Diatome and Free Ride (couplé); 31 to 4 Meadow Court and Khalife (couplé); 8 to 1 Tom Rolfe; 22 to 1 Anilin; 23 to 1 Emerald and Ardaban (couplé); 24 to 1 Blabla; 33 to 1 Ragazzo; 55 to 1 Marco Visconti; 60 to 1 Demi Deuil; 70 to 1 Sigebert; 75 to 1 Oncidium and Timmy Lad; 80 to 1 Carvin; 90 to 1 Soderini; 120 to 1 Francilius.

SEA BIRD chestnut colt 1962	Dan Cupid ch. 1956	Native Dancer gr. 1950	Polynesian br.1942	Unbreakable b. 1935	Sickle
					Blue Grass
				Black Polly b.1936	Polymelian
					Black Queen
			Geisha gr. 1943	Discovery ch. 1931	Display
					Ariadne
				Miyako gr. 1930	John P. Grier
					La Chica
		Vixenette ch. 1944	Sickle br. 1924	Phalaris br. 1913	Polymelus
					Bromus
				Selene b. 1919	Chaucer
					Serenissima
			Lady Reynard b. 1939	Gallant Fox b. 1927	Sir Gallahad
					Marguerite
				Nerva b. 1928	Fair Play
					Zephyretta
	Sicalade b. 1956	Sicambre br.1948	Prince Bio b.1941	Prince Rose b.1928	Rose Prince
					Indolence
				Biologie ch. 1935	Bactériophage
					Eponge
			Sif br. 1936	Rialto ch. 1923	Rabelais
					La Grêlée
				Suavita b. or. br. 1928	Alcantara II
					Shocking
		Marmelade b. 1949	Maurépas ch. 1937	Athelstan b. 1922	Teddy
					Dedicace
				Brocéliande ch. 1925	La Farina
					Reine Mab
			Couleur b. 1939	Biribi br. or gr. 1923	Rabelais
					La Bidouze
				Colour Bar br. 1930	Colorado
					Lady Disdain

Bon Mot 1966

Bon Mot

The Field

The appearance on Longchamp on the afternoon of the 1966 Prix de l'Arc de Triomphe was totally altered compared with the day of Sea Bird's victory. A magnificent new grandstand had been built. This stand was complemented by the new weighing room and paddock, which had been completed the previous year, and all combined to make Longchamp one of the most modern and luxurious racecourses in the world. It was certainly comparable with, if not better than, the new facilities at Ascot. All the foreign visitors commented on the excellence of the viewing facilities for the spectators, and with the paddock and unsaddling enclosure situated directly behind the stand, the amount of walking that the spectator had to do in order to see the horses parading was cut to a minimum. In fact it was difficult to find fault in the layout.

Unfortunately there was an outbreak of swamp fever in France in 1966 which prevented any French horses competing in England or Ireland after 15 May. There was also a severe coughing epidemic, which caused several of the best French three-year-olds to miss the Prix de l'Arc de Triomphe, and many French trainers and turfistes expected one of the five foreign challengers to be victorious in the race. The result was that Lt. Col. John Hornung's Aunt Edith, trained in England by Noel Murless, started clear favourite at 15 to 4; she was the first foreign challenger to do so since Ribot had won Longchamp's greatest race for the second consecutive year in 1956.

Aunt Edith had made a profound impression on the French racing public the previous autumn, when Lester Piggott had guided her to an effortless 8-length victory in the Prix Vermeille over Dark Wave with those good fillies Long Look and Tadolina third and fourth. Although the Prix de Diane heroine and runner-up, Blabla and Cantilêne, and the Irish Oaks winner, Aurabella, had all fallen in the Prix Vermeille, the authoritative manner in which Aunt Edith had disposed of her rivals had left no doubt in the minds of those present that she was the champion filly of her generation in Europe. Prior to this success, Colonel Hornung's filly had finished a highly promising runner-up in the Musidora Stakes at York in May on her racecourse début and had then won the Nassau Stakes at Goodwood in July. In 1966 Aunt Edith only ran three times before competing at Longchamp. She beat Fighting Charlie and Mintmaster 4 lengths and 1½ lengths in the Yorkshire Cup (14f) on 12 May. Fighting Charlie had won the 1965 Ascot Gold Cup and went on to win it for the second consecutive year in June with Mintmaster third, so this Yorkshire Cup victory established Aunt Edith in the highest class. She next ran in the Hardwicke Stakes at Royal Ascot but she failed to give Prominer 7 lbs and finished a disappointing fifth nearly 15 lengths behind. She had moved into a challenging position three furlongs out but found absolutely nothing when Piggott asked her to make her effort. No clear explanation has ever been found for this lack-lustre display. Aunt Edith returned to Ascot for the King George VI and the Queen Elizabeth Stakes in July. On this occasion she took the lead early in the straight and ran on strongly to defeat Sodium half a length with Prominer, who failed to quicken at the distance, 2 lengths away third. The three-year-old Sodium, who had finished fourth to Charlottown in the Derby at Epsom, had reversed the form in the Irish Sweeps Derby and beaten him a length. In September Sodium was again to defeat Charlottown in the St Leger. Thus Aunt Edith had established her superiority over one of the best representatives of the 1963 generation. Provided that her inexplicable failure in the Hardwicke Stakes had been an uncharacteristic aberration, she appeared a worthy favourite for the 1966 Prix de l'Arc de Triomphe and there were high hopes that she would give her trainer, Noel Murless, and her jockey, Lester Piggot, their first success in the race.

The English four-year-olds were represented at Longchamp by Louis Freedman's, I Say, who had been third to Sea Bird and Meadow Court in the 1965 Epsom Derby. I Say was successful in three of his five races in 1966 before his arrival in France. Unquestionably his best performance had been to win the Coronation Cup in June. In the lead from the start, he had no difficulty in repulsing the challenges of Prominer and Atilla by 2 lengths and 3 lengths. Atilla had won the 1965 Gran Premio del Jockey Club and was to take the Grosser Preis von Baden later in 1966, so the form bore the closest inspection. In the Hardwick Stakes I Say found the concession of ten pounds to Prominer too much and was beaten 6 lengths and 2½ lengths by him and Rehearsed with the disappointing Aunt Edith 5½ lengths behind him fifth. In his preparatory race before the Prix de l'Arc de Triomphe, the Red Deer Handicap at Ascot on 26 September, I Say had a facile 6 lengths victory over the useful handicapper Early to Rise. On his very best form I Say could conceivably run into a place on 9 October, but his front running tactics were definitely better suited by a track like Epsom than one such as Longchamp.

The three-year-old filly, Parthian Glance, made up the English trio of challengers for the 1966 Prix de l'Arc de Triomphe. Like her dam, Tenacity, who had run behind Puissant Chef in 1961, Parthian Glance was a late developing filly, but she had easily won her last three races. At Royal Ascot she was not extended to beat Varinia, who had been third to Valoris and Berkeley Springs in the Oaks, by 3 lengths in the Ribblesdale Stakes. She next comfortably defeated Predicament a length in the Yorkshire Oaks. Finally she was most impressive when trouncing Predicament and Varinia 2½ lengths and 3 lengths in the Park Hill Stakes at Doncaster. Indeed it was difficult to gauge exactly how good Parthian Glance was, and her starting price of 21 to 1 clearly reflected that she was a possible outsider.

There were, in reality, two challenges from Ireland: Prominer trained by Paddy Prendergast and Valoris, who had been trained by Vincent O'Brien until her defeat in the Prix Vermeille. Prominer's success in the Hardwicke Stakes had been most impressive, but in great measure had been a result of the favourable weight advantage he enjoyed. He was, moreover, held by both I Say on their running in the Coronation Cup and by Aunt Edith on the form of the King George VI and the Queen Elizabeth Stakes. On 23 September he deadheated with Red Vagabonde for La Coupe de Maisons-Laffitte (10f) a head in front of Manderley with another Arc outsider, Corfinio, unplaced. Prominer may have found the ten furlongs a little on the sharp side for him, but this shared victory did little to enhance his prospects on Sunday 9 October.

Valoris had been an extremely good filly in the spring. She won the Irish 1000 Guineas from Loyalty and then decisively beat Berkeley Springs and Varinia 2½ lengths and 3 lengths in the Oaks at Epsom. On that running there was little to choose between her and Parthian Glance. In the Irish Guinness Oaks, when starting at 7 to 2 on, Valoris ran so deplorably that she could only finish a moderate fifth to Merry Mate. She was tried in blinkers in the Prix Vermeille in September, but she faded out of contention even before the straight was reached. It seemed, therefore, that she had gone over the top and after the race Lester Piggott reported she was not the same filly she had been in the spring. Vincent O'Brien immediately announced her retirement to stud after this comprehensive defeat, but her owner, Charles Clore, decided to give her one more chance. He sent her to E. Fellows at Chantilly for her final preparations for the Prix de l'Arc de Triomphe, for which she started at 95 to 1, a fair estimate of her prospects on her recent displays.

There were two challengers from Italy on Sunday 9 October; the four-year-old Marco Visconti, who had finished tenth to Sea Bird in 1965, and the three-year-old colt Ciacolesso. Marco Visconti had won his first two races in 1966 over 11 furlongs and 12 furlongs in Rome in April. On 5 June he was most unlucky to go down by three-quarters of a length to Ben Marshall in the Premio Presidente della Repubblica at the same Capannelle Course, for he was ridden most injudiciously. Lester Piggott, therefore, was engaged to ride Marco Visconti in the Gran Premio di Milano on 19 June. The combination had no difficulty in disposing of Chio by 4½ lengths with Ciacolesso sixth and Appiani unplaced. Chio had been runner-up to Appiani in the Derby Italiano, and was to beat Astese 4 lengths in the St Leger Italiano on 18 September, so Marco Visconti's victory established him as the best horse in Italy. Unfortunately he did not have a preparatory race before the Prix de l'Arc de Triomphe which cannot have helped his cause and he was difficult to restrain. If his jockey, Camici, could control Marco Visconti's natural impetuosity, there was no question that his chance against the weaker field in 1966 was much brighter than it had been against Sea Bird and Reliance the previous year, which was reflected in his starting price of 13 to 1.

The other Italian invader, Ciacolesso, had won both the Premio Ostia Antica and the Premio Alfredo Doria very easily before he defeated Fouquet 2 lengths in the Premio Parioli. He finished a very game and courageous fourth to Appiani in the Derby Italiano as he had damaged a foot a few days prior to the race and had to run with a felt pad between his hoof and his plate. After his defeat in the Gran Premio di Milano, he could only manage ninth place behind Atilla in the Grosser Preis von Baden and doubts were raised about his stamina. Ciacolesso did score an easy 6 length victory in the Premio Isola Pescatori (11f) against moderate opponents at San Siro, but many good Italian judges felt that his owner, Luigi Mantovani, would have been better advised to have run him in the Prix du Moulin de Longchamp.

On paper the best of the French defenders was Mme François Dupré's three-year-old colt, Danseur, a full brother to La Sega, the heroine of the 1962 Poule d'Essai des Pouliches, the Prix Saint-Alary, and the Prix de Diane. Runner-up on his seasonal début, he easily won his next two races at Longchamp in the spring. On 5 June at Chantilly he scored his third consecutive victory when he beat A Tempo and Cadmus three-quarters of a length and half a length in the Prix du Lys. In the Grand Prix de Paris the Prix Hocquart and Prix Greffulhe winner, Hauban, and Bon Mot looked likely to fight out the finish early in the straight, but as soon as Yves Saint-Martin pushed Danseur through a gap on the rails, he sprinted home a decisive 3-length winner from Hauban with Bon Mot a neck away third and Vasco de Gama and A Tempo unplaced. It was a most convincing performance, but it is sad to relate that his owner, François Dupré, died after a long illness in Jamaica the day after this victory. A short time afterwards Danseur contracted the cough, which severely interrupted his training programme. While on paper Danseur held the best credentials of the French runners, the dislocation to his training and his lack of a preparatory race inevitably made his odds of 5 to 1 for the 1966 Prix de l'Arc de Triomphe seem somewhat ungenerous.

Danseur was re-opposed on 9 October by Walter Burmann's Bon Mot, who in his first seven outings as a three-year-old had won two races and had been placed in the highest company on the other five occasions. In the spring Bon Mot won the Prix des Marronniers and La Bourse, both over twelve furlongs, at Longchamp. In the Prix Hocquart on 8 May he was far from disgraced when third to Hauban and Sea Hawk. As a result Hauban started favourite for the Prix du Jockey Club at Chantilly but he ran pounds below his best and finished unplaced. Saint-Martin, riding Nelcius, took the lead shortly after turning into the straight. Throughout the final furlong, Bon Mot challenged Nelcius very strongly but failed to catch him by a neck, with Behistoun 1½ lengths away third, Sea Hawk fifth, Taneb sixth and Hauban in rear. Bon Mot again ran very well when third to Danseur and Hauban in the Grand Prix de Paris. In his last race before the Prix de l'Arc de Triomphe, the Prix Royal Oak on 18 September, the firm going was all in favour of Martin Fabiani's Vasco de Gama, who beat Bon Mot half a length with A Tempo 3¼ lengths away fifth. Walter Burmann's colt was ideally suited by soft going, which he was to encounter at Longchamp on 9 October since there had been persistent rain in Paris throughout the previous week. Vasco de Gama, however, detested soft going and every extra millimetre of rain added another pound on his back. His only other victory in four outings before the Prix de l'Arc de Triomphe had been in the Grand Prix du Printemps at Saint Cloud on 30 May when the going had been very firm. Thus there appeared a distinct likelihood that in the prevailing conditions Bon Mot would reverse the Prix Royal Oak form with Vasco de Gama on 9 October.

Alec Weisweiller's Behistoun had run third to Soleil and Hauban in the 1965 Grand Critérium. In the Prix Greffulhe he again ran Hauban to a head, the same distance that had separated them the previous autumn. In the Prix Lupin, Behistoun snatched victory in the last strides by a neck and three-quarters of a length from Barbare and Soleil. Soleil, who had not stayed ten and a half furlongs, had earlier won the Poule d'Essai des Poulains from Barbare. Behistoun, therefore, was considered Hauban's biggest danger in the Prix du Jockey Club, but proved no match for either Nelcius or Bon Mot. He again had to be content with third place when beaten 2 lengths and half a length by Sea Hawk and Diatome in the Grand Prix de Saint Cloud, with Taneb fourth and Sigebert fifth. In his preparatory race for the Prix de l'Arc de Triomphe, Behistoun finished a disappointing fifth to Pasquin and Carvin in the Prix du Prince d'Orange. It was difficult, therefore, to enthuse about his chance of winning on 9 October, although on his very best form he might run into a place.

None of the other five French three-year-olds in the field for the 1966 Prix de l'Arc de Triomphe seemed to possess credentials one would normally associate with a potential winner. André Rueff's Lionel had won two handicaps at Longchamp and Saint-Cloud over twelve furlongs in the early summer, but he sprang into prominence when he beat Fantomas, the Prix du Cadran winner, and Corfinio a head and 2 lengths in the Grand Prix de Deauville, with Carvin half a length away fourth. If he were to maintain that improvement, Lionel could be considered a possible outsider. A Tempo had shown his best form in the spring when he beat Nelcius in the Prix de Guiche and he had been runner-up to Danseur in the Prix du Lys. But afterwards his form had deteriorated and he was decisively beaten 4 lengths by Arjon in the Prix Henry Delamarre on 2 October. Cadmus had won over a mile at Saint-Cloud and over ten furlongs at Longchamp in the spring. After his defeat by Danseur and A Tempo in the Prix du Lys, he ran Silver Shark to a head in the Prix d'Ispahan (9½f). On 11 September at Longchamp he ran Taneb, who was receiving 10 lbs, to 2 lengths. While he was likely to reverse that form with Taneb, it seemed probable that Cadmus would find several too good for him in the Prix de l'Arc de Triomphe.

Ernest Masurel's Taneb had put up his best performance when fourth to Sea Hawk, Diatome and Behistoun in the Grand Prix de Saint-Cloud. In the Prix Greffulhe and Prix du Jockey Club, however, he had failed to reach a place and even Cadmus seemed likely to gain his revenge on him in Longchamp's most prestigious race. Henry Aubert's Eclat, had won two small races, one over ten furlongs at Longchamp and one over eleven furlongs at

Saint-Cloud in June. His best performance had been to run Nelcius to 2 lengths in the Prix de Chantilly on 4 September, but his proximity to the Prix du Jockey Club winner was undoubtedly flattering and it would have caused much surprise were he to win on 9 October.

There were four French fillies in the line-up for the 1966 Prix de l'Arc de Triomphe. Fine Pearl, twice successful in three outings as a juvenile, easily won the Prix Finlande at Le Tremblay on 1 June. She then impressively beat Bergame and Sweet Sauce, the third in the Poule d'Essai des Pouliches, by 1½ lengths and 2½ lengths in the Prix de Diane. Fine Pearl did not run again till she finished a short neck and three-quarters of a length behind Haltilala and Bubunia in the Prix Vermeille, with Si Sage three-quarters of a length away fourth and the disappointing Valoris and Fandine in rear. Since it was later discovered that Fine Pearl was in season that afternoon and the fact she may have been in need of the race after her long absence from the racecourse, her connections felt she would give a much better account of herself in the Prix de l'Arc de Triomphe for which she was to start at 11 to 1.

Although Si Sage had finished fastest of all in the Prix Vermeille, there seemed no reason to suppose she would gain her revenge on Fine Pearl. On 4 September she was a little over half a length behind Bubunia in the Prix de la Nonette. Her only victory in 1966 had been in the Prix de Malleret, when she had received four pounds from Bubunia and beaten her a neck with Chérie Noire 2 lengths back third. Although Deforge chose to ride Si Sage in preference to Marco Visconti, it seemed likely that she would find several of her rivals too good for her on 9 October.

Neither Chérie Noire, whose sole success in 1966 had been gained over a mile at Saint-Cloud in the early spring, but who was clearly inferior to Fine Pearl and Si Sage, nor Fandine, who had won a small race at Clarefontaine in September, could be said to have even a remote chance in the 1966 Prix de l'Arc de Triomphe. Their respective starting prices of 95 to 1 and 140 to 1 could not even be called generous.

The best of the four older French runners were unquestionably Carvin and Sigebert. Carvin had been slow to find his form in 1966. He had been unplaced in the Prix Ganay and the Grand Prix de Saint-Cloud. In August he won the Grand Prix de Vichy for the second consecutive year but against moderate opposition. He was only fourth in the Grand Prix de Deauville, but he showed a glimmer of his good three-year-old form when he ran Pasquin to half a length in the Prix du Prince d'Orange. Carvin, however, had always shown his best form in the autumn so could not be totally ruled out of reaching a place. Mme Jean Stern's Sigebert, however, had run most consistently in 1966. He only failed by half a length to hold Diatome in the Prix Boïard at Saint-Cloud in March with Carvin unplaced. He then ran Diatome to three-quarters of a length in the Prix Ganay, with Prominer and Carvin again in rear. In May Sigebert won the Prix d'Harcourt and was then third to the surprise winner, Tajubena, and Diatome in the Prix Dollar on 9 June. His only real failure was when he finished fifth in the Grand Prix de Saint-Cloud, but he comfortably beat Corfinio and Red Vagabonde 1½ lengths and a short head in the Prix Gontaut-Biron at Deauville. The principal doubt about Sigebert was his stamina; it was generally felt that he was better suited by ten or eleven furlongs and that the stiff mile and a half at Longchamp would test his stamina to its limit. Nevertheless, at his generous starting price of 50 to 1 he looked an outsider with the best value on account of his consistency.

The chances of the other two older French runners on 9 October was not all apparent. Corfinio won two small races in the summer, one over ten furlongs at Saint-Cloud, the other over twelve furlongs at Maisons-Laffitte. Whenever Corfinio competed in the highest company, however, he always found several of his rivals too good for him. In spite of the fact that Red Vagabonde deadheated with Prominer in La Coupe de Maisons-Laffitte, there was little to choose between him and Corfinio on their running in the Prix Gontaut-Biron. In his four outings prior to that he had been unplaced and almost certainly was better over a mile and a mile and a quarter. It would have been a major surprise if either Red Vagabonde or Corfinio were even able to reach a place in the 1966 Prix de l'Arc de Triomphe.

At the beginning of the year it had been expected that François Dupré's Reliance would carry all before him with the retirement of Sea Bird, but a series of mishaps caused him to retire to stud without racing as a four-year-old. That had left Diatome to carry the flag for the 1962 generation, but unfortunately after his defeat by

Sea Hawk in the Grand Prix de Saint Cloud, he injured a tendon and had to be taken out of training. Sea Hawk also broke down after this victory and never raced again. Hauban contracted the cough so badly after the Grand Prix de Paris that he also had to miss the Prix de l'Arc de Triomphe. Deprived of so many good horses, it is hardly surprising that the French expected a foreign victory on 9 October. A week prior to the race one of their principal hopes was the Prix du Jockey Club winner Nelcius, who had so easily won the Prix de Chantilly in early September. His final gallop on Monday 2 October was very impressive and both his trainer, Miguel Clément, and his owner, Paul Duboscq, were more than hopeful about his chance. Disaster, however, then struck. First Duboscq's sugar plantation in Jamaica was ravaged by a hurricane and he immediately had to leave Paris for the West Indies. Secondly two days after the gallop Clément found that Nelcius had succumbed to the cough and he was forced to withdraw his charge the next morning.

The Final Preparations

At the beginning of the week it had been reported that Danseur, who now seemed the best of the home contingent, had done a good gallop with the Prix de la Nonette winner Bubunia. His trainer, François Mathet, a stickler for the truth no matter how unpalatable, immediately wrote a letter to the French racing press; it was published on Friday and in it Mathet was at pains to stress that the earlier report was incorrect and that, in fact, Bubinia had gone much better than Danseur in the gallop. This frankness was much to be commended, but it also unahppily meant that the Grand Prix de Paris winner would not be at his best two days later.

The only favourable news about the French runners concerned Bon Mot who had gone really well in his final workout at the beginning of the week. His chance was also enhanced by the persistent rain that fell in the Paris area for the next six days. It rained again early in the morning of Sunday 9 October and the afternoon was wet and grey. The Stewards called in all the jockeys taking part in the Prix de l'Arc de Triomphe and they were instructed to keep a straight line for the first two furlongs in order to eliminate any possible interference.

The Race

In the preliminaries the favourite Aunt Edith looked well, but unquestionably the pick of the paddock were Sigebert and I Say. Bon Mot also was one of the better-looking colts in the parade. Danseur, as with all Mathet-trained runners, was turned out in magnificent condition, but his eye was dull and he appeared listless. The filly Parthian Glance seemed to have gone in her coat, which did not bode well for her chance.

There were no untoward incidents at the start and as the field broke from the stalls Prominer was the first to show from Vasco de Gama, I Say, Fandine, Sigebert, Danseur, Bon Mot and Aunt Edith. On settling down Marco Visconti, who refused to be restrained, took up the running from Ciacolesso with Sigebert, Vasco de Gama, I Say, Carvin and A Tempo close up behind them. They were followed by a group consisting of Aunt Edith, Cadmus and Bon Mot. At the top of the hill, Ciacolesso and I Say just headed Marco Visconti, who was closely tracked by a bunch containing Sigebert, Vasco de Gama, Carvin, A Tempo and Aunt Edith, who were followed by Bon Mot and Fandine. There was little change in the order coming down the hill, though Lionel and Behistoun could both be seen improving their positions from the rear. I Say and Ciacolesso still disputed the lead ahead of Marco Visconti together with Sigebert, who was going very easily, as the field approached the straight. Immediately behind these leaders A Tempo, Aunt Edith, Bon Mot, Carvin, Lionel and Behistoun were all poised to launch their challenges once the straight was reached. Both I Say and Ciacolesso quickly began to weaken as soon as the field had negotiated the turn. Flavien immediately shot Sigebert past them and rapidly went 2 lengths clear. It was soon evident that both Aunt Edith and Marco Visconti were in trouble and that it was not to be their day. It was left to A Tempo tracked by Bon Mot to set off in pursuit of Sigebert who was making the best of his way home.

Two furlongs out, Sigebert still led A Tempo by 1½ lengths with Bon Mot the same distance behind in third place on the outside. Walter Burmann's colt produced a fine burst of acceleration to catch A Tempo a little more than a furlong out. He then seemed to hang fire and it appeared that he would not catch Sigebert. However, his young jockey, Freddie Head, got him going again 150 yards out with a couple of

Bon Mot (10) beating Sigebert (rails) and Lionel (17) half a length and 2 lengths with A Tempo (16) half a length away and Behistoun fifth

strokes of his whip. By then Sigebert was approaching the limit of his stamina and, as his stride began slowly to shorten, Bon Mot made relentless if unspectacular progress. He caught the tiring Sigebert fifty metres from the line. Sigebert struggled gamely but finally went under by half a length with Lionel, who had finished like an express up the rails, 2 lengths behind third. A Tempo, who ran on at one pace, was half a length behind in fourth place, with the same distance separating him and Behistoun, whose very late run had enabled him to deprive Carvin of fifth place. Chérie Noire was seventh, the disappointing Aunt Edith eighth and I Say ninth. To the great surprise and pleasure of the French, the first seven runners home were all trained and bred in France.

For Bon Mot's trainer, Willie Head, it was a second success in the Prix de l'Arc de Triomphe, as he had trained Le Paillon to win in 1947. His grandson and Bon Mot's jockey, Freddie Head, however, was not then nineteen, so he became the youngest jockey to win Longchamp's most important race since Boullenger had won on Oroso. Immediately after the race Freddie Head thought that Bon Mot would have won even if the track 'had been drier'. He said that he had tracked Aunt Edith and when, on the final turn, he saw that she was not going well, he 'went for a gap on the outside and set off in chase of Sigebert'. Léon Flavien, Sigebert's jockey, disagreed with this opinion: 'If the going had been less heavy, I'm sure Sigebert would have held his lead over Bon Mot,' he told trainer Max Bonaventure. The latter had achieved a fine feat of training, as he also trained the third horse, Lionel. Indeed if there were a hard-luck story, it concerned Lionel who, turning into the straight, was boxed in on the rails and then was almost brought to his knees. However he still managed to regain his balance and produce a remarkable burst of speed in the last 400 metres to take third place.

Piggott reported that turning into the straight he had still thought the favourite, Aunt Edith, was in with a winning chance, but that when he asked her to go, she pulled out nothing. He added that, 'though I think the ground was a little softer than she cares for, she must have gone over her best'. Despite Aunt Edith's defeat, Lt. Col. John Hornung, her owner, was frequently congratulated that afternoon on the news of the engagement of his niece, Tanya Wilson, to a 25-year-old business executive, Didier Machenaud.

There was apparently little excuse for the poor showing of Danseur, who clearly was some way below his best and in addition did not relish the conditions underfoot. François Mathet had compensation for his defeat, however, for he trained Farhana and Silver Shark to win the Prix de l'Abbaye de Longchamp and the Prix du Moulin de Longchamp respectively for his patron, the Aga Khan, besides winning the first race, the Prix de la Cascade, with Arpad Plesch's Caishot. Neither of the best three-year-old fillies, Fine Pearl and Parthian Glance, had made any show at all. Duncan Keith, I Say's jockey, thought he would have been much closer on faster ground and Prominer's lack-lustre display was also blamed on the going. Nevertheless Bon Mot had proved the best horse on the day and it is idle to speculate what might have been had the going been different.

The Rest of the Season

Bon Mot's owner, Walter Burmann, had been a very successful breeder of pedigree cattle for many years. He had turned his attention to thoroughbred breeding approximately ten years earlier, and had horses in training in Ireland, France and England. In 1960 he had won the Irish Derby with Chamour; four years later his filly, Ancasta, whom he had bred, had won the Irish Oaks. He had also bred Bon Mot, and undoubtedly his sound knowledge of genetics and his ability to sift good from bad stock had helped to make him a successful breeder. Both he and his trainer, Willie Head, decided to refuse an invitation for Bon Mot to compete in the Washington D.C. International Stakes at Laurel Park as they felt he had had a hard season. The Prix de l'Arc de Triomphe form was, nevertheless, upheld at Laurel Park, for Behistoun beat Anilin and Assagai 2¼ lengths and 1¼ lengths, with Vasco de Gama fourth. Anilin had beaten Salvo by inches in the Preis von Europa at Cologne, with Carvin 5 lengths away third, in late October. A Tempo disappointed in the Prix du Conseil Municipal when unplaced to Pasquin, though another failure behind Bon Mot, Eclat, finished fourth. On the same day Marco Visconti paid another compliment to Bon Mot, however, when he beat Ciacolesso and the Irish St Leger winner, White Gloves, 1 length and three-quarters of a length in the Gran Premio del Jockey Club at San Siro, with Corfinio fourth.

Pedigree and Stud Record

Bon Mot remained in training in 1967. He had four races before he contracted the cough which caused him to be retired to stud. On his reappearance he clearly needed the race and only managed fourth place by the fluke winner Claquesous, Behistoun and Nelcius in the Prix Boïard at Saint-Cloud in March. In the Prix Ganay on 9 April he was beaten half a length and 1½ lengths by Behistoun and Great Nephew, but by this time the distance was almost certainly too sharp for him. He put up his best performance in the Prix de Lutece in May when he led throughout and easily beat Parthian Glance and Lionel 2½ lengths and 3 lengths.

The firm going was all against him in his final race when fourth to Busted in the King George VI and the Queen Elizabeth Stakes at Ascot. Although on balance his four-year-old career had been disappointing, when Bon Mot retired he had won four races and had been runner-up three times, third on the same number of occasions, once fourth and only once out of the frame. He had also won a total of 1,940,770 francs for his owner.

Bon Mot's sire, Worden, was a very useful colt who, like his son, was at his best on soft going. He won the Washington D.C.

International Stakes, the Prix du Conseil Municipal and the Premio Roma. He also finished third to La Sorellina and Silent in the 1953 Prix de l'Arc de Triomphe the same year as he occupied the same place behind Pinza and Aureole in the King George VI and the Queen Elizabeth Stakes, in which he had also been third the previous year behind Tulyar and Gay Time. The best of his progeny before Bon Mot had been the 1959 Prix de Diane winner, Barquette, and the 1962 Grand Prix de Paris victor Armistice. He was also the grandsire of Vasco de Gama and the maternal grandsire of Behistoun.

Bon Mot's dam, Djebel Idra by Phil Drake, was a big, attractive filly, who was acquired by Walter Burmann as a yearling from la Comtesse de Chambure. She was placed on her only outing as a juvenile and ran five times as a three-year-old. She won a maiden race at Phoenix Park in April 1960, but then competed in the highest class. She failed to reach a place in the Irish 1000 Guineas, the Irish Oaks and the Irish St Leger. In her only other race she was beaten approximately ten lengths when out of the frame to Never Too Late II in the Epsom Oaks. If she had been a little disappointing on the racecourse, she more than compensated her owner by her success in the paddocks. Four of her five foals were winners, of which the best undoubtedly was Bon Mot. Her filly Tidra by Prince Taj won two races including the Prix Saint-Alary, and she was runner-up to Gazala in the 1967 Prix de Diane. Tidra herself became a very useful broodmare: she is the dam of Tennyson by Val de Loir who won the 1973 Grand Prix de Paris and was runner-up to Roi Lear in the Prix du Jockey Club, and of the useful colt Turville, who won three races and ran third in the Prix du Jockey Club and the Prix Lupin. Djebel Idra was also the dam of Tycoon by Tamerlane, whose six victories included the Prix Perth. Her filly, Belidra by Prince Taj, gained one victory and was herself the dam of the ill-fated Bonami by Le Fabuleux, who was runner-up to My Swallow in the 1970 Grand Critérium. Certainly Walter Burmann could never have regretted Djebel Idra's purchase as a yearling.

Djebel Idra's dam, Djebellica by Djebel, ran in the colours of le Comte de Chambure. In all she won three races: the Acomb Stakes in England and a race over twelve furlongs as a three-year-old at Saint-Cloud, after she had run second to Belle of All, the 1951 1000 Guineas winner, in the Coronation Stakes at Royal Ascot. Her finest performance, nevertheless, was when she beat the Irish 1000 Guineas winner, Queen of Sheba, a neck in the Irish Oaks with the Prix de Diane winner Stratonice 6 lengths behind in third place. Her best performance as a juvenile was when she finished third to Pharsale and Ksarinor in the Prix de la Forêt. Although, along with Stratonice, she disappointed behind Orberose when unplaced in the Prix Vermeille, Djebellica had had a very successful career on the racecourse. As a broodmare she was equally successful for she produced seven winners, including Bon Mot's dam, Djebel Idra, the best of which were Cambremont by Sicambre and the filly Torbella by Tornado. Cambremont won three races including the 1965 Poule d'Essai des Poulains in which he beat Grey Dawn, and he finished third to Sea Bird and Diatome in the Prix Lupin. Torbella won both the Dewhurst Stakes and the Prix de Malleret, and was runner-up to Amante in the 1958 Irish Oaks. She herself was a successful broodmare and her produce included Carlemont, one of whose two victories was the Sussex Stakes at Goodwood, and Bella Carlotta (Prix Finlande), and dam of Beauvallon (Horris Hill Stakes), and Princess Tora (Prix de la Calonne), and grandam of Princess Lida (Prix Morny and Prix de la Salamandre); Avaray (Prix de la Jonchère) was another of Torbella's foals. Djebellica also produced Vezelay by Verso II, whose two victories included the Prix du Palais Royal. She was the grandam of Wenduyne by Moutiers, who won the 1969 Irish 1000 Guineas, and Critique, the runner-up to Recitation in the 1980 Grand Critérium, and winner of the Cumberland Lodge Stakes and the Hardwicke Stakes also traces to Djebellica.

Djebellica's dam, Nica by Nino, was also a very good filly; her seven victories in three seasons of racing included the Prix du Conseil Municipal and the Prix de Maréchaux. In addition to Djebellica she got four other winners, the best of which unquestionably was Montenica by Djebel, whose four victories included the 1947 Prix de Diane. Montenica also ran the 1000 Guineas winner Imprudence to half a length in the Poule d'Essai des Pouliches and was possibly an unlucky third to Procureuse in the Prix Vermeille as she tracked a below-par Imprudence and it was too late when her jockey realised he had made a tactical error. Nica's Verso II filly Vice Versa ran third in the Epsom Oaks in 1949 and herself was the grandam of

Crivelli by Alizier (St Leger Italiano 1963). Her Donatello II filly Donica won twice including the Prix de Royallieu, and she was the dam of Don Icario (Grand Critérium d'Ostende). Nica was also the grandam of Picounda whose three victories included the Prix de Malleret but who in addition was runner-up to Douve in the Prix de Diane and to Wild Miss in the Prix Vermeille, and ran third to Ribot and Beau Prince in the 1955 Prix de l'Arc de Triomphe.

Nica's dam, Canalette by Cannobie, was unbeaten in three races as a juvenile and was one of the best two-year-old fillies of her generation for she won both the Prix la Rochette (1000 m) and the Prix Robert Papin. She only won one race as a three-year-old, the Prix la Rochette (2200 m) but she was also an excellent broodmare getting five winners. In addition to Nica the two best were Canot, who won the Critérium de Saint Cloud, the Prix Henry Delamarre, the Prix d'Harcourt and the Prix Jean Prat, and who was also runner-up in the Prix du Jockey Club and the Grand Prix de Paris and ran third to Eclair au Chocolat in both the Prix Royal Oak and Prix dc l'Arc de Triomphe in 1938, and MacKann, who won the Poule d'Essai des Poulains the following year, the Queen Anne Stakes at Royal Ascot and the Prix Daphnis.

Canalette's dam, Hallebarde by Admiral Crichton, was another prolific broodmare producing nine individual winners of thirty-nine races, including Hallebardier by Badajoz (Prix la Rochette, 2200 m, and Grand Prix de Marseille), Le Debardeur by Badajoz (Grand Prix du Printemps), Bachelette II by Badajoz (Prix Chloë), and La Belladeuse, the dam of La Savoyarde by Filibert de Savoie and Chuchoteur by Chubasco, who both won the Prix des Sablons. The family traces back to Trampoline by Tramp, the dam of Glencoe by Sultan, the winner of the 2000 Guineas, the Ascot Gold Cup and the Goodwood Cup. Considering, therefore, from what an illustrious female line Bon Mot came, it was hardly surprising that he turned out as good as he was.

Bon Mot spent four seasons at stud in France and in 1972 was exported to Japan and recently has been sent from there to the United States. The best of his progeny was unquestionably Lassalle, the winner of the Ascot Gold Cup and the Prix du Cadran. Other good winners that he has sired include Panpryl (Prix de Sandringham), Robertino (Prix Herod), Gargano (Grand Handicap d'Evry), and Calisson (Prix du Lac). In Japan his son Nisso Pross was a stakes winner in 1980 and in that country his progeny have amassed more than $2 million in prize money. Nevertheless the only top-class horse that he has sired is Lassalle, so he has been less successful as a stallion than as a racehorse. He has recently emerged as a potential important broodmare sire. Son of Love (St Leger 1979) Nice Havrais (Prix de Fontainebleau and runner-up in the Grand Critérium), and Sharpman (Prix Omnium, Prix des Yearlings, runner-up in the Poule d'Essai des Poulains and Prix Lupin and third in the Prix du Jockey Club), were all out of broodmares sired by Bon Mot.

PRIX DE L'ARC DE TRIOMPHE 2400 metres
at Longchamp on Sunday 9 October 1966. For three-year-olds and upwards. Going: Heavy.
1,000,000 francs and 78 per cent of the entry fees to the winner; 300,000 fr. and 12 per cent of the entry fees to the 2nd; 200,000 fr. and 6 per cent of the entry fees to the 3rd; 100,000 fr. and 4 per cent of the entry fees to the 4th. 100,000 fr. to the breeder of the winner; 30,000 fr. to that of the 2nd; 20,000 fr. to that of the 3rd. Value to the winner: 1,095,862 francs.

BON MOT	ch.c	by Worden – Djebel Idra by Phil Drake	3.55½k	Walter Burmann	F. Head	1
SIGEBERT	b.c.	by Alizier – Senones by Prince Bio	5.60k	Mme Jean Stern	L. Flavien	2
LIONEL	ch.c.	by Herbager – La Strada by Fervent	3.55½k	André Rueff	H. Samani	3
A TEMPO	b.c.	by Tombeur – Lucky Shoe by Pot O'Luck	3.55½k	G. L. de Mola	R. Jallu	4
BEHISTOUN	br.c.	by O'Grady – Beotie by Worden	3.55½k	Alec Weisweiller	G. Serpereau	5
CARVIN	b.c.	by Marino – Coraline by Cor de Chasse	4.60k	Jean Boutillier	H. Mathelin	6
CHERIE NOIRE	br.f.	by Soleil Levant – Mefair by Medium	3.54k	Jacki Clérico	A. Gibert	7
AUNT EDITH	ch.f.	by Primera – Fair Edith by Hyperion	4.58½k	Lt. Col. John Hornung	L. Piggott	8
I SAY	br.c.	by Sayajirao – Isetta by Morland	4.60k	Louis Freedman	D. Keith	9
CADMUS	b.c.	by Supreme Court – Covert Side by Abernant	3.55½k	Mme André Magnus	J. Taillard	10
RED VAGABONDE		by Red God – La Vagabonde by Vandale	5.60k	Harry Kartz	W.B. Pyers	
CORFINIO		by Tissot – Cesaproba by Scai	4.60k	Ugo Tudini	R. Poincelet	
PROMINER		by Beau Sabreur – Snob Hill by Rockefella	4.60k	J.R. Mullion	D. Lake	
MARCO VISCONTI		by Antonio Canale – Magonza by Traghetto	4.60k	Alarico Palmieri	E. Camici	
ECLAT		by Herbager – Fée du Nil by Free Man	3.55½k	Henry Aubert	M. Garcia	
DANSEUR		by Tantième – La Danse by Ménétrier	3.55½k	Mme François Dupré	Y. Saint-Martin	
VASCO DE GAMA		by Bel Baraka – Princesse Bala by Prince Bio	3.55½k	Martin Fabiani	M. Philipperon	
CIACOLESSO		by Barba Toni – Ca' Lion by Traghetto	3.55½k	Luigi Mantovani	C. Ferrari	
TANEB		by Tapioca – Ténébreuse by Téléferique	3.55½k	Ernest Masurel	J.C. Desaint	
VALORIS		by Tiziano – Vali by Sunny Boy	3.54k	Charles Clore	W. Williamson	
SI SAGE		by Sicambre – Tête Sage by Sayani	3.54k	Comte Edouard Decazes	J. Deforge	
FANDINE		by Lavandin – Fantasque II by Wild Risk	3.54k	Mme Charles Levanti	A. Jézéquel	
PARTHIAN GLANCE		by Parthia – Tenacity by Wilwyn	3.54k	Mme W.H.D. Riley-Smith	R. Hutchinson	
FINE PEARL		by Fine Top – Seed Pearl by Tourment	3.54k	Baronne Geoffroy de Waldner	J. Massard	

24 runners
BON MOT trained by W. Head; bred by F.W. Burmann.
Distances: ½ length; 2 lengths; ½ length; 1½ lengths; head; ½ length; 1½ lengths; 3 lengths. Time: 2 mins 39.80 seconds.
Pari-Mutuel: Win: 6fr. 30. Places: 2 fr. 30; 12fr. 40; 5fr. 30.
S.P.: 15 to 4 Aunt Edith; 5 to 1 Danseur; 53 to 10 BON MOT; 11 to 1 Fine Pearl; 12 to 1 Behistoun; 13 to 1 Marco Visconti; 16 to 1 Lionel; 18 to 1 Si Sage; 20 to 1 Vasco de Gama; 21 to 1 Parthian Glance; 28 to 1 Carvin; 29 to 1 I Say; 32 to 1 Corfinio; 33 to 1 Cadmus; 37 to 1 Taneb; 50 to 1 Sigebert and A Tempo; 55 to 1 Prominer and Eclat; 95 to 1 Chérie Noire and Valoris; 130 to 1 Red Vagabonde; 140 to 1 Fandine; 170 to 1 Ciacolesso.

BON MOT chestnut colt 1963	Worden ch. 1949	Wild Risk b. 1949	Rialto ch. 1923	Rabelais b. 1900	St. Simon
					Satirical
				La Grêlée ch. 1918	Helicon
					Grignouse
			Wild Violet b. 1935	Blandford b. 1919	Swynford
					Blanche
				Wood Violet ch. 1928	Ksar
					Pervenchères
		Sans Tares ch. 1939	Sind b. 1933	Solario b. 1922	Gainsborough
					Sun Worship
				Mirawala br. 1923	Phalaris
					Miranda
			Tara ch. 1932	Teddy b. 1913	Ajax
					Rondeau
				Jean Gow b. 1920	Neil Glow
					Jane Shore
	Djebel Idra b. 1957	Phil Drake b. 1952	Admiral Drake b. 1931	Craig An Eran b. 1918	Sunstar
					Maid of the Mist
				Plucky Liège b. 1912	Spearmint
					Concertina
			Philippa br. 1943	Vatellor br. 1933	Vatout
					Lady Elinor
				Philippa of Hainault br. 1932	Teddy
					Pride of Hainault
		Djebellica ch. 1948	Djebel b. 1937	Tourbillon b. 1928	Ksar
					Durban
				Loïka ch. 1926	Gay Crusader
					Coeur à Coeur
			Nica ch. 1934	Nino ch. 1923	Clarissimus
					Azalée
				Canalette ch. 1922	Cannobie
					Hallebarde

Topyo 1967

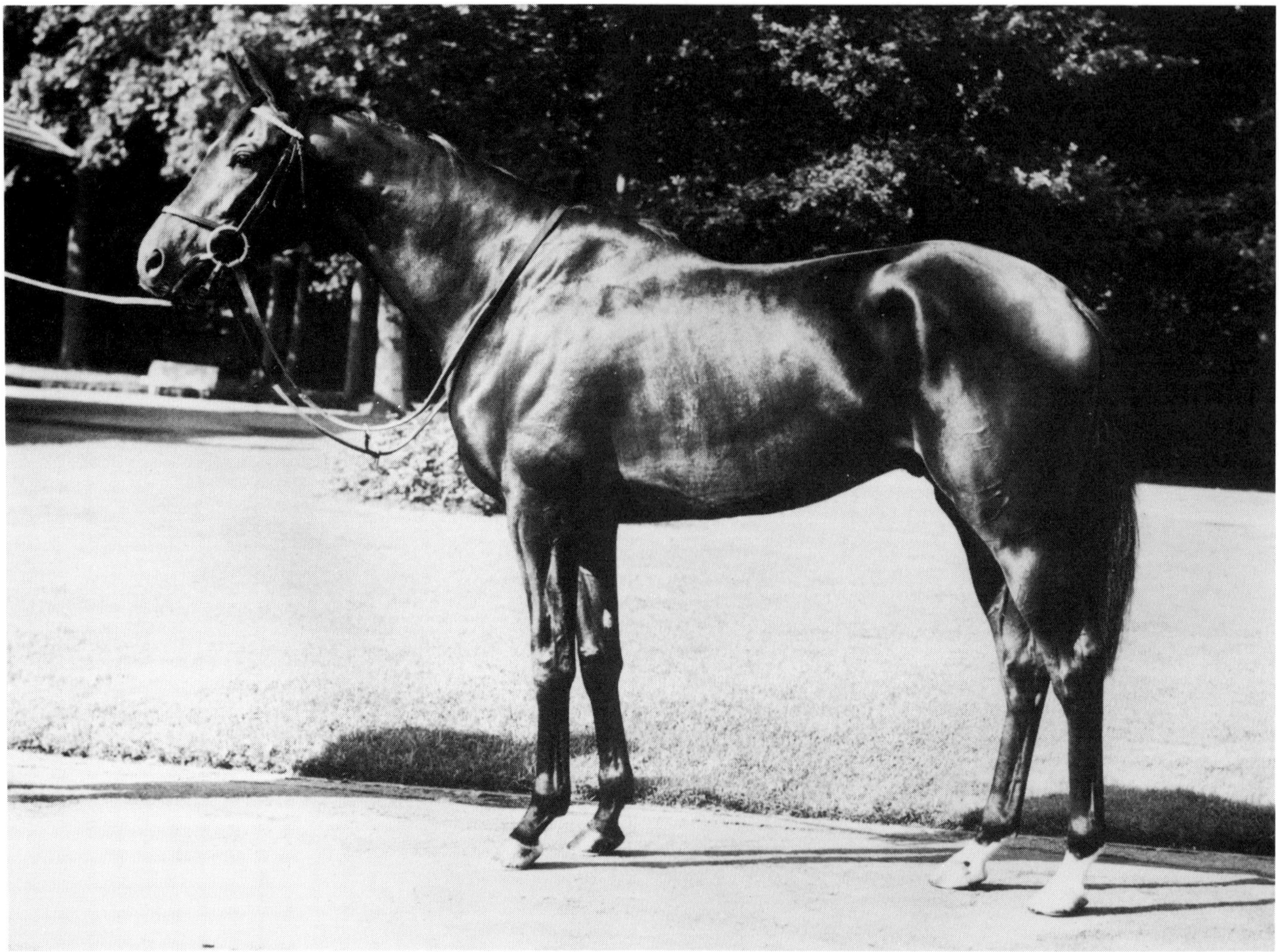

Topyo

The Field

There were thirty runners for the 1967 Prix de l'Arc de Triomphe, which was a record for the race. The field comprised sixteen three-year-olds, eleven four-year-olds and three six-year-olds. One third of the runners were challengers from abroad: Salvo, Ribocco, Sucaryl and Dart Board represented England; two came from Italy, Ciacolesso and Astese; two were from the Soviet Union, Anilin and Actach; while White Gloves was the sole challenger from Ireland.

Charles Engelhard's Ribocco and Gerald Oldham's Salvo appeared to be the best of the five English runners. As a juvenile Ribocco had won three of his four outings including the Observer Gold Cup at Doncaster and the Washington Singer Stakes at Newbury. He had also put up a meritorious performance when runner-up to the very fast Bold Lad in the Champagne Stakes. In the spring of 1967, however, he was a severe disappointment. Clearly needing the race when only fourth in the Craven Stakes at Newmarket, Ribocco was easily beaten by the moderate French Vine in the Dee Stakes at Chester in May. He was tried in blinkers for the first time in the Lingfield Derby Trial but with catastrophic results; he finished a remote and dismal fifth to another moderate colt, Heave Ho. Many felt, therefore, that he had failed to train on, but in fact Ribocco had taken an immediate aversion to the starting stalls in January, and it was not until the end of March that he was persuaded to enter them on the gallops. In spite of the presence of Lester Piggott in the saddle, Ribocco was allowed to start at the long odds of 22 to 1 for the Epsom Derby. Piggott brought him wide when he made his challenge in the straight. The 2000 Guineas winner, Royal Palace, had taken the lead over two furlongs out but Ribocco almost drew level with him approaching the distance. George Moore had kept enough in reserve, however, to enable Royal Palace to draw 2½ lengths ahead of Ribocco at the winning post with Dart Board 2 lengths away third. It transpired that this transformation in Ribocco's performance was a result of the much warmer weather following the Lingfield Derby Trial. It had been a very cold spring and Mr Engelhard's small Ribot colt did not thrive in such conditions. Even more remarkable was the report of the vet who examined him ten minutes after the race. He reported that Ribocco showed no signs of having had a race and it was discovered he had a much slower than normal heart-beat. On 1 July in the Irish Sweeps Derby at the Curragh Ribocco easily beat the highly regarded Sucaryl and Dart Board by three-quarters of a length and three lengths. In his next race, the King George VI and the Queen Elizabeth Stakes at Ascot, Ribocco was set to concede 2 lbs more than weight for age to Busted and Salvo. Neither he nor Salvo were able to match Busted's brilliant burst of speed which carried him to a 3-length victory over Gerald Oldham's horse. Ribocco, however, was a very creditable third, only a neck behind Salvo but over 6 lengths ahead of Sodium and Nelcius. Ribocco was beaten a head and 6 lengths by Dart Board and Tumbler in the March Stakes at Goodwood in August, his preparatory race for the St Leger. This display, however, should be ignored for the runner-up almost slipped the field and though Dart Board went after him, Piggott, not wishing to subject his mount to an unnecessarily hard race, had clearly accepted the situation even before the straight was reached. The St Leger itself was a completely different story. Ribocco swooped on the Princess of Wales's Stakes winner, Hopeful Venture, a hundred yards out and almost contemptuously disposed of him by 1½ lengths. The Derby Italiano winner, Ruysdael II, and Dart Board were a further half a length and 1½ lengths behind third and fourth. With this victory Ribocco completely redeemed his reputation, which had been so tarnished in the spring and could lay claim to be the second best three-year-old colt in England; his only superior was Royal Palace. He was thus a worthy representative of his generation in England to take on the best in France in the Prix de l'Arc de Triomphe.

Strictly on their running behind Busted, however, there was little between the St Leger winner and Gerald Oldham's Salvo. In theory Ribocco should just come out the best at Longchamp, but Salvo had shown continual improvement throughout the summer. In 1966 as a three-year-old he had won the Craven Stakes, the Vaux Gold Tankard and the Newbury Autumn Cup. He had ended the season by running the Russian champion Anilin to a head in the Preis von Europa at Cologne. In April 1967 Salvo only failed by a head to beat the previous year's Derby winner, Charlottown, in the John Porter Stakes at Newbury. He then had five lengths to spare over the useful stayer, Mintmaster, in the Yorkshire Cup. In the Hardwicke Stakes at Royal Ascot Salvo met Sodium, who had won the Irish Sweeps Derby and the St Leger in 1966. Sodium gave Gerald Oldham's horse a hefty bump at the furlong pole, but Salvo gamely kept on and beat him by half a length. After his second in the King George VI and the Queen Elizabeth Stakes, Salvo met Luciano, the hitherto unbeaten winner of the Deutsche Derby and considered the best German horse since Orsini, in the Grosser Preis von Baden. He met his match, however, in Salvo who beat him three-quarters of a length with the remainder of the field trailing 9 lengths in arrears. Like Ribocco, Salvo was unquestionably the second best of his generation in England; only Busted was clearly better. It was also, however, equally difficult to prognosticate with any degree of accuracy which of the two, Salvo or Ribocco, would emerge as the best of the English challengers on Sunday 8 October.

The other trio from England appeared to have much less chance in the Prix de l'Arc de Triomphe. Michael Sobell's Dart Board had finished behind Ribocco in three classics. There were some who considered him unlucky in running at both Epsom and Doncaster, but in reality it was difficult to envisage him reversing those places with the small Ribot colt. Apart from his slightly fortuitous victory at Goodwood in the March Stakes, he had only won one other race in 1967, the Brighton Derby Trial in May.

R.J. Sigtia's Sodium had been in the top flight of his generation in 1966. Besides winning the Irish Sweeps Derby and the St Leger, defeating Charlottown on each occasion, he had also finished runner-up to Aunt Edith in the King George VI and the Queen Elizabeth Stakes and had been fourth in the Derby at Epsom after a somewhat unlucky run. With the exception of the Hardwicke Stakes when runner-up to Salvo, Sodium was a dismal disappointment as a four-year-old. On his reappearance in the Coronation Cup he only managed to beat one other runner. He started favourite for the Eclipse Stakes at Sandown but trailed in 12 lengths behind Busted, and then he took no part in the finish of the King George VI and the Queen Elizabeth Stakes. In his and Sucaryl's final race before travelling to Paris, the David Robinson Stakes at Ascot on 25 September, he finished last of the five runners, 13 lengths behind the Murless-trained colt.

Only Royal Palace was considered Sucaryl's superior among the three-year-olds at Noel Murless's stable. Although he may have been a little flattered by his proximity to Ribocco at the Curragh, it was an excellent performance as it was Sucaryl's first race as a three-year-old. The weights for the News of the World Handicap at Goodwood were published before the Irish Sweeps Derby; Sucaryl, therefore, was very leniently treated and he won with all the ease that was anticipated. In the Prix de la Côte Normande, however, he could only finish fifth, 6½ lengths behind the very easy winner Topyo. Sucaryl's limitations were further exposed when he was beaten three-quarters of a length and 1½ lengths by Samos III and Rixtrag in the Prix Royal Oak with Pétrone 2¼ lengths away fifth and Phaeton sixth. His 1½-length victory over Mariner in the David Robinson Stakes at Ascot showed his well-being, but in reality Sucaryl was only the Murless second string for the 1967 Prix de l'Arc de Triomphe. He also trained Stanhope Joel's brilliant four-year-old Busted in 1967 and his scintillating victories in the Eclipse Stakes over Great Nephew and Appiani II and in the King George VI and the Queen Elizabeth Stakes had established him as the best four-year-old in Europe. In early September Busted contemptuously disposed of Fiasco by 4 lengths in the Prix Henri Foy at Longchamp. The manner of his victory left little doubt in the minds of spectators that they had seen the winner of the Prix de l'Arc de Triomphe. It was immensely regrettable, therefore, that Busted broke down shortly before the race. Stanhope Joel had invited so many of his friends to Paris to see him run that he immediately decided to buy Sucaryl from Lady Sassoon, for whom he had run all summer, as a replacement. There was, however, one further element not in Sucaryl's favour. His dam, Sweet Angel by Honeyway, had a reputation for producing foals of great ability but with a frequent disinclination to fight when strongly challenged.

The challenger from Ireland, White Gloves, appeared a total outsider. His three successes in 1966 had included the Irish St Leger and he had run third, a length and three-quarters of a length behind Marco Visconti and Ciacolesso, in the Gran Premio del Jockey Club in October, with Astese fifth. On his reappearance in April 1967 he won the Ballymoss Stakes and later that month he beat Signal Infesta in the Desmond Stakes at the Curragh. Nevertheless he was a distant fifth to Salvo in the Hardwicke Stakes. Although only 2 lengths and a head behind Charlottown and Nelcius in the Coronation Cup at Epsom, White Gloves was very flattered by this bare result for Charlottown was an exceptionally easy winner. Indeed White Gloves's starting price of 210 to 1 could be said to reflect his chance on Sunday October 8th.

Luigi Mantovani sent two of his string to compete in the 1967 de l'Arc de Triomphe, the four-year-olds Ciacolesso and Astese. Ciacolesso had done nothing in 1967 to suggest that he would fare better in Paris than he had done the previous year. In the late summer he ran in France and won the Prix Guillaume le Conquérant (10f) at Deauville and a minor event over 11½ furlongs at Longchamp on 10 September. He only managed ninth place behind Lionel in the Grand Prix de Deauville, however, and he was beaten a neck, 2½ lengths and 1½ lengths by Carmarthen, Fiasco and Fin Bon in the Prix du Prince d'Orange. His stable companion, Astese, had been a useful three-year-old when his victories included the Gran Premio Citta di Napoli, and the Premio Roma in which he beat Marco Visconti and Ciacolesso 1½ lengths and 1 length. As a four-year-old, however, he ran disappointingly; the only time he reached the frame in five outings in Italy and France was when he finished 4½ lengths third to A Tempo, who was conceding him 4 lbs, in the Prix de Reux (15f) at Deauville. He was a distant tenth in the Grand Prix de Deauville and last to Samos III in the Prix la Rochette at Longchamp on 6 September. It was generally thought, therefore, that both Ciacolesso and Astese would find the opposition much too tough in the 1967 Prix de l'Arc de Triomphe.

A much better case could be made out for one of the two Soviet entries Anilin. Since he had finished such a creditable fifth to Sea Bird and Reliance in 1965, Anilin was unbeaten in all but one of his nine outings when he finished runner-up to Behistoun in the Washington D.C. International Stakes. He had been giving Salvo 6 lbs more than weight for age when he beat him in the Preis von Europa in 1966. He had only two races in 1967 before coming again to Paris: he easily won the Budapest Prize in East Berlin on 28 August and in September he beat his three-year-old stable-companion Actach in the Steinradler Rennen. Actach had won the Prize of Moscow and the Grand Prize of Central and Eastern Europe in East Berlin earlier in the season, in addition to three other races in Moscow. Clearly something of an unknown quantity, Actach was overshadowed by Anilin who had already proved himself against international competition and whose starting price of 12 to 1 was not an overestimate of his chance.

There were eleven French three-year-olds in the field for the 1967 Prix de l'Arc de Triomphe and undoubtedly the best of these were Roi Dagobert and the filly, Silver Cloud, both of whom were trained by Etienne Pollet. The Comtesse de la Valdène's Roi Dagobert had been hobdayed as a yearling and so had not been entered for the classics. On his début as a juvenile he won the Prix de Blaison. He did not have the best of runs in the Critérium de Maisons-Laffitte but finished fastest of all and only went down by a length to the very good filly Gazala. He was made favourite for the Grand Critérium but was rather surprisingly beaten a neck by his stable-companion, Silver Cloud, with Taj Dewan a short neck away third, Frontal fifth, Carmarthen sixth and Topyo seventh. On his reappearance on 2 April in the Prix Greffulhe (10½f), Roi Dagobert took the lead entering the straight at Longchamp but was all out to hold Phaeton and Pétrone by half a length and a neck with Topyo and Frontal 2½ lengths and half a length away fourth and fifth. The Prix Noailles, however, which was run in heavy going and over an extra half furlong at Longchamp on 23 April, was a completely different story. The Comtesse de la Valdène's colt again took the lead shortly after the straight was reached but then proceeded to trounce Grandier and Phaeton by 6 lengths and the same in exceptionally impressive fashion. Although probably a better horse in soft going, it was his vastly superior stamina which was principally responsible for the way in which he totally humiliated his rivals. Over the shorter distance of the Prix Lupin on 14 May and on good going, Roi Dagobert was unable to utilise this stamina to the same extent; although he was under strong pressure 400 metres out, he courageously battled on to beat Astec and Fin Bon a short head and half a length. Unfortunately he met with a setback after the Prix Lupin and did not race again till Sunday 8 October. He was, however, generally recognised as the best of his generation in France and there was no question that he would not be able to make full use of his greatest asset, his stamina, in the Prix de l'Arc de Triomphe. His absence of nearly five months from the racecourse cannot have helped his cause but, even so, he was made favourite at 7 to 2 for Longchamp's most prestigious race, which went to underline the immensely favourable impression he had made on those who saw his victory in the Prix Noailles.

Claude Puerari's Silver Cloud had won the Prix de la Massellière in addition to her success in the Grand Critérium from three outings as a two-year-old. She was a little disappointing in the spring: she only finished fourth in the Prix de la Grotte (7½f) at Longchamp on 16 April, and sixth to Gazala in the Poule d'Essai des Pouliches two weeks later. She showed something of her two-year-old brilliance when she ran Gazala and Tidra to half a length and 2 lengths in the Prix de Diane at Chantilly in June. On 3 September Silver Cloud beat Casaque Grise half a length in the Prix de la Nonette. This performance was made to look even better when Casaque Grise beat Percale and Heath Rose a length and a short neck in the Prix Vermeille a fortnight later, with Amoricana sixth and Gazala unplaced. Silver Cloud had sidestepped the Prix Vermeille as the Prix de l'Arc de Triomphe was her principal autumn objective. She certainly seemed back to her best and sure to run prominently even if Roi Dagobert was the stable's major hope.

Baron Louis de la Rochette's Astec, who had run Roi Dagobert so close in the Prix Lupin, looked the best of the other French three-year-old colts. He had started the season by winning the Prix Djebel at Maisons-Laffitte on 31 March, and had then narrowly failed to catch Carmarthen when beaten a short neck in the Prix Daru on 16 April. In the Prix du Jockey Club at Chantilly, Astec was lying third behind the leader, Minamoto, entering the straight. He made his challenge with a furlong to go and had three-quarters of a length to spare over Minamoto on the line with the fast-finishing Taj Dewan 2 lengths away third, half a length ahead of Frontal in fourth place. Phaeton finished seventh, Pétrone eighth, Topyo ninth, and Carmarthen amongst the backmarkers. Astec found the distance of the Grand Prix de Paris too far for him and was beaten 2½ lengths and 4 lengths into third place by Phaeton and Pétrone. Astec had only one more outing before the Prix de l'Arc de Triomphe in the Prix de Chantilly at Longchamp on 19 September. He was only able to manage fourth place, however, 5 lengths behind the subsequently disqualified In Command and Taj Dewan. His trainer, A. Lieux, was not unduly depressed nevertheless; he felt that Astec had been most unlucky in running as he had been hemmed in on the rails at a vital stage of the race and he expected his charge to give a prominent display on Sunday 8 October.

On paper Mme George Courtois's, Taj Dewan, also appeared to hold sound credentials for the Prix de l'Arc de Triomphe. He had won the Prix Montenica at Maisons-Laffitte in March. He had fought back so gallantly when Royal Palace, looking the assured winner, passed him in the 200 Guineas that only a short head separated the pair at the winning post. In the Prix du Jockey Club he had finished like an express train to snatch third place, and he was far from disgraced when he ran Taneb and Nelcius to a short head and 2½ lengths in the Grand Prix de Saint-Cloud with the 1966 Epsom Derby winner, Charlottown, only sixth. Although he was beaten 2 lengths by In Command in the Prix de Chantilly, he was almost certainly in need of the race, which was awarded to him anyway in the stewards' room for the interference he had suffered from the winner. Nevertheless Taj Dewan was permitted to start at the generous odds of 33 to 1 as several good French judges felt he was better suited by a straight course and by ten furlongs.

Baron Guy de Rothschild's Frontal had only finished fifth to Roi Dagobert in the Prix Greffulhe. On 7 May he comfortably won the Prix Hocquart (12f) at Longchamp by 1½ lengths and 1 length from Minamoto and Carmarthen with Phaeton only fifth. If judged on that race alone, Frontal could not be disregarded on Sunday 8 October, but he was only fourth in the Prix du Jockey Club and in his only other race, the Prix de la Côte Normande, he had filled the same position 6½ lengths behind the winner, Topyo, and only a short head in front of Sucaryl. Surprisingly he was to start joint third favourite with Ribocco at 10 to 1 for the Prix de l'Arc de Triomphe; these odds seemed rather flattering to him, especially when compared with those of his Deauville conqueror, Topyo, 82 to 1.

Mme Léon Volterra's Topyo had begun his three-year-old career by running Grandier to a neck in the Prix de Guiche and he was then fourth to Roi Dagobert in the Prix Greffulhe. On 4 May he won the Prix la Force (10f) at Longchamp but was a disappointing ninth in the Prix du Jockey Club. The distance of the Prix d'Ispahan when he was fourth to Caldarello was clearly too short for him. He was, however, most impressive in the Prix de la Côte Normande. He won unchallenged by 4 lengths from Nashwake, to whom he was giving 9 lbs, with Frontal fourth, Sucaryl fifth and the Poule d'Essai des Poulains hero, Blue Tom, only eighth. Lester Piggott rode him in the Prix de Chantilly, but he was a disappointing seventh behind In Command and Taj Dewan. Certainly an erratic performer, there were some who considered Topyo a possible outsider if he could reproduce the form he showed in the Prix de la Côte Normande.

Phaeton and Pétrone had finished 2½ lengths in front of Topyo in the Prix Greffulhe. On 1 May Pétrone had won the Prix d'Iéna at Longchamp but Mme Jean Stern's Phaeton had proved no match for Roi Dagobert in the Prix Noailles. He finished just in front of Pétrone when sixth in the Prix du Jockey Club and beat him 2 lengths in the Grand Prix de Paris when both gained their revenge on Astec. Both Phaeton and Pétrone had only one more outing before the Prix de L'Arc de Triomphe in the Prix Royal Oak at Longchamp in September, when they finished a disappointing fifth and sixth to Samos III. Not merely did both require a longer distance than twelve furlongs, but their poor showing in the Prix Royal Oak did nothing to enhance their chance on 8 October.

A much better prospect seemed to be Mme R. B. Strassburger's Carmarthen who had beaten

Astec in the Prix Daru and had run third to Frontal in the Prix Hocquart. Although he ran very disappointingly in the Prix du Jockey Club, he put up a very creditable performance when he beat Fiasco and Fin Bon a neck and 2½ lengths in the Prix du Prince d'Orange with Ciacolesso fourth.

Major Lionel Holliday's three-year-old filly Heath Rose had won the Imperial Produce Stakes at Kempton on her sole outing as a two-year-old. She was then sent to France to be trained by Topyo's trainer Mick Bartholomew Jr as a three-year-old. Heath Rose needed the race when beaten into third place by Armoricana in the Prix de Minerve at Le Tremblay on 29 July. She then gave a fine performance when only failing by a neck to give the useful English filly, Whirled, 9 lbs in the Prix de la Calonne (8f) at Deauville. On 11 September Heath Rose beat Chèrie Noir a length over nine and a half furlongs at Longchamp. She gave her best display of the season when she attempted to make all the running in the Prix Vermeille but was caught in the final stages and beaten a length and a short neck by Casaque Grise and Percale with Armoricana nearly 8 lengths away sixth and Gazala unplaced. On this running Heath Rose was only a few pounds behind Silver Cloud and her odds of 80 to 1 represented possible each way value.

The other three-year-old filly in the 1967 Prix de l'Arc de Triomphe, Armoricana, had begun her three-year-old campaign in South-West France where she won twice over nine furlongs at Mont de Marsan. In addition to the Prix Minerve, she won a ten-furlong handicap at Saint-Cloud on 27 June, but she had no chance with either Heath Rose or Silver Cloud on her running in the Prix Vermeille and was likely to be outclassed on Sunday 8 October.

Neither of the two French six-year-olds, Dédini and Pot aux Roses had the remotest chance in the Prix de l'Arc de Triomphe. Omar Sharif's Dédini had won the Prix de Porte Maillot (7f) in June but was essentially a miler. He was only running as it was to be the last race of his career. Pot aux Roses had been very flattered when 2½ lengths behind A Tempo and Nelcius in La Coupe de Maisons-Laffitte; in reality his sole claim to fame was to have beaten the mighty Sea Bird's moderate half-brother, Syncom, 2 lengths in a minor event at Maisons-Laffitte in May.

Paul Duboscq's Nelcius, the hero of the 1966 Prix du Jockey Club, had been very unfortunately withdrawn at the eleventh hour from the Prix de l'Arc de Triomphe later that year. He did not prove to be as good a four-year-old as he had been the previous season. Beaten into third place by the fluke winner, Claquesous, in the Prix Boïard in March, Nelcius could only manage seventh place behind Behistoun in the Prix Ganay. He ran Charlottown to 2 lengths in the Coronation Cup at Epsom, and might possibly have been closer had the field not crawled at such a farcical tortoise pace for the first five furlongs. In the Grand Prix de Saint-Cloud Nelcius took the lead at the distance and looked certain to win until Ernest Masurel's Taneb came flying in the final hundred metres and beat him a short head in the last stride. Taj Dewan was 2½ lengths away third and Charlottown only sixth. However, in the King George VI and the Queen Elizabeth Stakes at Ascot Nelcius dropped back with a mile to go and finished a dismal last to Busted. Clearly he was not himself that afternoon; his connections maintained he was suffering from sunstroke, but whatever the cause of this poor performance, it was undoubtedly stones behind his best form. He had one preparatory race before the Prix de l'Arc de Triomphe when he was beaten a head by A Tempo in La Coupe de Maisons-Laffitte (10f) on 22 September. All in all it was difficult to enthuse about Nelcius's chance on Sunday 8 October.

A Tempo, fourth to Bon Mot the previous year, again had a busy season in 1967. He won the Prix Edmond Blanc (8f) at Saint-Cloud on 4 March but then had to wait till his success over Nelcius in September before another victory came his way. After finishing fourth to Behistoun in the Prix Ganay, he lost his form in midsummer. Nonetheless he ran a fine race in the Grand Prix de Deauville when only beaten half a length by Lionel with Arjon a short head away third and Ciacolesso and Astese ninth and tenth. Still, A Tempo normally found one or two opponents too good for him when competing in the highest class and the very best he could hope for was a minor place in the Prix de l'Arc de Triomphe in 1967.

Lionel, third to Bon Mot the previous season, had made no show in the Prix Boïard or in the Prix Ganay, and was easily beaten 2½ lengths and 3 lengths by Bon Mot and Parthian Glance in the Prix de Lutèce on 7 May. He was a distant fourth to Salvo in the Hardwicke Stakes at Royal Ascot, but when he ran the Prix du Cadran runner-up, Pompon Rouge, to a short neck in the Prix de la Pelouse at Chantilly in July, he showed he was coming back to his best. His second consecutive victory in the Grand Prix de Deauville suggested that he might prove the best of the French four-year-olds in the field for the 1967 Prix de l'Arc de Triomphe, but that certainly did not guarantee that Lionel would even do as well as he had done the previous October.

Ernest Masurel's Taneb had run Great Nephew to half a length in the Prix de Pâques at Le Tremblay in the spring but had failed in the Prix Ganay, the Prix d'Harcourt and the Prix Dollar. His best performance in 1967 was his victory over Nelcius in the Grand Prix de Saint Cloud. He had no preliminary race before the Prix de l'Arc de Triomphe after this success which cannot have been to his advantage, as was the fact that he had always run his best races at Saint-Cloud and not Longchamp.

Arjon, the 1966 Prix Henry Delamarre winner, had a reputation of only being at his best in the autumn, although he had been runner-up to Ilix in the Deutsches Derby the previous season. He only had two outings in the Prix de Pâques and in the Prix de Reux at Deauville before he finished third in the Grand Prix de Deauville. Some were of the opinion that Arjon might have finished even closer to Lionel had he had more luck in running. There was certainly little to choose between him and A Tempo, but that did not seem to hold out much hope for him in the Prix de l'Arc de Triomphe.

Mrs Riley-Smith's Parthian Glance remained in France after the 1966 Prix de l'Arc de Triomphe and was trained in 1967 by Mick Bartholomew Jr. However, she failed to reproduce her three-year-old form which had made her the best staying filly of her generation in England. Her best performance in four outings before 8 October had been to run Bon Mot to 2½ lengths in the Prix de Lutèce on 7 May. In a small race at Longchamp over a mile on 19 September many considered her very unlucky, for she was almost brought down. Nevertheless it was difficult to see her taking a hand in the finish of her second Prix de l'Arc de Triomphe.

The ex-provincial horse Carnaval appeared to have an even remoter chance. In the spring he won over ten furlongs at Lyon but his best performance had been to run Dan Kano to a neck in the Grand Prix de Vichy. He did admittedly finish third to Busted and Fiasco in the Prix Henry Foy, but the contemptuous ease with which Stanhope Joel's champion disposed of his rivals meant that their proximity to Busted in no way reflected their ability or lack of it.

The Final Preparations

For the first year for a long time there was no runner from François Mathet's stable in the Prix de l'Arc de Triomphe. Danseur, who had easily won the Prix de Cadran, unfortunately broke down in the Ascot Gold Cup and was retired to stud. The English champion three-year-old, Royal Palace, who had missed the St Leger on account of injury, was also an absentee as he was aimed at the Champion Stakes. Charlottown also had been retired to stud after his failure in the Grand Prix de Saint-Cloud. Both Minamoto and Bon Mot were withdrawn from the field because they fell victim to the cough, as did the Prix de Diane and Poule d'Essai des Pouliches heroine, Gazala, who was scratched four days before the race. On 3 October four horses were found to be coughing at Harry Wragg's Abingdon Place stables in Newmarket, so it was a relieved trainer who dispatched an unaffected Salvo to Paris on Thursday 5 October, the same day as Ribocco left England. Anilin and Actach had arrived in Paris the previous day. They were accompanied by Eugene Dalmatov, the director of the Moscow Hippodrome and the Voshod Stud. It was his first visit to Paris and he described it as 'most enjoyable'. Naturally he was invited to the pre-race dinner at Maxim's. He also revealed that Anilin had not run in the spring in the Soviet Union because he had been standing at stud before going to East Berlin in August. In his final gallop at Chantilly Anilin left the watchers in no doubt about his well-being as he left Actach as if his stable-companion was standing still.

Roi Dagobert did his last serious work on Tuesday 3 October over 6½ furlongs with the sprinter Polytune and was most impressive. The same morning Topyo was noted going very well in his final gallop. Astec also satisfied his connections and appeared to be in peak condition. Neither Pétrone nor Taj Dewan impressed the watchers of their final work. Indeed many French trainers felt that, on account of Roi Dagobert's long absence from the racecourse, either Salvo or Ribocco might well become the first English horse to win the Prix de l'Arc de Triomphe since Migoli. Charles Engelhard, in fact, turned down an enormous offer for his Ribot colt only a short time before the race.

The going on Sunday 8 October was heavy, which was in Roi Dagobert's favour but it was a beautifully sunny and fine day. Inevitably there was a huge crowd at Longchamp with many visitors from England and Ireland. In spite of the fact that there were no runners from the United States, the large number of Americans present illustrated how pre-eminent a position the Prix de l'Arc de Triomphe held in the international racing calendar. 'Bull' Hancock, one of the world's most successful breeders, and his wife, John D. Schapiro, who represented Maryland, Bill Perry and Jack Kennedy from the New York Jockey Club, and Captain and Mrs Moseley, Bill Brady and Jimmy Stewart from Hollywood Park were amongst the throng of personalities in the American racing world who had made the trip across the Atlantic.

The Race

In the preliminaries none of the runners looked better than Ribocco, though the big and well-proportioned Roi Dagobert was much admired. Topyo's trainer, Bartholomew, also revealed that Mme Léon Volterra's colt had been almost brought to his knees in the Prix de Chantilly and added that he would not be 'surprised if he runs well'.

Phaeton was slowly away from the number one stall and lost quite a lot of ground. The first to show in front was Pot aux Roses closely tracked by Heath Rose on the outside followed by Frontal, Sucaryl, Roi Dagobert, Nelcius, Taj Dewan and Anilin. After 400 metres Heath Rose took over the lead from Frontal, Sucaryl, Nelcius, Taj Dewan, Anilin and Topyo. This group was tracked by Arjon, Carmarthen and Pétrone, while both Ribocco and Silver Cloud were at the back of the field a little in front of Phaeton. Approaching the top of the hill by the Porte d'Auteuil the order was little changed, except that Silver Cloud had improved her position into the middle of the field. Heath Rose continued in the lead during the descent to the entrance to the straight and, as the field turned for home, she held a definite advantage over a group consisting of Sucaryl, Frontal, Roi Dagobert and Topyo, with Nelcius, Anilin, Taj Dewan, who was coming under pressure, Silver Cloud and Carmarthen on their heels. Ribocco, meanwhile, had made progress on the rails and was in the middle of the field with Salvo close on his tail. As soon as Heath Rose reached the straight, George Moore dashed Sucaryl into second place and for a brief moment his challenge looked menacing but it was short-lived, and 300 metres out he began to fall back to join Frontal, Anilin and Nelcius who had all begun to weaken and drop out of contention. Pyers on Topyo now successfully challenged the gallant Heath Rose and went into the lead hotly pursued on the outside by Roi Dagobert and Silver Cloud. Lester Piggott, meanwhile, had managed to manoeuvre Ribocco off the rails and began to make up ground very rapidly on the wide outside, while Salvo, who had been tracking him throughout, was beginning a very promising run on the rails. The first to weaken was Silver Cloud followed by Heath Rose a hundred metres out. It seemed only Roi Dagobert or the fast-finishing Ribocco could catch Topyo who had got first run. The lack of a race for five months then began to tell on the French champion, for although he ran on gallantly, the burst of acceleration he showed in the Prix Noailles was missing and seventy yards out Ribocco caught and passed him. As Heath Rose had weakened a few seconds earlier, she veered off the rails and left a gap for the totally boxed-in Salvo. Although he was 3 lengths behind Topyo with a hundred metres to go, Gerald Oldham's colt now produced a devastating burst of acceleration. Ribocco appeared to have used up too much of his energy passing Roi Dagobert and, though catching him with every stride, seemed to be held by Topyo in the last twenty-five metres, but Salvo was travelling faster than any of the three. The gap, however, had come seconds too late and at the winning post Topyo had a neck to spare over Salvo, who snatched second place from Ribocco by a short head. Roi Dagobert was another half a length away fourth, a neck in front of Heath Rose, who was 2 lengths ahead of Silver Cloud. She was followed by Parthian Glance, Arjon, Sucaryl and Nelcius. As soon as the field passed the post, Taillard on Heath Rose was the first to offer his congratulations to Pyers.

Topyo, at odds of 82 to 1, became then the longest-priced winner in the history of the Prix

Topyo (26) beating Salvo (virtually hidden on rails), Ribocco (8) and Roi Dagobert (25) a neck, a short head and half a length with Heath Rose (rails) fifth, Silver Cloud (30) sixth and Parthian Glance (15) seventh

de l'Arc de Triomphe and inevitably the post-mortems began immediately. In reality the moral winner of the race was Roi Dagobert. To produce a colt who had not seen a racecourse since 14 May and finish less than a length behind the winner in a race as competitive as the Prix de l'Arc de Triomphe was a great feat of training on Etienne Pollet's part and said much for the ability of the colt himself. Nevertheless on the day and in the circumstances Topyo had proved the better horse. Lester Piggott said that Ribocco had taken a long time to find his rhythm and that he had to urge him all the way to keep up with the field. In addition he felt that the holding going was against his mount. Certainly at the entrance to the straight Ribocco had been given a great deal to do, but it may well have been the fault of the colt himself rather than Piggott for he had shown himself to possess a mind of his own in the spring. If there had been an unlucky loser it was certainly Salvo, who had fully 9 lengths to make up on Topyo from 300 metres out and then found his way blocked by Heath Rose until she veered off the rails. Hutchinson, who rode Salvo, had had Ribocco in his sights throughout the race and entering the straight had been tucked in on his heels and so had had to make up even more ground on the leaders. In hindsight he might have tried to improve his position a little earlier, but who can say what trouble he would have met if he had managed to extricate himself from the rails? Immediately after the race

Hutchinson reiterated the view of most spectators that if he had found a gap 'only twenty yards earlier, I would have won'. Perhaps, however, the most sporting reflection came from David McCall, Charles Engelhard's racing manager, who commented: 'You would have to live a long time to see a better horserace than that.'

None of these thoughts and comments on what might have been should detract from the excellent feat of training achieved by Mick Bartholomew Jr. He had saddled not merely the winner of the Prix de l'Arc de Triomphe but also Heath Rose and Parthian Glance, who had finished fifth and seventh and who started at odds of 80 to 1 and 130 to 1 respectively. Even after Topyo's disappointing performance in the Prix de Chantilly neither he nor Bill Pyers, his 34-year-old Australian jockey, lost confidence in him and had felt sure he would regain his Deauville form in the Prix de l'Arc de Triomphe. Bill Pyers, for whom it was a first victory in the race, related afterwards that he had tracked Sucaryl all the way from the start and that when he weakened 300 metres out he shot Topyo past him and 'that was that'. He further added, 'I am the happiest man on earth. This is the best moment of my career.' Madame Léon Volterra felt that he was the only jockey who understood Topyo and that Lester Piggott, who rode him in the Prix de Chantilly, did not.

His happiness, however, was short-lived. The following Wednesday a French lawyer Jean-Marc Denis read of Pyers's success. He remembered that the jockey had been the defendant in a case in which he had represented a Mme Monique Geraud de Seguin, who had been injured in a car accident with Pyers in July the previous year. Pyers had not been present at the trial and was sentenced in absentia to a year in gaol. The lawyer naturally wished to know why Pyers was still at liberty. On Thursday 12 October the jockey was detained at his home in Lamorlaye and was first held at the gendarmerie at Senlis, but a little while later was transferred to the Santé prison. The lawyer successfully blocked all money due to him from his victories on the racecourses in France, including that which he should have collected for his victory in the Prix de l'Arc de Triomphe. On Tuesday 17 October a Paris court decreed that the jockey had to remain in custody in the Santé gaol pending an appeal against his sentence. In all, Pyers spent several months in prison as a result of this affair and at the Newmarket Houghton sales Bartholomew said one of the reasons he had come to England was to find a replacement for him as stable jockey. It was the first time that a jockey's success in the Prix de l'Arc de Triomphe resulted in him serving a prison sentence.

It is also rather surprising to relate that Topyo's success was the first time that the famous red and white colours of Mme Léon Volterra had been carried to victory in Longchamp's most prestigious race. In fact it had not been a particularly lucky race for either Mme Volterra or her husband. The latter only once had one of his horses placed – in 1934 when Casterari was runner-up to Crapom. In the year of her husband's death Mme Volterra's Amour Drake was third to Coronation and Double Rose, and the following year, 1950, L'Amiral was third to Tantième and Alizier. But since then none of her runners had reached the frame until Topyo's victory in 1967. Initially she accepted an invitation to run Topyo in the Washington D.C. International Stakes, but changed her mind and he was retired for the season.

The Rest of the Season

Charles Engelhard accepted an invitation for Ribocco to run at Laurel Park nevertheless. It was not, perhaps, the wisest decision since the cold weather, in which the Ribot colt did not thrive at all, had already set in. Ribocco lost ground at the start, an almost insuperable disadvantage on this tight circuit, and eventually only managed seventh place behind Fort Marcy and Damascus. Afterwards Piggott reported that he had never been running at all during the race.

Anilin, who had one of his most infrequent off days on Sunday 8 October, made his final appearance in the Preis von Europa a fortnight later. He totally outpointed Luciano by 4 lengths, with Taneb 5 lengths back in third place. Nevertheless the form of the Prix de l'Arc de Triomphe did not work out as well as in most years. Parthian Glance was only sixth to Bagdad in the Prix du Conseil Municipal with Carmarthen and Arjon further behind. Ciacolesso was beaten 2½ lengths and 3 lengths by Ruysdael and Carlos Primera in the Gran Premio del Jockey Club. In the Premio Roma Pétrone beat Ciacolesso by 2 lengths but was the same distance adrift of the winner, Carolos Primera. In the Champion Stakes Reform beat

Taj Dewan very comfortably by 2 lengths with Royal Palace, who was clearly not at his best, 1½ lengths away third. In the Grand Prix de Marseille Parthian Glance was only third to Monsieur Boussac's three-year-old Alfaro, and Le Potelet.

Pedigree and Stud Record

Topyo remained in training as a four-year-old but failed to win in four outings. He was beaten into fourth place in the Prix Boïard at Saint-Cloud by Traghetto, Bagdad and Pétrone. He was then last to Taj Dewan, Roi Dagobert and Pétrone in the Prix Ganay at Longchamp. In the Grand Prix de Saint-Cloud he could only manage eighth place behind Hopeful Venture, Minamoto and Vaguely Noble. Topyo gave his best performance on his final appearance on a racecourse when he ran Royal Palace and Felicio to three-quarters of a length and a short head in the King George VI and the Queen Elizabeth Stakes at Ascot with the Irish Sweeps Derby winner, Ribero, fourth. Admittedly both Felicio and Topyo were very fortunate to get so close to Royal Palace, because he broke down so badly in the final furlong. In December 1968 Mme Volterra sold Topyo to Japan where he took up stud duties.

The 1967 Prix de l'Arc de Triomphe winner did not meet with any great success as a stallion in Japan, but that may have been partly as a result of his premature death in 1974. He never figured amongst the top twenty stallions in Japan between 1972 and 1979 and none of his progeny won a Group I race. His offspring won a total of seventy-seven races worth 528,923,900 yen.

Topyo's sire, Fine Top by Fine Art, was a useful if not quite top-flight performer on the racecourse. He won sixteen races from thirty-six outings in four seasons. At three he won the Prix de Guiche and the Prix Eugène Adam and finished fourth to Guersant in the Poule d'Essai des Poulains. As a four-year-old his victories included the Prix Boïard, the Prix de la Forêt and the Prix du Rond-Point. He was also successful in the Prix de la Forêt as a five-year-old when he won the Prix du Point-du-Jour and the Prix du Chemin de Fer du Nord. Before Topyo, Fine Top had already established his reputation as a sire for he had got Sanctus (Prix du Jockey Club and Grand Prix de Paris 1963), Fine Pearl (Prix de Diane 1965) and Vinteuil (Grand Critérium, San Siro 1962). Fin Bon won the 1966 Prix Robert Papin and was third to Roi Dagobert in the Prix Lupin. His and Topyo's successes were principally responsible for Fine Top being placed third in the list of sires in France in 1967. He was also the sire of Top Dream who won the Prix de la Côte Normande, like Topyo.

Topyo's dam, Deliriosa by Delirium, won two minor races as a juvenile and was placed twice but as a three-year-old she only managed two places in moderate events. Her first foal, Berto by Umberto, won one small race as a three-year-old, but her next foal, Solios by Soleil Levant, only achieved five places. There was no doubt that Topyo was the best of her offspring.

Deliriosa's dam, La Fougueuse by Admiral Drake, met with no success on the racecourse but was a very good broodmare, for she produced seven winners in addition to Deliriosa. The best of these was Lilya by Clarion, whose victories included the Prix du Moulin de Longchamp, the Prix d'Astarté and the Handicap de la Tamise. Eranhild by Vatellor, the winner of the Prix Reiset and third to Charlottesville and Kirkes in the 1960 Grand Prix de Paris, and La Vigoureux by Tambourine, who won eight races between 1969 and 1972 worth 333,918 francs, were the most notable of her other foals.

La Fougueuse's dam, La Foux by Vatout, never ran but she also was a successful broodmare. She produced four winners, the outstanding of which was Mistral by Admiral Drake who won the Poule d'Essai des Poulains, the Prix Greffulhe and the Prix Lupin in 1945 and later became a good stallion.

La Foux's dam, Quick Change by Hurry On, ran three times unplaced in England where she was bred. Captain Jefferson Davis Cohn imported her into France in 1929. At stud she got six winning foals of nine races amongst which her filly Quick Arrow by Casterari was the dam of Steady Aim, the winner of the Oaks in 1946.

Quick Change in fact arrived in France only two years after her dam, Broderie by Tracery, had been imported by M. H. Corbière at the age of twelve. Broderie ran only four times as a three-year-old but was unplaced in selling company on each occasion. She produced one winner of a small race in England before her departure for France, where she succeeded in getting three individual winners of nine races between 1930 and 1936. Her full sister, Queen's Guild, won six races between 1919 and 1923 in

England, which might have influenced M. Corbière to purchase Broderie.

Broderie traced back to the Old Orange Girl by Kingston, the half sister to Western Australian by Melbourne, the winner of the Derby, the St Leger and the 200 Guineas in 1853 and of the Ascot Gold Cup the following year.

When Topyo retired to stud he had won four races; the Prix de l'Arc de Triomphe, the Prix de la Côte Normande, the Prix la Force and Prix la Flèche and had been placed eight times. He had won a total of 1,447,727 francs. He certainly ranks as one of the most fortunate winners of the Prix de l'Arc de Triomphe, although he was the best horse on the day in the circumstances. Nevertheless in any handicap of the winners of the Prix de l'Arc de Triomphe since 1945, it is certain that he would figure amongst the bottom weights.

PRIX DE L'ARC DE TRIOMPHE 2400 metres
at Longchamp on Sunday 8 October 1967. For three-year-olds and upwards. Going: Good.
1,000,000 francs and 78 per cent of the entry fees to the winner; 400,000 fr. and 12 per cent of the entry fees to the 2nd; 200,000 fr. and 6 per cent of the entry fees to the 3rd; 100,000 fr. and 2 per cent of the entry fees to the 4th. 100,000 fr. to the breeder of the winner; 40,000 fr. to that of the 2nd; 20,000 fr. to that of the 3rd. Value to the winner: 1,119,788 francs.

TOPYO	b.c.	by Fine Top – Deliriosa by Delirium	3.55½k	Mme Léon Volterra	W.B. Pyers	1
SALVO	ch.c.	by Right Royal – Manera by Macherio	4.60k	Gerald Oldham	R. Hutchinson	2
RIBOCCO	b.c.	by Ribot – Libra by Hyperion	3.55½k	Charles Engelhard	L. Piggott	3
ROI DAGOBERT	b.c.	by Sicambre – Dame d'Atour by Cranach	3.55½k	Comtesse de la Valdène	Y. Saint-Martin	4
HEATH ROSE	b.f.	by Hugh Lupus – Cherished by Chanteur	3.54k	Major Lionel Holliday	J. Taillard	5
SILVER CLOUD	ch.f.	by Dan Cupid – Beactive by John's Joy	3.54k	Claude Puerari	J. Massard	6
PARTHIAN GLANCE	b.f.	by Parthia – Tenacity by Wilwyn	4.58½k	Mme W.H.D. Riley-Smith	A. Barclay	7
ARJON	br.c.	by Janitor – Astral II by Watling Street	4.60k	Paul Hilger	M. Philipperon	8
SUCARYL	b.c.	by St Paddy – Sweet Angel by Honeyway	3.55½k	Stanhope Joel	G. Moore	9
NELCIUS	br.c.	by Tenareze – Namagua by Fontenay	4.60k	Paul Duboscq	A. Gibert	10
ANILIN		by Element – Analogichnaya by Agregat	6.60k	Haras Voshod	N. Nasibov	
DEDINI		by Dedicate – The Runner by Olympia	6.60k	Omar Sharif	J.L. Durry	
POT AUX ROSES		by Spoutnik – Rosée by Patachon	6.60k	Anatole Saint-Jean	M. Jathan	
CARNAVAL		by Fast Fox – Cambre by Sicambre	4.60k	Charles Giraudon	Y. Parenti	
CIACOLESSO		by Barba Toni – Ca' Lion by Traghetto	4.60k	Luigi Mantovani	C. Ferrari	
ASTESE		by Barba Toni – Moleca by Traghetto	4.60k	Luigi Mantovani	J. Lindley	
TANEB		by Tapioca – Ténébreuse by Téléferique	4.60k	Ernest Masurel	J.C. Desaint	
A TEMPO		by Tombeur – Lucky Shoe by Pot O'Luck	4.60k	G.L. de Mola	M. Larraun	
WHITE GLOVES		by High Hat – Gallamoud by Mahmoud	4.60k	Mme M.A. Moore	L. Ward	
LIONEL		by Herbager – La Strada by Fervent	4.60k	Mme André Rueff	H. Samani	
SODIUM		by Psidium – Gambade by Big Game	4.60k	R.J. Sigtia	F. Durr	
PETRONE		by Prince Taj – Wild Miss by Wild Risk	3.55½k	Edward Constantine	R. Poincelet	
TAJ DEWAN		by Prince Taj – Devinette by Elbé	3.55½k	Mme Georges Courtois	F. Head	
ASTEC		by Prince Taj – Rush Floor by Guersant	3.55½k	Baron Louis de la Rochette	A. Jézéquel	
FRONTAL		by Le Haar – Faireale by Borealis	3.55½k	Baron Guy de Rothschild	J. Deforge	
DART BOARD		by Darius – Shrubswood by Straight Deal	3.55½k	Michael Sobell	A. Breasley	
PHAETON		by Sicambre – Pasquinade by Vandale	3.55½k	Mme Jean Stern	L. Flavien	
CARMARTHEN		by Devon – Kuwait by Persian Gulf	3.55½k	Mme R.B. Strassburger	G. Thiboeuf	
ACTACH		by Steinadler – Ambitzia by Agregat	3.55½k	Haras Voshod	A. Zekachev	
ARMORICANA		by Bel Baraka – Ardmore by Norseman	3.55½k	Jean Bardon	V. Dupays	

30 runners.
TOPYO trained by C. Bartholomew Jr; bred by Mme Léon Volterra.
Distances: Neck; short head; ½ length; neck; 2 lengths; 1 length; 2½ lengths; 1 length; 2 lengths. Time: 2 mins 38.20 seconds.
Pari-Mutuel: Win: 82fr. 60. Places: 17fr. 70; 4fr. 80; 3fr. 30.
S.P.: 7 to 2 Roi Dagobert; 5 to 1 Ribocco; 10 to 1 Salvo and Frontal; 12 to 1 Anilin and Actach (couplé); 14 to 1 Astec; 19 to 1 Silver Cloud; 22 to 1 Lionel and Phaeton; 27 to 1 Sucaryl; 28 to 1 Pétrone and Carmarthen; 30 to 1 Dart Board; 32 to 1 Arjon; 33 to 1 Taj Dewan and A Tempo; 35 to 1 Nelcius; 45 to 1 Carnaval; 60 to 1 Ciacolesso, Astese and Taneb; 80 to 1 Heath Rose; 82 to 1 TOPYO; 110 to 1 Pot aux Roses and Armoricana; 120 to 1 Sodium; 130 to 1 Parthian Glance; 140 to 1 Dédini; 210 to 1 White Gloves.

TOPYO bay colt 1964	Fine Top br. 1949	Fine Art b. 1939	Artist's Proof ch. 1926	Gainsborough b. 1915	Bayardo
					Rosedrop
				Clear Evidence bl. 1920	Tracery
					Honora
			Finnoise b. 1930	Finglas b. 1923	Brûleur
					Fair Simone
				Unfortunate b. 1911	Marlow
					Cimb
		Toupie b. or br. 1943	Vatellor br. 1933	Vatout b. 1926	Prince Chimay
					Vashti
				Lady Elinor b. 1919	Teddy
					Madame Royale
			Tarentella br. 1934	Blenheim br. 1927	Blandford
					Malva
				Andalusia b. 1925	Alcantara II
					Confection
	Deliriosa b. 1956	Delirium ch. 1945	Panorama ch. 1936	Sir Cosmo br. 1926	The Boss
					Ayn Hali
				Happy Climax br. 1921	Happy Warrior
					Clio
			Passed Out ch. 1935	Solario b. 1922	Gainsborough
					Sun Worship
				Ambrosia b. 1929	Manna
					Rosemead
		La Fougueuse b. 1950	Admiral Drake b. 1931	Craig An Eran b. 1918	Sunstar
					Maid of the Mist
				Plucky Liège b. 1912	Spearmint
					Concertina
			La Foux br. 1935	Vatout b. 1926	Prince Chimay
					Vashti
				Quick Change br. or bl. 1924	Hurry On
					Broderie

Vaguely Noble 1968

Vaguely Noble

The Field

There was a great air of excitement and expectancy at the Newmarket December Sales in 1967 when Brook Holliday's champion two-year-old, Vaguely Noble by Vienna, was led into the sales ring. As a result of the death duties that were incurred on the estate of his father, Major Lionel Holliday, Brook Holliday had found it necessary to dispose of one of the most promising juveniles his father had bred for a great many years. On his début Vaguely Noble had run the more experienced Sweet Thanks, to a neck in the Sandgate Stakes at Newcastle in late August. Vaguely Noble was expected to win the Faversham Stakes at Doncaster in September. The going was fast that afternoon and he was beaten three-quarters of a length by the unraced Saraceno; this disappointment may have been occasioned either by the ground or because Vaguely Noble needed more time. He next ran in the Sandwich Stakes at Ascot in soft going. The official distance of his victory over The Moorings was 12 lengths but many felt that the judge had erred on the side of conservatism. To gauge his true merit Vaguely Noble's final race as a juvenile was in the Observer Gold Cup over a mile at Doncaster. The opposition was formidable: it included the Champagne Stakes winner and runner-up, Cheb's Lad and Lorenzaccio, who had also been placed in the Prix Robert Papin and the Prix Morny;

Riboccare, the winner of the Chesham Stakes and runner-up to Remand in the Royal Lodge Stakes; Hibernian, the Beresford Stakes victor; the highly regarded but unraced Connaught from Noel Murless's stable; and Doon, who had won his last three races impressively. Two furlongs out Williamson on Vaguely Noble found himself boxed in but, riding at his most forceful, he found a passage through and took the lead a furlong from the winning post. From that moment the race turned into a procession and Vaguely Noble, who put more and more daylight between himself and his rivals with every stride, had an astonishing 7 lengths to spare over his nearest pursuer, Doon, on the line. It was a much more impressive display than Ribocco's the previous season and notwithstanding Noblesse's brilliant performance in the same race in 1962, no winner of the Observer Gold Cup had given such a humiliating drubbing to his opponents and in such scintillating fashion. Like Major Holliday's winner in 1963, Pushful, Vaguely Noble was not engaged in the English classics, which may have been another of the reasons for his appearance a little over a month later in the sale ring at Newmarket.

The bidding opened at 80,000 guineas, an immensely high figure at that time and double the previous highest price for a horse in training, 37,500 guineas, paid by M. Edmond Blanc when he bought Flying Fox in 1900. In no time the figure had reached 136,000 guineas at which price it was knocked down to Dr R. A. Franklyn, a Californian plastic surgeon from Hollywood. Incredibly he had only bought his first racehorse in August, four months earlier, but his colours had never been carried in public on the day he acquired Vaguely Noble. He was not destined to race in the United States but was first sent to Ireland to Paddy Prendergast. A short while later the underbidder at Newmarket, Nelson Bunker Hunt, the oil multi-millionaire, came to an agreement with Dr Franklyn to buy a half-share in the colt. It was as a result of his influence that he was then sent to France to be trained by Etienne Pollet, as his principal objective in 1968 was the Prix de l'Arc de Triomphe.

Vaguely Noble made his début in France in the Prix de Guiche over an extended nine furlongs at Longchamp on 21 April. He took command at the distance and beat Semillant and the Prix Boïard winner Taglietto with the utmost ease by 3 lengths and the same. Unable to run in the Prix du Jockey Club as he had not been entered, Vaguely Noble contested the Prix du Lys (12f) at Chantilly the same afternoon. He totally outclassed his rivals; his nearest pursuer, Nightshade, was toiling 8 lengths behind him as he passed the winning post. His first real test, and the first time he met older horses, came in the Grand Prix de Saint-Cloud on 7 July, for which there was a high-class field of twenty. Vaguely Noble's pacemaker, Admiral's Boy, took over the lead after two furlongs and led the field into the straight, but Deforge, his jockey, kept the favourite well back at the rear. He began to make up ground rounding the turn but on the outside. Half-way up the straight Vaguely Noble was still headed by Minamoto, Pola Bella, Taglietto, Hopeful Venture and Topyo, and it began to appear that he had been given too much to do, and thus it transpired. Hopeful Venture managed to reach Minamoto inside the final twenty-five metres and went on to win by a neck. Vaguely Noble, however, who had improved steadily and relentlessly throughout the final two furlongs, found that the post came too soon and was still 2 lengths behind in third place, half a length ahead of Pola Bella. Samos III was 3¾ lengths away sixth, a length and a half ahead of Topyo in eighth place and such good horses as Lionel, Gazala and Grandier even further back. Although Her Majesty the Queen's Hopeful Venture had won the Hardwicke Stakes at Royal Ascot and had been runner-up to Ribocco in the St Leger in 1967 and was certainly no sluggard, the result must have been a disappointment for Vaguely Noble's owners. There were several of the opinion that Deforge had given his mount too much to do in such a large field. He had ridden him in all his three races in France but was never to do so again, a reflection perhaps of his owner's opinion of the way he handled Vaguely Noble at Saint-Cloud. Bill Williamson, who had partnered him to his two successes in England as a juvenile, was engaged to ride him in his next race, the Prix de Chantilly, and in the Prix de l'Arc de Triomphe itself. In the Prix de Chantilly he beat the useful Felicio on a tight rein by 4 lengths, with Danoso a length away third. He was clearly going to take a great deal of beating in the Prix de l'Arc de Triomphe for which he started favourite at the apparently generous price of 5 to 2.

The previous October had, however, also seen the emergence of another potential champion in Paris: Sir Ivor by Sir Gaylord who was owned by Raymond Guest, the United States Ambassador to Ireland. It seemed more than logical, therefore, that Sir Ivor should have been trained by Vincent O'Brien at Baldoyle. On Sunday 15 October, 1967, he was sent to Longchamp to take on the best French two-year-olds in the Grand Critérium. Although only sixth on his début in the Tyros Stakes (6f) at the Curragh, he then had beaten the future Irish 2000 Guineas winner, Mistigo, in the Probationers Stakes (7f) in late July and on 10 September he had trounced the highly rated Candy Cane in the most impressive fashion by 3 lengths in the National Stakes (7f). The Grand Critérium was run on heavy going on a cold and rather wet autumn afternoon. Lester Piggott on Sir Ivor was always going easily and with a pulverising burst of speed Raymond Guest's colt went clear of his rivals 200 metres out. At the winning post he was 3 lengths clear of the hitherto unbeaten Pola Bella with Timmy My Boy third. Pola Bella, a daughter of the brilliant Bella Paola, had earlier numbered the Critérium de Maisons-Laffitte amongst her victories and was unquestionably the best juvenile filly in France. This success stamped Sir Ivor as a very promising classic hope and certainly raised the hopes of his owner that he would land his bet of £50,000 to £500 each way, which he had made earlier that season with William Hill, the famous bookmaker, for the 1968 Epsom Derby.

Sir Ivor was sent to Italy by Vincent O'Brien to escape the rigours of the winter in Ireland, and he had no difficulty in winning the 2000 Guineas Trial at Ascot on his reappearance. In the mile classic at Newmarket, however, he was opposed by the unbeaten English two-year-old champion, Petingo, whose victories had included the Gimcrack Stakes and the Middle Park Stakes. In the latter Lester Piggott had ridden him to victory. Although Petingo was trained by his father-in-law, Sam Armstrong, Piggott had had no hesitation during the winter in opting to ride Sir Ivor in the 2000 Guineas. This judgment could not be faulted: Joe Mercer sent Petingo into the lead coming down into the Dip, but Sir Ivor, whom Piggott had kept at the back of the field in the early stages, again produced a devastating burst of speed and most impressively defeated Petingo by 1½ lengths, with Jimmy Reppin 2½ lengths away third.

Sir Ivor next started 5 to 4 on favourite for the Epsom Derby. Since there was some doubt about his getting the mile and a half, as he was by Sir Gaylord, Piggott held Raymond Guest's colt up. When Jim Joel's Connaught went on five furlongs out, Sir Ivor was only eighth. Connaught was still about 5 lengths clear with a little under two furlongs to go and looked to have slipped his field. But the favourite had made steady progress to take second place and when Piggott asked him for his effort at the distance, the response was another scintillating burst of acceleration, which settled the issue beyond all doubt. He beat Connaught 1½ lengths with Mount Athos 2½ lengths away third. Immediately after he dismounted Piggott declared that he was the best Derby winner he had ever ridden, while Vincent O'Brien rated him superior to Ballymoss.

A very serious rival to Vaguely Noble in the Prix de l'Arc de Triomphe had thus firmly staked his claims. Sir Ivor, however, then proceeded to tarnish his reputation by losing his next three races. Piggott was claimed for Charles Engelhard's Ribero, a full brother to Ribocco, in the Irish Sweeps Derby on 29 June. Liam Ward, therefore, took the mount on Sir Ivor who started at 3 to 1 on. It was an immense shock therefore when Sir Ivor produced nothing when Ward asked him to make his effort a furlong out and Ribero strode away to beat him 2 lengths with Val d'Aoste the same distance away third. Almost certainly the tremendous amount of ground that he had made up on Connaught at Epsom had had a more lasting effect on him than had appeared at the time. It was all the more surprising, therefore, that but seven days later he should take on Royal Palace and Taj Dewan in the Eclipse Stakes at Sandown. It is customary for champions in the United States to run much more frequently than their European counterparts, and possibly it was his owner who encouraged this extremely bold policy. Nevertheless in the United States champions do meet defeat more often as a result and for a trainer to keep a horse in peak condition for a protracted period of time is well-nigh impossible. In the event Sir Ivor ran with his usual courage but was beaten a short head and three-quarters of a length by Royal Palace and Taj Dewan. That afternoon it looked as if he had gone slightly over the top and that he would benefit greatly from a couple of months rest.

Sir Ivor had one preliminary race before the Prix de l'Arc de Triomphe in the Prix Henry Delamarre in late September in which he failed by half a length to give the useful French three-year-old Prince Sao 9 lbs. In the Prix de la Côte Normande, Prince Sao had been runner-up to Semillant, a short head in front of Felicio, and in addition he had been third to Timmy My Boy and Zeddan, the Poule d'Essai des Poulains winner, in the Prix Eugène Adam. His victory in recept of 9 lbs over Sir Ivor, therefore, was not such a catastrophe as some felt it was immediately after the race. Sir Ivor had been off the course for nearly three months and almost certainly required the race to bring him to his peak. Certainly Vincent O'Brien was not despondent and was looking forward to the clash with Vaguely Noble. If there were a genuine doubt about the Epsom Derby hero, it was that he might find the stiff twelve furlongs at Longchamp beyond his stamina.

It must not be thought, however, that the 1968 Prix de l'Arc de Triomphe merely concerned these two colts. The 1965 generation of fillies had thrown up two very good fillies, La Lagune and Pola Bella, and one who was exceptional, Mme R. B. Strassburger's Roselière, who was unbeaten as a three-year-old before she ran at Longchamp on 6 October. She totally outclassed the opposition when she won the Prix Pénélope (10½f) by 5 lengths from La Plagne on 13 April. In the Prix de Diane at Chantilly on 16 June Roselière ran on gamely to account for Pola Bella by a length, with La Lagune out of the first ten. Earlier that spring Pola Bella had won the Poule d'Essai des Pouliches and the Prix de la Grotte. She was later a good fourth in the Grand Prix de Saint-Cloud and in the Prix de la Nonette she gave 7 lbs and a head beating to the Nassau Stakes winner Hill Shade, all of which performances reflected credit on Roselière. This trio clashed again in the Prix Vermeille at Longchamp in September: Roselière confirmed her superiority in a most convincing fashion beating Pola Bella and La Lagune 4 lengths and 2 lengths with the Irish Guinness Oaks winner, Celina, in the rear. On this performance Roselière represented a very distinct threat to Vaguely Noble and Sir Ivor; she started third favourite for the Prix de l'Arc de Triomphe at 6 to 1 and it would have come as no major surprise if she were to defeat them both.

La Lagune had been a very good filly in the spring. Unbeaten in two outings as a juvenile, she had effortlessly won the Prix Vanteaux (9½f) by 5 lengths on 21 April. In the Oaks at Epsom her 5-length trouncing of Glad One, the runner-up to Front Row in the Irish 1000 Guineas, had been even more impressive. Her failure in the Prix de Diane only a fortnight later was almost certainly because the race had come too soon for her. In her only other outing before 6 October in the Prix Vermeille, she showed a glimpse of her brilliance in the spring but was so comprehensively defeated by Roselière that the best she could hope for was one of the minor honours in the Prix de l'Arc de Triomphe.

For the first time in many years it appeared that there might be a possibility of Germany winning Longchamp's most prestigious race. In 1967 Luciano had proved himself to be the best German colt since Orsini; he won six races including the Deutsches Derby and the Deutsches St Leger. As a three-year-old he only met defeat twice: once at the hands of Salvo by three-quarters of a length in the Grosser Preis von Baden, and then when runner-up to Anilin in the Preis von Europa. As a four-year-old Luciano was unbeaten in four races. He won the Grosser Preis von Nordrhein-Westfalen at Düsseldorf, but unquestionably his best victory was when he beat the useful English four-year-old Chicago by 3 lengths with Samos III only fourth in the Grosser Preis von Baden. Chicago had won three races earlier that year in England including the Henry II Stakes at Sandown, and he was to beat Park Top 4 lengths in the Cumberland Lodge Stakes at Ascot on 26 September. Certainly Luciano could not be left out of the reckoning on 6 October even if he was taking on Europe's best and was something of an unknown quantity, and his starting price of 15 to 2 reflected the confidence of his connections.

Mme Courtois's Dhaudevi also had done little wrong as a three-year-old: in six outings he had been successful four times and runner-up twice. In the Grand Prix de Paris he had beaten the Prix du Jockey Club winner, Tapalque, by 2½ lengths, though the latter might have finished closer if his jockey, Yves Saint-Martin, had not lain so far out of his ground entering the straight. Nevertheless Dhaudevi gained another impressive success when he beat Torpid and Levmoss 4 lengths and 3 lengths in the Prix Royal Oak. Levmoss had defeated Canterbury and Park Top in the Oxfordshire Stakes at Newbury in August which made this victory by Dhaudevi even more praiseworthy. The only doubt about him was that twelve furlongs might be a little on the short side for him.

Baron Guy de Rothschild's Luthier had only been fourth to Sir Ivor in the Grand

Critérium in October 1967, but had proved himself to be a very useful colt the following spring. Although only fifth to Val d'Aoste on his reappearance in the Prix Greffulhe, Luthier then ran out a decisive 3-length winner from Soyeux in the Prix Noailles on 28 April with Val d'Aoste only sixth. In the Prix Lupin Luthier took the lead with 200 metres to go and easily held the challenge of Felicio by 1½ lengths, with Soyeux another 2¾ lengths back in fourth place. Baron Guy de Rothschild's colt, therefore, started at odds-on for the Prix du Jockey Club, but he produced nothing when asked to make his challenge two furlongs out and finished a disappointing 5½ lengths sixth to Tapalque. The feeling at Chantilly that afternoon was that Luthier did not truly stay twelve furlongs. He reverted to a mile on his only other outing before the Prix de l'Arc de Triomphe and had no trouble in defeating Locris 4 lengths in the Prix Jacques le Marois at Deauville in August. The firmly held and widespread conviction that twelve furlongs was too far for him, however, was the reason why he was permitted to start at odds of 11 to 1 on 6 October, despite the fact that he was undoubtedly a very high-class colt.

Daniel Wildenstein was doubly represented in the 1968 Prix de l'Arc de Triomphe by the four-year-old Pétrone and the year younger Felicio. As François Mathet was running Pola Bella in the Prix du Moulin de Longchamp, there were no runners from his stable in the Prix de l'Arc de Triomphe. Thus Pétrone's trainer, Maurice Zilber, was able to engage Yves Saint-Martin for him. That was certainly one of the principal reasons why these two stable-companions were coupled in the betting at the seemingly ungenerous odds of 5 to 1. Pétrone began his four-year-old campaign by finishing half a length and 2½ lengths behind Taglietto and Bagdad in the Prix Boïard on 23 March at Saint-Cloud. In the Prix Ganay he ran Taj Dewan and Roi Dagobert to a short neck and 3 lengths with Carmarthen over 10 lengths away sixth. Nevertheless he could only manage third place, 6 lengths and a neck behind Carmarthen and Franc Castel in the Prix d'Harcourt (10½f) at Longchamp on 5 May. Pétrone was not disgraced when beaten a length and 2½ lengths by Pardallo and Samos III in the Ascot Gold Cup nor when he was third in the Grand Prix de Deauville, a short head and a length behind Soyeux and Blazer with Tit Bits unplaced. Pétrone gained his first success of the season when he beat the Prix Dollar winner, Grandier (gave 9 lbs), 8 lengths in the Prix Henri Foy with Arjon 3½ lengths away fourth and Carmarthen (gave 9 lbs) another 5 lengths back seventh. In the Prix du Prince d'Orange Pétrone demonstrated that this runaway victory was no fluke for, only receiving 2 lbs from Grandier, he beat him by 3 lengths. Certainly Pétrone had come to his best at the right time, but on a line through Grandier, who had finished out of the first ten in the Grand Prix de Saint Cloud, he still came out several pounds behind Vaguely Noble, Sir Ivor and Roselière.

Daniel Wildenstein's Felicio had run consistently throughout the season but surprisingly had not won a single race in 1968. On paper his best performance had been to run Royal Palace to three-quarters of a length in the King George VI and the Queen Elizabeth Stakes at Ascot in July with Topyo third and Ribero fourth. His proximity to Royal Palace was, however, more than flattering as Jim Joel's four-year-old champion broke down on his near foreleg in the straight and was never to race again. Nevertheless Felicio had been runner-up, 2½ lengths behind Timmy My Boy, in the Prix de Fontainebleau (8f) at Longchamp on 15 April and had occupied the same position behind Luthier in the Prix Lupin. He made no show in either the Poule d'Essai des Poulains or in the Prix du Jockey Club and was only third, 4 lengths and a short head behind Semillant and Prince Sao in the Prix de la Côte Normande at Deauville. It seemed very likely, therefore, that Felicio would find several of his rivals too good for him on 6 October and certainly the odds against the Wildenstein pair did not represent value.

Mme Strassburger's Carmarthen, who had won the Prix du Prince d'Orange in 1967, was a heavy going specialist and had been very lightly raced in 1968. On his seasonal reappearance in the Prix Ganay the going had been good and he had only finished sixth to Taj Dewan. The going for the Prix d'Harcourt on 5 May was heavy. Carmarthen led throughout to beat Franc Castel very convincingly by 6 lengths with Pétrone in receipt of 7 lbs a neck away third. His only other outing before the Prix de l'Arc de Triomphe was four months later in the Prix Henri Foy on 1 September when the going was good. He was beaten 16½ lengths into seventh place by Pétrone who was receiving 9 lbs, but with so many factors against him Carmarthen's chances on 6 October should not be judged on that run alone. In the Prix de l'Arc de Triomphe he was to have the heavy going that suited him so much and he was sure to give a much better account of himself then he had done in the Prix Henri Foy. Indeed, if judged on his facile victory in the Prix d'Harcourt, his odds of 90 to 1 seemed as generous as those of Pétrone at 5 to 1 seemed cramped.

There was one challenger from England for the 1968 Prix de l'Arc de Triomphe, Charles Engelhard's Ribero. Like his full brother Ribocco, Ribero had not shown his best form as a three-year-old until June. His enormously surprising victory over Sir Ivor and Val d'Aoste by 2 lengths and the same made him a couple of pounds better than the Prix du Jockey Club winner Tapalque. Unfortunately he unseated his jockey on the way to the start of the King George VI and the Queen Elizabeth Stakes and covered nearly a mile before he was eventually recaptured. It was to his credit, therefore, that he was able to finish fourth to Royal Palace, Felicio and Topyo. In the Prix Kergorlay at Deauville on 4 August, however, he ran totally without distinction finishing at least 11 lengths behind Pardallo in eighth place. Ribero redeemed his reputation somewhat in the St Leger at Doncaster: he took the lead a furlong and a half out and ran on courageously to hold the fast-finishing Canterbury by a short head. On the other hand, the field for the St Leger in 1968 appeared to be sub-standard and there was little confidence in England that Ribero would repeat his Curragh victory over Sir Ivor or even reach a place at Longchamp on Sunday 6 October.

There were three other French three-year-olds in the field for the 1968 Prix de l'Arc de Triomphe: Madame René Gribe's Soyeux, Marcel Boussac's Danoso and G.L. de Mola's Tit Bits, the best of which was Soyeux. He had run consistently throughout the season finishing fourth in the Prix Lupin, the Prix du Jockey Club and the Prix Royal Oak, in addition to his second place behind Luthier in the Prix Noailles. His best performance had been to win the Grand Prix de Deauville by a short head and a length from Blazer and Pétrone, but on balance it appeared likely that Soyeux would find several of his opponents too good for him in Longchamp's most prestigious race.

It was very difficult, however, to give Danoso much chance on Sunday 6 October. Unraced as a juvenile he had been a very impressive winner over 10½ furlongs at Longchamp in April but was then beaten into second place in the Prix la Force in May. Such was his reputation at home that he started second favourite in the Prix du Jockey Club but he made absolutely no show at all at Chantilly. Admittedly he was only beaten 4 lengths and a neck by Vaguely Noble and Felicio in the Prix de Chantilly at Longchamp in September, but of course Vaguely Noble won in a half-canter.

G.L. de Mola's Tit Bits, who won the Prix des Marronniers in April, appeared totally outclassed in the Prix de l'Arc de Triomphe. In May he beat Riboccare and Apipe a short neck and 2 lengths in the Prix d'Ièna (12f) at Longchamp but was relegated to third place for interference to the runner-up. Tit Bits subsequently ran with no distinction at all in five outings including the Prix du Jockey Club and the Grand Prix de Deauville; his lowly sixth place in the Grand Prix de Clairefontaine one would usually associate with a moderate pacemaker for one of the favourites for the Prix de l'Arc de Triomphe.

There was again a challenger from the Soviet Union, the three-year-old Zbor who had won five races from seven outings at distances ranging from nine to fourteen furlongs in Moscow. He was, however, beaten into second place by his stable-companion Zagreb, who was withdrawn from the Prix de l'Arc de Triomphe a few days before the race, in a minor event at Longchamp in September. There was no way that Zbor could be considered another Anilin and it would have been a great surprise if he even managed to reach the first six.

Comtesse Margit Batthyany's Samos III, the 1967 Prix Royal Oak winner, was without a victory in 1968. In the Prix de Barbeville (15½f) in April he ran Pardallo and Lingot to short necks; he was also runner-up to Pardallo in both the Prix Jean Prat and the Ascot Gold Cup. In the Prix Kergorlay he was only third to Pardallo and he was only sixth to Hopeful Venture in the Grand Prix de Saint-Cloud, when Vaguely Noble finished 4 lengths in front of him. Samos III certainly needed a distance in excess of twelve furlongs to show his best and his slightly disappointing performances over longer distances in 1968 appeared to rule him totally out of contention in the Prix de l'Arc de Triomphe.

The field was completed by Paul Hilger's, Arjon. He had won the Prix de Bois Roussel (11f) in June and had beaten Taneb in the Prix de la Pelouse (12f) at Chantilly in July. He was, however, a well-beaten fourth to Pétrone in the Prix Henri Foy, and whenever he competed in the highest class, Arjon invariably found several of his opponents too good for him.

The Prix du Jockey Club winner, Tapalque, had run so disappointingly when last to Dhaudevi in the Prix Royal Oak that his trainer François Mathet immediately scratched him from the Prix de l'Arc de Triomphe. The third at Chantilly and the Curragh, Val d'Aoste, was another absentee after his lack-lustre display in the Prix Henry Delamarre. The Ascot Gold Cup winner, Pardallo, was withdrawn a few days before the Prix de l'Arc de Triomphe as he was unsuited by heavy or soft going. It had been intended to run the Marchese Mario Incisa della Rocchetta's Ruysdael, who had won the Derby Italiano, the Gran Premio d'Italia and the Gran Premio del Jockey Club in 1967, but he did not satisfy his connections in his preparation and was withdrawn in September. The Prix Ganay winner, Taj Dewan, also missed the Prix de l'Arc de Triomphe as his autumn objective was the Champion Stakes.

The Final Preparations

Vaguely Noble's jockey, the Australian Bill Williamson, flew to France on Monday 1 October and partnered his mount on his final gallop at Les Aigles the following morning. Etienne Pollet afterwards expressed himself more than satisfied with its outcome and was very optimistic about his charge's prospects in the Prix de l'Arc de Triomphe. The same day Luthier did his final work more than satisfactorily with Taj Dewan. La Lagune did her final gallop with Carmarthen; the four-year-old colt came out much the best and François Boutin, La Lagune's trainer, was in two minds about running the Oaks winner, being finally persuaded to do so by her owner, Henry Berlin. George Bridgland was extremely pleased with Roselière's condition, and a few days before her participation in the Prix de l'Arc de Triomphe felt sure that she would run very well. It was, in fact, reported that Mme François Dupré, who owned Pola Bella, had made a big offer for Roselière which was turned down.

The Race

In the preliminaries on Sunday 6 October the powerful Vaguely Noble looked magnificent and he and Sir Ivor stood out on looks. Unfortunately the German challenger Luciano appeared very nervous and fretful during the parade and pulled hard all the way to the start where Samos III also began to play up a little. As the runners were being loaded into the stalls, Damoso gave Ribero a kick. His jockey, Brian Taylor, dismounted but seemed satisfied that there was nothing serious wrong.

Deforge sent his mount Luthier immediately into the lead from Roselière as soon as the stalls opened, and they were followed by Pétrone, the Russian challenger Zbor and Dhaudevi. They were tracked closely by Vaguely Noble, Arjon and Danoso, with Soyeux and the hard-pulling Luciano just behind, and Sir Ivor, Carmarthen and La Lagune among the backmarkers. Shortly before the Petit Bois, Zbor was pushed up to dispute the lead with Luthier ahead of Roselière, Pétrone, Vaguely Noble and Arjon. After the Petit Bois Luthier again took a definite lead ahead of Roselière, and Sir Ivor began to improve his position into the first half-dozen. At the Porte d'Auteuil, the top of the hill, Luthier still led from Roselière tracked by Luciano and Soyeux together, and Vaguely Noble, Arjon, Zbor, Sir Ivor and Dhaudevi who were closely bunched. Felicio, La Lagune and Danoso were still in touch, but Tit Bits was already falling well behind the others. Luthier and Roselière remained at the head of affairs coming down the hill but Zbor quickly faded out of the leading group, and well before the straight was reached Soyeux began to drop out of contention. On the turn both Arjon and Luciano were clearly done with, and Luthier led Roselière into the straight, closely tracked by Vaguely Noble and Sir Ivor, who was followed by Dhaudevi, Felicio and the improving La Lagune and Carmarthen. Williamson immediately made his effort on Vaguely Noble and smoothly caught the two leaders. Deforge had slowed the pace in the middle stages on Luthier in order to try to conserve Luthier's stamina, so it was essential that Williamson made his move as soon as the straight was reached. Ridden with hands and heels only, Vaguely Noble quickly opened up a gap of 3 lengths and it was only Sir Ivor who was able to go after him. Luthier, whose stamina limit was reached,

Vaguely Noble (10) has 3 lengths advantage over Sir Ivor (12)

was back-pedalling rapidly whilst Roselière was plugging on at one pace. The burst of speed that Vaguely Noble had produced to go clear of the field was quite breathtaking. Sir Ivor at the same time unleashed a run in pursuit of the favourite that rapidly took him 2 lengths clear of Roselière but it was evident that he was making no impression on the leader. Two hundred metres out it was apparent that he was not going to catch Vaguely Noble who still had not been touched with the whip: the pair continued to put daylight between them and the remainder of the field, but so easily was Vaguely Noble going that 100 metres from the line Williamson was able to glance casually behind at his floundering rivals.

At the winning post he was 3 lengths clear of Sir Ivor, who was 4 lengths ahead of his nearest pursuer. Indeed from the beginning of the straight there had seemed no doubt about the first two places and the race had then only concerned the two minor honours. Roselière had run on gamely all the way up the straight but was deprived of third place by the four-year-old Carmarthen close to home. Carmarthen, who had been among the backmarkers entering the straight, eventually began to get into top gear with two furlongs to go and had finished so fast that he was 1½ lengths ahead of Roselière on the line, with La Lagune who also finished well another half a length back fifth and Dhaudevi a nose behind her sixth. They were followed by Felicio, Samos III, Luthier and Soyeux. Not one of their opponents, however, had ever threatened either Vaguely Noble or Sir Ivor in the straight, and, well as he ran, Sir Ivor had never looked a serious danger to the winner. Doctor Franklyn's gamble of paying 136,000 guineas for the Vienna two-year-old ten months before had, therefore, paid handsome dividends.

Unquestionably Vaguely Noble's performance compared favourably with those of Sea Bird, Ribot and Exbury in the previous thirteen years. There had been seven individual winners of eleven classics in the field and he had treated them all with the utmost disdain. Immediately after the race Williamson stated that he knew 'I had won the race after half a mile'. Lester Piggott declared that it was the best race that Sir Ivor had ever run but that Vaguely Noble was 'a better horse'. He added, 'I was with the winner all the way but he just ran away from me in the straight.' Roselière's jockey, Y. Josse, said that when Vaguely Noble and Sir Ivor made their efforts, 'I felt as if Roselière was standing still. My filly ran well but there is a difference in class between her and the winner and runner-up.' J.L. Durry, Carmarthen's jockey, felt he might have been a little closer to Sir Ivor if his mount had got into top gear a little earlier, while Freddie Head, who had ridden Dhaudevi, thought he might have been closer than sixth if there had been a stronger pace. Nevertheless, considering the dead going, Vaguely Noble's time of 2 minutes 35·2 seconds was good. The fact that Pola Bella easily beat Agy in the Prix du Moulin de Longchamp merely paid tribute to Vaguely Noble's prowess. While his victory was yet another example of Etienne Pollet's mastery of the art of training thoroughbreds, it was not entirely a Gallic triumph. Vaguely Noble's breeder, Major Lionel Holliday, who had contributed so much to English breeding in the previous thirty years, had achieved posthumously his greatest success. The afternoon had in fact begun with another English victory when Be Friendly, owned by the doyen of English race commentators, Peter O'Sullevan, had landed the Prix de l'Abbaye de Longchamp. The only disappointment of the 1968 Prix de l'Arc de Triomphe was that the very good German invader, Luciano, had become so fretful before the race that, as his jockey Ron Hutchinson afterwards reported, he was unable to show his best form, though even if he had done so he would not have threatened either of the first two.

The Rest of the Season

Almost inevitably the form of the 1967 Prix de l'Arc de Triomphe worked out very well. Sir Ivor took the lead a furlong and a half out in the Champion Stakes at Newmarket on 19 October. So easily was he going that Lester Piggott was able to ease him before the winning post and yet he still beat Locris and Candy Cane 2½ lengths and 2 lengths with Taj Dewan a little over 2 lengths away fifth. Sir Ivor later ran in the Washington D.C. International Stakes at Laurel Park. With a furlong to go the two very useful American horses Czar Alexander and Fort Marcy were disputing the lead in front of Carmarthen while Lester Piggott on Sir Ivor appeared boxed in on the rails in fourth place. Fortunately Carmarthen, under pressure, began to drift away from the rails towards the centre of the course and a gap for Sir Ivor appeared. He again produced that brilliant acceleration he had shown at Epsom. Lester Piggott shot him through the gap and he caught Czar Alexander and Fort Marcy with fifty metres to go. He won going away by three-quarters of a length from Czar Alexander with Fort Marcy and Carmarthen a nose and 1¼ lengths away third and fourth. La Lagune was another 4 lengths back in fifth place, 5 lengths in front of Pétrone in sixth. After the race there was widespread criticism of Lester Piggott's tactics in the American racing press; it was quite unjustified for Sir Ivor had to be held up for a late run when running over a mile and a half. The Americans were just not used to the tactics of European jockeys. Nonetheless the outcome of the race merely reflected credit on Vaguely Noble. On 27 October another Prix de l'Arc de Triomphe failure, Samos III, beat the winner of the Cesarewitch Handicap, Major Rose, 2½ lengths in the Prix Gladiateur. Only Felicio, who did not reach the frame behind Zamazzan in the Prix du Conseil Municipal, failed to advertise the form of the Prix de l'Arc de Triomphe.

Pedigree and Stud Record

Vaguely Noble did not race again after his triumph on 6 October, but was retired to stud in the United States. His sire, Vienna by Aureole, who had been owned by Sir Winston Churchill, had been a good if not quite top-flight performer. As a three-year-old he won the Blue

Riband Stakes at Epsom and the Red Deer Stakes at Ascot and finished third to St Paddy and Die Hard in the 1959 St Leger. He won three races as a four-year-old, but undoubtedly his best two performances that season were to run Petite Etoile to a neck in the Coronation Cup at Epsom and St Paddy to a neck in the Hardwicke Stakes at Royal Ascot. As a five-year-old Vienna's two victories included the Prix d'Harcourt and he was runner-up to Misti in the Prix Ganay in front of Match and Taine. He was at stud in England from 1963 to 1968 when he was exported to France. As a result of Vaguely Noble's victories in 1968 Vienna was fourth in the list of sires in France in 1968. In 1967 he had been the leading sire of two-year-olds in England for, in addition to Vaguely Noble, his 1965 crop included Kursaal who won nine races. Nevertheless Vaguely Noble was unquestionably the best of all his progeny.

On his retirement to stud Vaguely Noble had won six of his nine races, had finished runner-up twice and third once. As a two-year-old he won £18,443 and as a three-year-old 1,544,150 francs. Vaguely Noble's dam, the attractive Noble Lassie by Nearco, was a useful filly as a three-year-old. As a juvenile she ran six times in the highest class competing in the Queen Mary Stakes and the National Breeders Produce Stakes when she was unplaced and in the Lonsdale Produce Stakes at Epsom when she finished fourth. As a two-year-old she almost certainly found the distances of these races too short. As a three-year-old she ran eight times and won twice and was placed twice. In August she won a twelve-furlong handicap at Ripon, but unquestionably her best performance had been when she led throughout to win the Lancashire Oaks over a mile and a half at Haydock in May 1959. Noble Lassie's success as a broodmare was, of course, ensured by Vaguely Noble. Her first foal, Glitter Gold by Gratitude, was placed twice in minor events at Haydock and Stockton in 1963, was then sold for 2,600 guineas at the Newmarket December Sales and was placed once in France in 1964. Her second foal, Attractive by Acropolis, won a thirteen-furlong handicap at Alexandra Park in June and was placed second in a maiden race at Catterick from four outings as a three-year-old. Noble Lassie gave birth to dead twins in 1963 and was barren in 1964. In 1965 she produced Vaguely Noble but was barren to Proud Chieftain in 1966 and her 1967 colt by No Argument was a maiden. Her 1970 foal, Regal Lady by Relko, won two small races at Ayr and Nottingham over thirteen furlongs in 1973 from seven outings at two and three. Regal Lady became the dam of the very useful filly Britannia's Rule by Blakeney, who beat Formulate in the Lupe Stakes at Goodwood before she ran third to Scintillate and Bonnie Isle in the 1979 Epsom Oaks. Another of Noble Lassie's foals, Vive La Reine, a full sister to Vaguely Noble, won a twelve-furlong event in France as a three-year-old but more importantly became the dam of the very good two-year-old R.B. Chesne, whose victories included the Laurent Perrier Champagne Stakes at Doncaster, the Washington Singer Stakes at Newbury and the Sandwich Stakes at Ascot in 1978. Unfortunately injury curtailed his three-year-old campaign when he only ran once, a very creditable length runner-up to the 2000 Guineas hero Tap on Wood in the Kiveton Park Steel Stakes (7f) at Doncaster in September 1979.

Noble Lassie's dam, Belle Sauvage, was bred by Lady MacDonald-Buchanan who sold her to Major Holliday for 3,700 guineas at the First October Sales at Newmarket in 1950. As a juvenile she showed no signs of merit in six outings but as a three-year-old in 1952 she won the Chorley Maiden Plate (12f) at Manchester in July, and was placed in a maiden fillies plate at Thirsk over a mile and in a ten-furlong handicap at Doncaster from eight outings. She proved to be a much better broodmare. Her second foal, Pandour by Petition, was runner-up to Aggressor in the Solario Stakes and second in the Foxhall Stakes at Goodwood in addition to finishing sixth in the Gimcrack Stakes in 1957. As a three-year-old he won the Dee Stakes at Chester, a maiden race at Windsor and a handicap at Lingfield. He did not win in ten attempts as a four-year-old but was placed eight times. As a five-year-old he won a minor event at Brighton and also was placed in the Chesterfield Cup at Goodwood and in the Arthur Lorraine Handicap at Sandown. He was exported to New Zealand where he became a successful stallion.

Belle Sauvage's foals, Master Gay by Narrator, and Countessa by Chanteur II, were only moderate; the former won an eight-furlong selling plate as a three-year-old at Worcester from sixteen outings, while the latter's sole success came in a six-furlong maiden event at Stockton. Belle Sauvage's 1959 foal Naturalist by Narrator, however, was a different proposition. As a juvenile he won the Carrs Stakes over six furlongs at Stockton and put up a very creditable display when fourth to Hidden Meaning, who was in receipt of 8 lbs in the Solario Stakes at Sandown. Although he did not win in six outings as a three-year-old, he was not disgraced when fourth to Pindari in the Lingfield Derby Trial and when again fourth to Pavot in the Queen's Vase at Royal Ascot. In all, six of Belle Sauvage's ten foals won either in the United Kingdom or on the continent.

Belle Sauvage's dam, Tropical Sun, was a more than useful filly. As a two-year-old she was placed on each of her three outings. As a three-year-old she won three of the five events in which she ran and was placed once. In May 1943 she won the Weyhill Plate (8f) at Salisbury and in June the George III Plate (10f) at Windsor. Later that month she ran Why Hurry and Ribbon to a neck and a length in the New Oaks Stakes at Newmarket. In August she won the Gosport Stakes (10f) at Salisbury and her only failure was when she was unplaced to Herringbone in the New St Leger Stakes at Newmarket in September. Tropical Sun was also successful at stud. Her winning progeny, in addition to Belle Sauvage, included Roaring Forties by Blue Peter who won the Tudor Stakes (8f) at Sandown and was third over ten furlongs in May 1948 at Newmarket to Riding Mill and Black Tarquin, and Brimstone by Dante, the winner of two races as a three-year-old over thirteen furlongs at Birmingham and fourteen furlongs at Salisbury before he was exported to Australia as a stallion, where he sired the winner of the Melbourne Cup, Baystone.

Tropical Sun's dam, Brulette, was an exceptionally good filly. Owned by Lt. Col. C. V. Birkin, she won over seven furlongs on her second racecourse appearance at Saint-Cloud in November 1930. On her début as a three-year-old she won the Prix Pénélope by 5 lengths and then took the Prix la Rochette (11f) at Longchamp very easily. In the Epsom Oaks she appeared boxed in a furlong out but in a very exciting finish beat Jour Course and Links Tor a length and three-quarters of a length. In the Prix Vermeille Brulette was beaten 3 lengths by her stable-companion Pearl Cap, who again defeated her into fourth place in the Prix de l'Arc de Triomphe. As a four-year-old she won the Prix du Cadran by a neck from Brûledur and was then bought for about five thousand guineas from the executors of Lt. Col. Birkin by Lord Woolavington. She left the stables of Frank Carter and was trained in England by Fred Darling at Beckhampton. Although beaten by Brown Jack in the Queen Alexandra Stakes at Royal Ascot, Brulette gained her revenge when she beat that illustrious stayer by 4 lengths in the Goodwood Cup. She was not as good as a five-year-old and was unplaced in four outings including the Ascot Gold Cup and the Goodwood Cup. At stud Brulette got seven individual winners: the best was undoubtedly Tropical Sun, but her colt Stockade by Big Game won five events from twelve outings and also ran third to Sayajirao in the Lingfield Derby Trial.

Brulette's dam, Seaweed by Spearmint, was an enormously important broodmare amongst whose progeny was Hotweed (Prix du Jockey Club). Lt. Col. Birkin sent Seaweed from England to his stud in France in 1924.

At stud Vaguely Noble has certainly been a great success, though his progeny have done much better in Europe than those that remain in the U.S.A. Dahlia won the 1973 Irish Guinness Oaks and the King George VI and the Queen Elizabeth Stakes as a three-year-old and again the following year. She also won the Benson and Hedges Gold Cup, the Prix Saint-Alary and the Prix Niel. In North America Dahlia competed in the autumn as a four-year-old and was nominated as grass horse of the year after her victories in the Canadian International at Woodbine and the Washington D.C. International Stakes. No other mare, not even her great rival Allez France, who beat her in the Prix de Diane, the Prix de l'Arc de Triomphe and in the Poule d'Essai des Pouliches, won as much prize-money as she. Almost as prolific a winner by Vaguely Noble was Exceller whose victories included the Grand Prix de Paris, the Prix Royal Oak, the Grand Prix de Saint-Cloud, the Coronation Cup, the Canadian International and other good races in California. Vaguely Noble's stock trained in France have included Mississippian (Grand Critérium, 2nd Prix Lupin, Poule d'Essai des Poulains and Observer Gold Cup); Gay Mecene (Grand Prix de Saint-Cloud, Prix Eugène Adam, 3rd King George VI and the Queen Elizabeth Stakes); Empery (Epsom Derby, 2nd Irish Sweeps Derby and 3rd Prix Lupin); Nobiliary (Prix Saint-Alary, Washington D.C. International Stakes, Prix de la Grotte, 2nd Epsom Derby); Ace of Aces (Sussex Stakes, Prix du Chemin de Fer du Nord, Oettingen Rennen, 2nd Prix de l'Abbaye de Longchamp); Vagaries (Prix Juigné and Grand

Prix d'Evry); and Corvaro. In England and Ireland his stock have included Noble Decree (Observer Gold Cup, 2nd 2000 Guineas); Inkerman (Joe McGrath Memorial Stakes and Sunset Handicap, California); Noble Saint (Yorkshire Cup); Valour (Grand Prix d'Evry); and the useful Duke of Marmalade. Those of Vaguely Noble's progeny who raced exclusively in the U.S.A. include Royal and Regal (Florida Derby and Bahamas Stakes), Regal and Royal (Jamaica Handicap) and Lemhi Gold.

While Vaguely Noble may not have been the equal of Sea Bird or Ribot, he certainly ranks amongst the best half-dozen winners of the Prix de l'Arc de Triomphe between 1945 and 1968. Only Exbury, Tantième and possibly Coronation at her best would have had a chance of defeating him at level weights, though it is by no means certain that they would have done so. He remains a tribute to his breeder Major Holliday and to his trainer Etienne Pollet, especially the latter, for he had the reputation of being a bad-tempered colt who had to be handled with care and patience.

PRIX DE L'ARC DE TRIOMPHE 2400 metres

at Longchamp on Sunday 6 October 1968. For three-year-olds and upwards. Going: Holding.
1,000,000 francs to the winner; 400,000 fr. to the 2nd; 200,000 fr. to the 3rd; 100,000 fr. to the 4th. 100,000 fr. to the breeder of the winner; 40,000 fr. to that of the 2nd; 20,000 fr. to that of the 3rd. Value to the winner: 1,156,600 francs.

VAGUELY NOBLE	b.c.	by Vienna – Noble Lassie by Nearco	3.55½k	Mme R.A. Franklyn	W. Williamson	1
SIR IVOR	b.c.	by Sir Gaylord – Attica by Mr Trouble	3.55½k	Raymond Guest	L. Piggott	2
CARMARTHEN	ch.c.	by Devon – Kuwait by Persian Gulf	4.60k	Mme R.B. Strassburger	J.L. Durry	3
ROSELIERE	br.f.	by Misti – Peace Rose by Fastnet	3.54k	Mme George Bridgland	Y. Josse	4
LA LAGUNE	b.f.	by Val de Loir – Landerinette by Sicambre	3.54k	H. Berlin	G. Thiboeuf	5
DHAUDEVI	ch.c.	by Dhaulagiri – Devivor by Worden	3.55½k	Mme Georges Courtois	F. Head	6
FELICIO	br.c.	by Shantung – Fighting Edie by Guersant	3.55½k	Daniel Wildenstein	W.B. Pyers	7
SAMOS III	ch.c.	by Sheshoon – Solotanzerin by Ticino	4.60k	Comtesse Margit Batthyany	A. Gibert	8
LUTHIER	br.c.	by Klairon – Flute Enchantée by Cranach	3.55½k	Baron Guy de Rothschild	J. Deforge	9
SOYEUX	ch.c.	by Fils de Roi – Soie Sauvage by Sériphos	3.55½k	Mme R. Gribe	H. Samani	10
ARJON		by Janitor – Astral II by Watling Street	5.60k	Paul Hilger	R. Jallu	
LUCIANO		by Henry the Seventh – Light Arctic by Arctic Prince	4.60k	A. Marcour	R. Hutchinson	
PETRONE		by Prince Taj – Wild Miss by Wild Risk	3.55½k	Daniel Wildenstein	Y. Saint-Martin	
DANOSO		by Val de Loir – Albara by Auriban	3.55½k	Marcel Boussac	R. Poincelet	
RIBERO		by Ribot – Libra by Hyperion	3.55½k	Charles Engelhard	B. Taylor	
TIT BITS		by Tombeur – Esmoura by Escamillo	3.55½k	G.L. de Mola	J. Massard	
ZBOR		by Balto – Zaria by Agregat	3.55½k	Haras Voshod	J. Pchoukov	

17 runners.
VAGUELY NOBLE trained by E. Pollet; bred by Major L.B. Holliday.
Distances: 3 lengths; 4 lengths; 1½ lengths; ½ length; nose; ¾ length; 1 length; short neck; ½ length. Time: 2 mins 35.20 seconds.
Pari-Mutuel: Win: 3fr. 50. Places: 2fr. 20; 3fr. 10; 11fr. 60.
S.P.: 5 to 2 VAGUELY NOBLE; 5 to 1 Felicio and Pétrone (couplé); 6 to 1 Roselière; 13 to 2 Sir Ivor and Dhaudevi; 15 to 2 Luciano; 11 to 1 Luthier; 31 to 1 La Lagune; 35 to 1 Soyeux and Ribero; 45 to 1 Danoso; 65 to 1 Zbor; 90 to 1 Carmarthen, Samos III and Tit Bits; 150 to 1 Arjon.

VAGUELY NOBLE bay colt 1965	Vienna ch. 1957	Aureole ch. 1950	Hyperion ch. 1930	Gainsborough b. 1915	Bayardo
					Rosedrop
				Selene b. 1919	Chaucer
					Serenissima
			Angelola b. 1945	Donatello II ch. 1934	Blenheim
					Delleana
				Feola br. 1933	Friar Marcus
					Aloe
		Turkish Blood b. 1944	Turkhan b. 1937	Bahram b. 1932	Blandford
					Friar's Daughter
				Theresina ch. 1927	Diophon
					Teresina
			Rusk b. 1935	Manna b. 1922	Phalaris
					Waffles
				Baby Polly b. 1924	Spearmint
					Pretty Polly
	Noble Lassie b. 1956	Nearco br. 1935	Pharos b. 1920	Phalaris br. 1913	Polymelus
					Bromus
				Scapa Flow ch. 1914	Chaucer
					Anchora
			Nogara b. 1928	Havresac II br. 1915	Rabelais
					Hors Concours
				Catnip b. 1910	Spearmint
					Sibola
		Belle Sauvage ch. 1949	Big Game b. 1939	Bahram b. 1932	Blandford
					Friar's Daughter
				Myrobella gr. 1930	Tetratema
					Dolabella
			Tropical Sun ch. 1940	Hyperion ch. 1930	Gainsborough
					Selene
				Brulette b. 1928	Brûleur
					Seaweed

Levmoss 1969

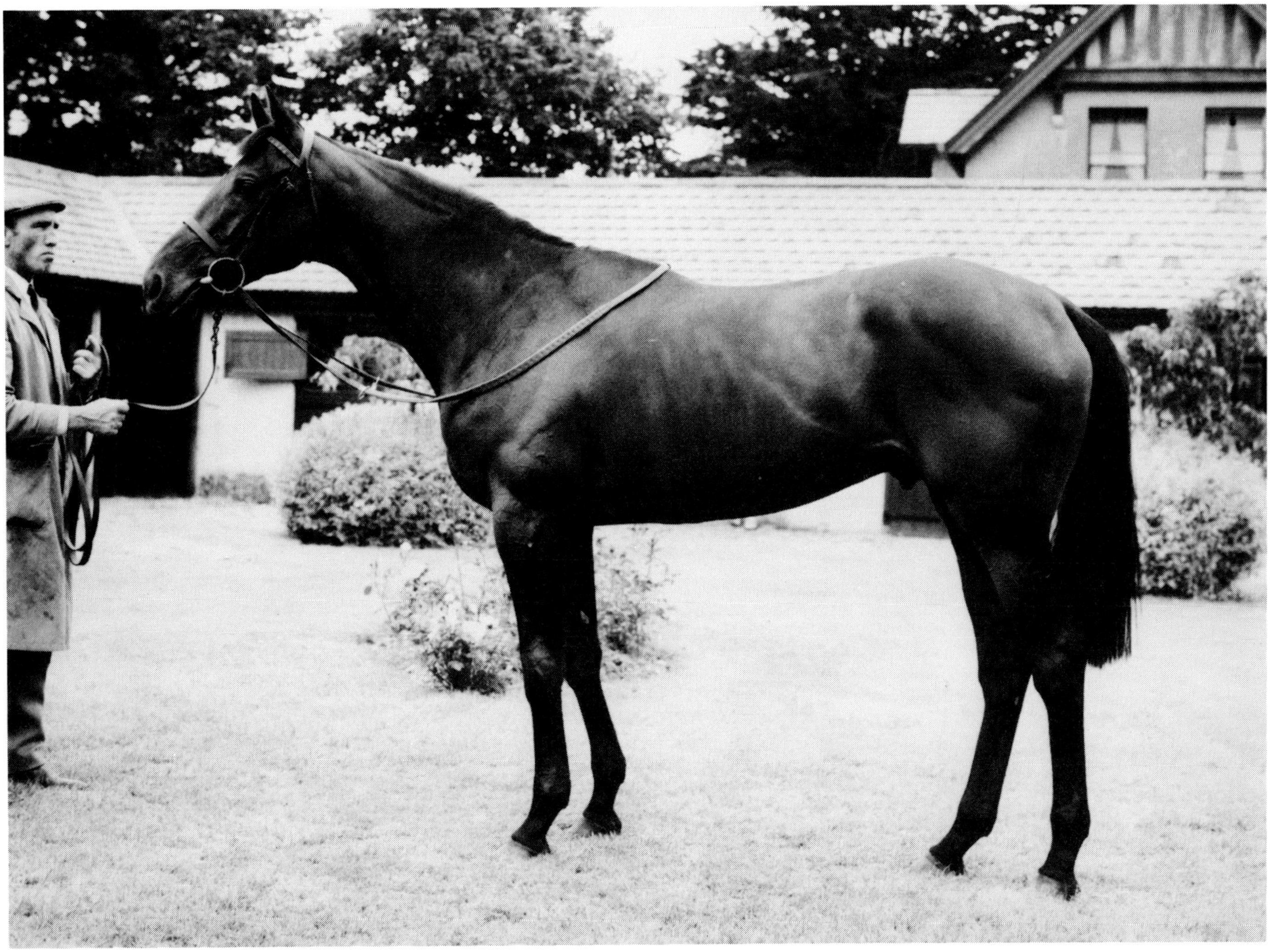

Levmoss

The Field

The retirement of both Vaguely Noble and Sir Ivor at the end of their three-year-old campaigns was a little disappointing but nevertheless understandable as there was little left for either of them to prove. One result was that there was a much bigger field for the Prix de l'Arc de Triomphe in 1969, including nine foreign challengers; six were from England: Park Top, Remand, Copsale, Blakeney, Ribofilio and Shoemaker; one from Italy, Bonconte di Montefeltro; one from Ireland, Levmoss and one from the U.S.S.R., Zbor. In addition, the Japanese five-year-old, Speed Symboli, and the Irish four-year-old, Candy Cane, who were trained in France in the autumn of 1969, had been running in their respective native lands at the beginning of the season.

The Italian challenger, Alarico Palmieri's three-year-old, Bonconte di Montefeltro, was justifiably held in very high regard by his connections. In the Premio Parioli at Rome on 13 April he made up 3 lengths inside the distance and beat Monarca going away by half a length. A month later Lester Piggott was engaged to ride Bonconte di Montefeltro in the Derby Italiano, also at Rome. He gained an even easier length and a half success over Olcio with Monarca the same distance away third. He then beat Gabord and Monarca on a tight rein by 3½ lengths in the Gran Premio d'Italia at San Siro on 8 June. It was after this victory that Lester Piggott agreed to ride Bonconte di Montefeltro in both the Prix Royal Oak and the Prix de l'Arc de Triomphe. This news made the London bookmakers install him as second favourite for Longchamp's great autumn race when they opened their ante-post lists. In the Prix Royal Oak the Italian colt was partnered by Piggott but could only manage to finish a disappointing fourth, a little over 2 lengths behind Le Chouan, Honeyville and Chaparral. Lester Piggott then asked Sr Palmieri to be released from his contract to ride the colt in the Prix de l'Arc de Triomphe in order to ride the Duke of Devonshire's outstanding mare Park Top. Naturally Sr Palmieri was reluctant to agree, but eventually, after consultation with the Italian Jockey Club, the Société d'Encouragement gave Piggot permission to ride the English mare. However this permission was only forthcoming on Tuesday 30 September, so which jockey was to ride Park Top was still in doubt only five days before the Prix de l'Arc de Triomphe. Her trainer, Bernard van Cutsem, had even taken the precaution of asking the young Sandy Barclay to take the ride should Piggott be unavailable and he had partnered her in one of her final gallops.

Lester Piggott had in fact never been beaten on Park Top before the Prix de l'Arc de Triomphe, though he had only ridden her in 1969. As a three-year-old, when she won four of her five races, Park Top's best performance had been to beat the Oaks runner-up St Pauli Girl half a length in the Ribblesdale Stakes at Royal Ascot. She was generally considered the best English filly of her generation. It was true she made no show behind Casaque Grise in the Prix Vermeille at Longchamp in September but her dullness of coat that afternoon made Bernard van Cutsem think that she may have gone over the top. As a four-year-old Park Top was a little disappointing. In six outings she only gained two victories in the Prix d'Hedouville at Longchamp on 8 September and in the Brighton Challenge Cup the month before. Probably her best performance was to run Levmoss and Canterbury to a neck and a head in the Oxfordshire Stakes at Newbury on 17 August. The distance of that race, one mile five furlongs and sixty yards, was almost certainly a little too far for her and in addition the going was soft, which blunted her devastating burst of finishing speed. The merit of the performance was emphasised when Canterbury ran Ribero to a short head in the St Leger.

It was intended that Park Top should be retired to stud after the Cumberland Lodge Stakes at Ascot in September 1968. She took the lead a furlong out and appeared to have the race at her mercy when she suddenly shortened her stride and was caught and beaten 4 lengths by Chicago, who later that autumn was to win the Gran Premio del Jockey Club and the Premio Roma. Such was the disappointment that Park Top had not finished her career on a winning note that in December the Duke of Devonshire decided to keep her in training for another season. Had Chicago not beaten her at Ascot, it is quite probable that her marvellous achievements as a five-year-old would never have occurred.

Park Top's five-year career began in a most unhappy way, for in February 1969 Bernard van Cutsem was informed that he had cancer of the jaw. This necessitated his travelling almost every week-day from Newmarket to London for treatment. But this cruel blow in no way deterred him from continuing training. After supervising the first and sometimes the second lots, he drove to London for his treatment and returned to Newmarket for evening stables. Such was the dedication of Park Top's most able trainer.

Park Top's first race in 1969 was the Prix de la Seine (12f) at Longchamp on 22 May when she was ridden for the first time in her career by Lester Piggott. She had no difficulty in beating the 1968 Ribblesdale Stakes winner, Pandora Bay, by 2½ lengths. Immediately after this victory Piggott agreed to ride the mare in the Coronation Cup at Epsom. There she produced an exhilarating burst of acceleration well inside the final furlong to beat Mount Athos and Connaught three-quarters of a length and 1½ lengths with Remand 6 lengths away fourth. It was these three who had occupied the minor places behind Sir Ivor in the previous year's Derby. Apart from having won the Coronation Cup virtually unextended, Park Top's time was a whole two seconds faster than Blakeney's in the Derby and Sleeping Partner's in the Oaks.

Partnered by Geoff Lewis in place of the suspended Lester Piggott, Park Top next took her revenge on Chicago by 1½ lengths in the Hardwicke Stakes at Royal Ascot. It had not been planned to run the mare in the Eclipse Stakes at Sandown, but the field cut up so badly as a result of the severe coughing epidemic that summer that it was decided to let her take her chance. Lewis was again engaged to ride when the decision was made to run Park Top as there was a distinct possibility that Piggott would have to ride Charles Engelhard's three-year-old Ribofilio. In fact he did not run, but Bernard van Cutsem and the Duke of Devonshire decided to retain Lewis, with the result that Piggott took the mount on Wolver Hollow. In the race Lewis found himself shut in on the rails on Park Top with two furlongs to go, although he had been given specific instructions to keep clear of the rails on the turn into the straight. When he finally extricated himself and brought Park Top round the entire field on the outside, Lewis found Lester Piggott had poached an unassailable lead on Wolver Hollow. Park Top was still 2½ lengths adrift at the winning post with Hogarth, the 1968 Derby Italiano winner, 3 lengths away third. In the King George VI and the Queen Elizabeth Stakes Park Top was reunited with Piggott and coming from last to first in the straight, she took the lead a furlong out and beat Crozier and Hogarth 1½ lengths and a neck with Felicio and Speed Symboli another 3 lengths and 4 lengths away fourth and fifth. The bare result, however, does not at all convey the measure of Park Top's total dominance over her rivals. In her final race before the Prix de l'Arc de Triomphe, the Duke of Devonshire's mare easily defeated Felicio and Pandora Bay by 2 lengths and a length in the Prix Foy at Longchamp on 7 September. Throughout 1969 Park Top had proved herself to be far and away the best of the older horses in training in either England or France and of the foreign challengers seemed to have much the best chance at Longchamp on Sunday 5 October.

J.J. Astor's four-year-old Remand had been an unlucky horse. He was unbeaten before the Epsom Derby. His three successes as a juvenile had included the Solario Stakes at Ascot, and on his sole outing before Epsom as a three-year-old he had beaten the highly regarded Connaught half a length in the Cheshire Vase. On Derby Day, however, he did not look at all well in the paddock and although he ran a gallant fourth to Sir Ivor, he was found to be running a temperature after the race. The virus that he had caught was so bad that he never reappeared on the racecourse in 1968. Certainly the sporting decision of his owner to run him in the Derby, despite his obviously distressed condition immediately before the race, contributed to the severity of the virus. He began his four-year campaign by winning the Westbury Stakes (10f) at Sandown from his pacemaker Hipster and Ribero, who probably did not train on as a four-year-old. Although favourite he could only manage fourth place, 8¼ lengths behind Park Top in the Coronation Cup at Epsom. Remand's ill-fortune continued in the Eclipse Stakes when he lashed out at the stalls, went temporarily lame and had to be withdrawn by the starter. He next ran third to Djakao and Arme d'Or, beaten 2½ lengths and the same, in the Grand Prix de Deauville with Speed Symboli 8½ lengths behind tenth. In his final race before the Prix de l'Arc de Triomphe he easily beat Crozier 4 lengths in a very fast time in the Cumberland Lodge Stakes at Ascot. This victory gave rise to hopes that he was coming back to his best, but on the other hand it was very difficult to envisage him gaining his revenge on Park Top.

Marcos Lemnos's six-year-old Copsale was trained at Newmarket by Sam Armstrong, but all his three victories in 1967 were gained in France. He won a mile and a half furlong event at Saint-Cloud and then decisively beat Duc d'Aquitaine 3 lengths in the Prix Maurice de Nieuil on the same course on 14 July. He had half a length to spare over Crossbow in the Prix de Reux (13f) at Deauville, but was beaten a neck by A Chara in the Prix Hubert de Pourtales in September, with Beaugency a little over 5 lengths away fifth. On balance, however, this honest and top-class handicapper was likely to meet several too good for him in the Prix de l'Arc de Triomphe.

The three English three-year-olds in the field on Sunday 5 October had finished first, second and fifth in the Epsom Derby. Arthur Budgett's Blakeney had only run twice as a juvenile, finishing fourth to Caliban in the Clarence House Stakes and winning the Houghton Stakes by half a length from Prince de Galles at Newmarket. He did not reappear until the Lingfield Derby Trial Stakes when he just failed to catch The Elk after his jockey, Lewis, managed to find himself in a lot of trouble three furlongs out. Many experienced race-readers were of the opinion that Blakeney ought to have won. Ernie Johnson was given the mount at Epsom. Blakeney was very fortunate to find a trouble-free run on the rails: he headed Shoemaker inside the final furlong and beat him a length. Prince Regent was another length back third with Ribofilio a length away fifth.

There was absolutely no question that the moral victor of the Epsom Derby was la Comtesse de la Valdène's Prince Regent. Although it was reported that he was bumped rounding Tattenham Corner, the fact remains that the meandering path followed by his jockey, Jean Deforge, was a major contribution to his virtually hopeless position as the field entered the straight. It was a clear example of the need for a jockey experienced in riding the Epsom track. In the final 2½ furlongs Prince Regent passed horse after horse but the post came too soon. He is widely regarded as one of the unluckiest Derby losers since the Second World War. This impression was totally confirmed in the Irish Sweeps Derby when he beat Ribofilio most decisively by a length. When Piggott sent Ribofilio into the lead a furlong out, Charles Engelhard's colt looked to have the race at his mercy, but he had no answer to the exhilarating turn of foot that Prince Regent produced with a hundred yards to go. The fact that the pair finished 5 lengths and a short head in front of the future Irish St Leger winner, Reindeer, and Blakeney, merely went to emphasise what a tremendous performance Prince Regent had put up.

Blakeney, who had never been able to get into a challenging position at the Curragh, was found to be suffering from sore shins immediately after the race with the result that he was unable to run in the Great Voltigeur Stakes. In the St Leger it was the Derby winner's turn to meet all kinds of interference in the straight so that he could only manage fifth place behind Intermezzo, Ribofilio and Prince Consort. On balance it seemed unlikely that Blakeney would confirm the Epsom Derby placings with Prince Regent, or for that matter with Ribofilio, on 5 October and his starting price of 60 to 1 reflected this opinion.

Charles Engelhard's Ribofilio was an enigmatic colt. As a two-year-old he had won the Chesham Stakes at Royal Ascot, the Champagne Stakes at Doncaster and the Dewhurst Stakes at Newmarket and had been placed top of the Free Handicap. His one failure had been when only sixth in the Prix Morny when Prince Regent had finished third a little over 4 lengths in front of him. After easily winning the Ascot 2000 Guineas Trial, Ribofilio started favourite for the Newmarket classic but he totally disgraced himself by refusing to gallop and Piggott eventually pulled him up. The mystery has never been cleared up and surprisingly he started favourite again at Epsom. For a colt whose preparation had been interrupted, he did not run at all badly to finish fifth, 8 lengths behind Blakeney. He was again preferred in the market to Prince Regent at the Curragh. Ribofilio easily beat his solitary opponent in the March Stakes at Goodwood, and for the fourth time started favourite in a classic, the St Leger. Like Blakeney, Ribofilio had a far from trouble-free run at Doncaster. When Piggott got him clear Gerald Oldham's Intermezzo was gone beyond recall. In fact no one would quote Lester Piggott's riding of Ribofilio in the 1969 St Leger as an example of his artistry at its best and most felt the colt to have been an unlucky loser. Endowed with undoubted ability, when he consented to show it, Ribofilio started at 11 to 1 on 5 October. Nevertheless the fact that Piggott deserted him for the first time in favour of Park Top made it seem likely that the most he could aspire to was one of the minor places.

The runner-up at Epsom, Peter Goulandris's Shoemaker, had won the Royal Stakes at Sandown and the Player-Wills Stakes at Leopardstown earlier that spring. However he was defeated by Harmony Hall in the Gordon Stakes at Goodwood and was a disappointing eighth to Intermezzo in the St Leger. He got up on the line to beat Yelapa a short head in the Prix Henry Delamarre at Longchamp on 28 September, but that was hardly a breathtaking performance and his odds of 70 to 1 on 5 October was a fair reflection of his likelihood of winning.

The 2 to 1 favourite for Longchamp's most prestigious race was la Comtesse de la Valdene's Prince Regent, trained by Etienne Pollet. As a juvenile he had won twice from four starts, a débutant's race at Chantilly over 5 furlongs and the Prix Saint-Roman at Longchamp. In the Prix Morny he had been third to Princeline and Ronda but he had disappointed when only fifth in the Prix de la Salamandre. On his reappearance as a three-year-old Prince Regent comfortably beat the subsequent Prix du Jockey Club winner, Goodly, by 2 lengths in the Prix Greffulhe on 7 April. He was all out to defeat Caliban and Belbury a head and a neck in the Prix Lupin on 18 May, but the heavy going that afternoon had been against him. His reputation was built mainly on his patent ill-fortune at Epsom and his devastating burst of speed at the Curragh. It seemed almost certain that Prince

Regent would emerge as the best of the quartet who had run in the Derby but whether his acceleration would match that of Park Top was a different question. There were some who considered that the 1969 Prix de l'Arc de Triomphe was going to concern only those two runners.

For the first time in its history there was a challenger for the Prix de l'Arc de Triomphe who had been one of the leading horses in Japan. Speed Symboli, a six-year-old by Royal Challenger, had won eighteen races in Japan before coming to be trained by Freddie Palmer in France in the summer of 1969. In November 1967 he had run a creditable fifth to Fort Marcy in the Washington D.C. International Stakes at Laurel Park. In his first race in Europe, the King George VI and the Queen Elizabeth Stakes at Ascot, Speed Symboli took the lead two furlongs out and eventually was only beaten 9 lengths into fifth place. In the Grand Prix de Deauville, however, he was beaten 13½ lengths into tenth place by Djakao. It would have been a major surprise if he were to finish even in the first three on 5 October.

The Russian four-year-old, Zbor, was running in his second consecutive Prix de l'Arc de Triomphe. In 1969 he had won two of the major middle distance races in the Soviet Union including the Krasnaya Presnya. Nevertheless he had been completely outclassed by Vaguely Noble and Sir Ivor the previous October and it seemed likely that he was again destined to play no part in the finish in 1969.

The invader from Ireland, Seamus McGrath's four-year-old Levmoss, had done little wrong in 1969. On 29 March at the Curragh he had run in the seven-furlong Gladness Stakes and not surprisingly had made no show over a distance patently too short for him. On 20 April he ran Zamazaan and Samos III to three-quarters of a length and the same in the Prix Jean Prat (15f) at Longchamp. In the Prix du Cadran on 25 May Levmoss took up the running soon after entering the straight and battled on gamely to repel the challenges of Zamazaan and Samos III by a neck and a head with Torpid 3 lengths away fourth. In the Ascot Gold Cup Levmoss again took the lead well over two furlongs out and scored a decisive 4-length victory over Torpid and Fortissimo. In his final outing before 5 October he was burdened with the welter weight of 10 st. 10 lbs in the Leinster Handicap (14f) at the Curragh but still proved much too good for his rivals. As a three-year-old Levmoss had steadily improved throughout the summer. After winning a small race at Phoenix Park in May, he had run fourth in both the Lingfield Derby Trial and in the Queen's Vase. He then put up his best performance as a three-year-old in the Oxfordshire Stakes at Newbury on 17 August, when he got up in the last strides to beat Canterbury and Park Top a neck and a head. Levmoss was only receiving 2 lbs more than the weight for age scale from the mare, while he actually was giving a pound to Canterbury as his jockey, Brian Taylor, could not do the weight. Levmoss put up another very creditable display when only beaten 4 lengths and 3 lengths by Dhaudevi and Torpid in the Prix Royal Oak. His final outing in the Leopardstown November Handicap (16f), when he carried 9st 4lbs to a 4-length victory over thirty-one rivals, was an indication of how much he might achieve in 1969. The ease with which he reversed the Prix Royal Oak form with Torpid in the Prix du Cadran and Ascot Gold Cup was an indication of the progress he made between three and four. Levmoss indeed had the distinction of being the only runner in the field for the 1969 Prix de l'Arc de Triomphe who had defeated Park Top. It was all the more surprising, therefore, that he was permitted to start at 52 to 1. It was generally held, however, that the distance would prove a little too sharp for him. The last Ascot Gold Cup winner to achieve success in the race had been Caracalla in 1946 and the only other had been Massine in 1924.

As we have seen, the French defenders were spearheaded by the three-year-old Prince Regent, but three of the older horses, Carmarthen, Grandier and Roselière, could all be given chances. Mme R.B. Strassburger's Carmarthen, who had finished third to Vaguely Noble and Sir Ivor, was lightly raced in 1969 as he had a distinct preference for heavy going. He had no difficulty in beating Coryas and Bagdad by 3 lengths and 1½ lengths in the Prix Exbury (10f) at Saint-Cloud on 22 March. In the Prix Ganay he took the lead 200 metres out and had a length and the same to spare over Grandier and Locris, with Luthier a head further back fourth and Roselière another 5½ lengths away sixth. This victory marked him as the best middle distance older horse in France, but he did not run again until the Prix du Prince d'Orange on 21 September on account of his preference for heavy going. After taking the lead 400 metres out he was caught close home by the Prix du Jockey Club winner, Goodly, and beaten a head with Candy Cane 2 lengths away third. Unfortunately he returned to the unsaddling enclosure lame with blood coming from his hock. Although the injury was not serious, he was on the easy list for a week and this interruption to his training at a vital stage before the Prix de l'Arc de Triomphe almost certainly put paid to his chance, quite apart from the fact that the going would not suit him.

Mme Pierre Ribes's Grandier again had had a successful season. After finishing runner-up to Carmarthen in the Prix Ganay, he beat Timmy My Boy 1½ lengths in the Prix d'Harcourt on 4 May at Longchamp, with Candy Cane a little under 1½ lengths away fifth and Roselière in rear. Grandier was only third, a neck and 2 lengths behind Semillant and Locris, in the Prix Dollar on 1 June, but he reversed the form in the Prix d'Ispahan when beating Locris a head on 29 June. Grandier was unlucky in running when only seventh to Felicio in the Grand Prix de Saint-Cloud but he gained another facile success when beating Apipe and Locris 1½ lengths and a head in the Prix Gontaut-Biron (10f) at Deauville on 16 August. Grandier had never won over a mile and a half, and the doubt that he might not stay that distance were the causes why his odds were a very generous 70 to 1; there were several good French judges who considered him as a potential outsider on 5 October.

Roselière, the best middle distance filly of her generation the previous season and fourth to Vaguely Noble in October 1968, had been but a shadow of herself in the spring and summer. Out of the frame in both the Prix Ganay and Prix d'Harcourt, she was a disappointing eighth to Felicio in the Grand Prix de Saint-Cloud. She beat Arme d'Or 2 lengths in the Prix de Pomone on 24 August at Deauville, but Arme d'Or was some way behind the best three-year-old fillies in 1969. Roselière's starting price of 11 to 1 for the 1969 Prix de l'Arc de Triomphe reflected much more her glories the previous year than her achievements as a four-year-old.

Lady Sassoon's four-year-old, Candy Cane, had had the misfortune to encounter Sir Ivor in Ireland as a juvenile. As a three-year-old he was an impressive 6-length winner of the Ballymoss Stakes in the spring but was most disappointing when favourite for the Irish 2000 Guineas; it later transpired he was ill. He only had one other outing in 1968 in the Champion Stakes in October when he finished a very creditable third 2½ lengths and 2 lengths behind Sir Ivor and Locris. He began his four-year-old career by running the three-year-old Seiko to a length in the Ballymoss Stakes at the Curragh and he was then fifth to Grandier in the Prix d'Harcourt. Candy Cane easily won the Royal Whip Stakes at the Curragh on 14 May and was then sent to France to be trained by François Boutin. Candy Cane won a small event over a mile and a half on 5 September at Chantilly but only managed third place behind Goodly and Carmarthen in the Prix du Prince d'Orange. It seemed probable that he would find several too good for him on 5 October though his chance was stones better than the seven-year-old Fiasco, who had been immensely flattered when runner-up to Busted in the Prix Henri Foy two years earlier, but whose accomplishments since then were utterly unnoteworthy. Fiasco was, in fact, completely outclassed in such an aristocratic equine field.

With the exception of Prince Regent, the French three-year-old colts seemed a little below the normal standard in 1969. Maurice Lehmann's Goodly was the most consistent. After running Prince Regent to 2 lengths in the Prix Greffulhe, he beat Belbury a short head in the Prix Noailles at Longchamp on 27 April. It was a better performance than the bare result shows for he took the lead at the entrance to the straight, but for some unknown reason faltered badly near the finish after looking to have the race in his pocket. In the Prix du Jockey Club he just got the better of Beaugency by a short head with Djakao, who had been well and truly boxed in a furlong out, a very fast-finishing third another three-quarters of a length behind. Goodly then ran Felicio to a head in the Grand Prix de Saint-Cloud with Grandier and Roselière only seventh and eighth. The race, however, was run at a pace slower than a hearse: the time of 2 minutes 47·2 seconds was ten seconds slower than the previous year, and it developed into a sprint in the straight with several justified hard-luck stories. In his final race before the Prix de l'Arc de Triomphe Goodly defeated the unfortunate Carmarthen in the Prix du Prince d'Orange. There seemed, however, no reason why he should beat Prince Regent and he was allowed to start at the generous odds of 37 to 1 on 5 October.

Mme Pierre Wertheimer's Beaugency, who had won both his starts as a juvenile including the

Prix Thomas Bryon at Saint-Cloud, had found the distance of the Prix de Guiche (9¾f) too short for him when half a length runner-up to Spring Stone. He just held on to beat Djakao a short neck in the Prix Hocquart on 11 May but then narrowly failed to catch Goodly at Chantilly. At the Curragh in the Irish Derby Beaugency was only able to finish a disappointing ninth to Prince Regent, and he was only fifth to A Chara and Copsale in the Prix Hubert de Pourtales in September, so it was not surprising he started at 59 to 1 on 5 October.

Baron Guy de Rothschild's Djakao followed his narrow defeats in the Prix Hocquart and Prix du Jockey Club by running Chaparral to 2½ lengths in the Grand Prix de Paris without ever looking like catching the winner. His best performance was reserved for the Grand Prix de Deauville when he easily beat Arme d'Or and Remand. After Prince Regent, Djakao was the most fancied of the French three-year-old colts, starting at 12 to 1 on 5 October.

Mme Weisweiller's Chaparral won both the Prix Darioletta and the Prix Berteux (15f) at Chantilly in the spring, before running a slightly disappointing fifth in the Prix de l'Espérance (15f) behind Nonancourt at Longchamp on 1 June. In the Grand Prix de Paris he never looked in danger after he took the lead early in the straight, and easily held Djakao by 2½ lengths. Nevertheless he was beaten a short head and a head by Le Couan and Honeyville in the Prix Royal Oak, and it seemed the best Chaparral could hope for on 5 October was one of the minor places.

Daniel Wildenstein's Yelapa had beaten Belbury 2½ lengths in the Grand Critérium and was rated the best French juvenile in 1968. On his reappearance in the Prix de Fontainebleau at Longchamp he was beaten a head by Le Mas Marvent. He still took his chance in the 2000 Guineas at Newmarket, but dropped out of contention soon after half way and he finished unplaced to Right Tack. Yelapa was found to be lame on his return to France and was given a long rest. He won a small 10-furlong event at Chantilly in early September and in the Prix Henry Delamarre, when runner-up to Shoemaker, he showed a glimmer of his two-year-old form, but his starting price of 47 to 1 on 5 October was a fair reflection of his chance.

Mme Léon Volterra's Belbury ran Goodly to a short head in the Prix Noailles but had flashed his tail when asked to make his effort. He ran with greater courage when only a head and a neck behind Prince Regent and Caliban in the Prix Lupin. In the Epsom Derby he was given far too much to do and in view of this, although he only finished ninth to Blakeney, it was a more creditable performance than the bare result conveys. In the Prix de Chantilly in September Belbury beat Harmony Hall and Le Mas Marvent. Harmony Hall had beaten Shoemaker in the Gordon Stakes at Goodwood, won the Princess of Wales's Stakes at Newmarket, and been awarded the Great Voltigeur Stakes at York on the disqualification of Intermezzo. Belbury thus was not totally out of contention on 5 October.

There were two three-year-old fillies in the field for the 1969 Prix de l'Arc de Triomphe, Marcel Boussac's Crepellana, and Arpad Plesch's Saraca. Crepellana was an impressive winner of a maiden race at Longchamp in the spring but in the Prix de Diane she appeared to have been given too much to do 400 metres out. Nevertheless she quickened in excellent fashion and collared Saraca well inside the final furlong and beat her and Glaneuse three-quarters of a length and 1½ lengths. Until that moment Saraca was the best three-year-old filly in France. She had been the best filly of her generation, as a juvenile winning her only two starts, including the Critérium de Saint Cloud by 8 lengths. On her reappearance in the Prix Pénélope she easily accounted for Belle Margot by 10 lengths. In the Prix Saint-Alary at Longchamp Saraca impressively defeated Koblenza, the Poule d'Essai des Pouliches winner, by 2 lengths. Saraca and Crepellana met again in the Prix Vermeille when Saraca got her revenge over the longer distance, but she still only had a short head advantage at the line. Certainly both could not be ruled out but since they were meeting the three-year-old colts and older horses for the first time, it was difficult to know if they were good enough to cope with the likes of Park Top and Prince Regent.

The Final Preparations

In his final testing work ten days before the Prix de l'Arc de Triomphe Prince Regent impressively accounted for Honeyville and Carmarthen, leaving them as if they were standing still. In her last gallop before leaving for Longchamp Park Top worked five furlongs with

Shaft and strode clear in the final furlong. On the Wednesday before the race Lester Piggott and the Duke of Devonshire had dinner with Bernard van Cutsem and Piggott told them he thought 'the mare will win – we shall certainly be in the first three'. Park Top left Cambridge Airport on Friday morning; Bernard van Cutsem had arranged for her to be stabled in a small yard in Maisons-Laffitte. To provide total security for her the Chief Inspector of Police at Newmarket had himself volunteered to accompany her.

After the final gallops of Goodly, Chaparral and Beaugency, Freddie Head decided to ride the Grand Prix de Paris winner, the mounts on Goodly and Beaugency being taken by Sandy Barclay and Jacques Taillard respectively. Blakeney arrived in Paris on Thursday, while Djakao was reported to be in excellent shape and there was a persistent tip for him at Chantilly.

The going on Sunday 5 October was firm and it was a beautiful sunny autumn afternoon. As usual there was a very cosmopolitan crowd at Longchamp including Americans, Russians, Japanese and inevitably a huge contingent of English and Irish, despite the £50 travel allowance limit imposed by England's socialist government. The Conservative Member of Parliament for Thanet, William Rees-Davies, had organised a weekend trip to Longchamp for the Racegoers' Society – 130 of whom had flown over for twenty-seven guineas. Naturally the British Ambassador to France, Sir Christopher Soames, was at Longchamp though, unlike the Duke of Marlborough's younger son, Lord Charles Spencer Churchill, he was not wearing morning dress which the smarter French racegoers wore at that time. The Earl of Suffolk, a frequent visitor to the Parisian tracks was in contrast hatless, while the Duke of Devonshire, wearing a slightly battered brown homburg, appeared more than a little anxious before the big race. Both the Marchioness of Tavistock and Patricia Rawlings also decorated the lawns of Longchamp that afternoon. The Aga Khan, who was to announce his engagement to the future Begum the following day, was another spectator.

The afternoon began with Lester Piggott winning the first three races. He guided Tower Walk to a three-quarters of a length victory in the Prix de l'Abbaye de Longchamp, won the Prix des Champs-Elysées on Park Top's galloping companion, Shaft, and then rode Vela to victory in the Critérium des Pouliches. The Duke of Devonshire felt that this glut of success could not enhance Park Top's chance and only reluctantly congratulated Piggott in the parade ring before he mounted his mare. Levmoss's jockey, Bill Williamson, commented after Vela's victory that 'Lester is using up all his luck'.

The Race

In spite of the large field there was no delay in loading the field into the stalls for the 1969 Prix de l'Arc de Triomphe. Immediately after the start Blakeney was the first to show from Bonconte di Montefeltro, but on settling down the Italian colt took over the lead from the English Derby winner, with Levmoss tucked in behind in third place followed by Chaparral. This leading bunch were closely tracked by Remand and Saraca close up on the rails. Ribofilio, Candy Cane and Shoemaker were in the middle of the field, but the two favourites Prince Regent and Park Top were on the rails at the back. The order remained almost unchanged till the top of the hill where Saraca began to lose her prominent position. Shoemaker improved as the runners began coming down the hill almost to join the leaders, and Grandier could be seen making significant progress from the mid-division. When the runners started to turn into the straight, Bonconte di Montefeltro drifted away from the rails and Williamson was able to slip Levmoss past him to take the lead nearly three furlongs out. At this stage both Piggott on Park Top and Lewis on Prince Regent, a couple of places in front of the mare, were trying to find a passage through beaten horses. Williamson was able to go clear on Levmoss as he entered the straight where Chaparral made a brief effort to challenge him, but was unable to maintain his run. Next Grandier and Candy Cane on the rails attempted to get on terms with the leader, but Levmoss continued his relentless gallop and was always holding them easily. Meanwhile Lewis on Prince Regent had managed to thread his way through the field but, on reaching the heels of Grandier and Candy Cane, he had no more to give and ran on at one pace. Park Top had been forced to the wide outside of the whole field to obtain a clear run. With a furlong to go Levmoss was clear of Grandier, Candy Cane, and Prince Regent with Chaparral and Belbury on the rails behind them. Blakeney and Remand together came next and, beginning her challenge almost

Levmoss (7) holds off Park Top (10) with Grandier (4) third

nearer the stands than the rails, was Park Top. She was at least seven lengths in arrears but she produced such an electrifying burst of acceleration that she swept past the six immediately in front of her with the speed one associates with a winner of the Prix de l'Abbaye de Longchamp. The winning post came, nevertheless, but metres too soon for her to catch Levmoss who had a rapidly diminishing three-quarters of a length to spare over her. The fact that she beat Grandier 3 lengths for third place merely illustrates how fast she was travelling in those final 200 metres. Candy Cane was a length behind in fourth place a short head in front of Prince Regent who beat Chaparral a neck. Then followed Remand, Belbury and Blakeney.

Lester Piggott and Park Top were the first back to the unsaddling enclosure but made no comment to the Duke of Devonshire. In the weighing room, however, he blamed himself saying: 'I left it too late, that's all there is to it and I should have won.' The French press were unanimous that he had made a tactical error and that the mare should have won. When talking to Bernard van Cutsem at Newmarket some considerable time after the race, he was of the opinion that Piggott judged himself harshly in the immediate aftermath. He said that in order to make full use of Park Top's fantastic burst of acceleration, she had to be held up and the fact that an opportunity to challenge had not presented itself earlier and that she had to come on the outside was one of those misfortunes that occur in racing. He added that if Piggott had tried to force her to close on the leaders earlier it would have blunted her speed, which everyone at Longchamp saw that afternoon.

Williamson, who became the first jockey to win the Prix de l'Arc de Triomphe in consecutive years on different horses since C.H. Semblat on Pearl Cap and Motrico in 1931 and 1932, was not of the opinion that Park Top was an unlucky loser. Immediately afterwards he said that, 'By half-way I felt my horse was going to win. He was going so well when Blakeney and Bonconte di Montefeltro weakened that I did not hesitate to push Levmoss on. I knew he could go on with his effort to the post. If Park Top had come up to me, I think I could have pulled out a bit extra.'

The controversy as to who should have won will never be resolved; there will be apologists for both Park Top and Levmoss each with sound cases. It is probable that Williamson won the race by the opportunist way in which he shot Levmoss into the lead three furlongs out, not knowing at that split second the trouble Lester Piggott would have trouble in extricating Park Top and getting her into a challenging position. There is no question, however, that both Levmoss and Park Top covered themselves with glory that afternoon.

For the others there were no valid excuses. Lewis felt that 'with a little better luck' Prince Regent would have been third but no nearer. Hutchinson reported that Ribofilio 'was never going at any stage' though he felt his draw, twenty-four, which forced him to make an immediate effort on the Engelhard colt had not improved his chance. Blakeney found the going too firm but still had run creditably. The only real hard luck story concerned Saraca who pulled a tendon so badly that she had to be retired immediately to stud. Crepellana also was slightly injured during the course of the race but not seriously enough to suggest she would have won had the incident not happened.

The Rest of the Season

Levmoss was retired to stud immediately after his greatest victory, but a week after the race Bernard van Cutsem felt that Park Top had recovered so well from her hard race that she should be allowed to take her chance in the Champion Stakes on 18 October. As so frequently happens the Champion Stakes comes too soon for many horses who have run well in the Prix de l'Arc de Triomphe. The French three-year-old filly, Flossy, trained by François Boutin, had the advantage of being a fresh horse and she beat Park Top and Lorenzaccio 2 lengths and 4 lengths with Connaught sixth and Shoemaker unplaced. Flossy had done a brilliant gallop with Candy Cane before the Prix de l'Arc de Triomphe, so the result was no big surprise to his connections. Park Top had four outings in 1970 winning La Coupe at Longchamp in May and the Cumberland Lodge Stakes at Ascot in September. She may have been a little unlucky to be runner-up to Caliban in the Coronation Cup at Epsom as she was not suited by the very firm going and returned to Newmarket with very sore legs which necessitated a very long rest and an abandonment of the plan to run her again in the King George VI and the Queen Elizabeth Stakes and the Prix de l'Arc de Triomphe.

For once the form of the Prix de l'Arc de Triomphe was not conclusively franked in the remainder of the 1969 season. Candy Cane was third to Karabas and Le Mas Marvent in the Prix du Conseil Municipal in which Belbury finished eighth and Fiasco was unplaced. No runners in the Prix de l'Arc de Triomphe took part in the Washington D.C. International which was won by Lester Piggott on Karabas. This success was some compensation for both Bernard van Cutsem and Piggott for Park Top's defeat. Remand ran in the Gran Premio del Jockey Club at San Siro but could only manage fourth place, 3¾ lengths behind Glaneuse.

Pedigree and Stud Record

Levmoss was the first victory in the Prix de l'Arc de Triomphe for his Irish trainer, Seamus McGrath, and only the second for an Irish-trained horse, the first having been the Vincent O'Brien-trained Ballymoss in 1958. McGrath, who had ridden as an amateur when he was young, had held a trainer's licence since 1942. He had saddled Panaslipper, who had come second to Phil Drake in the 1955 Epsom Derby and had gone on to win the Irish Derby, which McGrath was also to win with Weaver's Hall in 1973. In 1956 his charge, Roistar, had finished third to Lavandin and Montaval in the Derby, and the following year he had trained Silken Glider, who had only failed by a short head to catch Her

Majesty the Queen's Carrozza in the Oaks. Sweet Mimosa, a full sister to Levmoss, was also to win the Prix de Diane in 1970. Nevertheless Levmoss's victory in the Prix de l'Arc de Triomphe was unquestionably the biggest success McGrath had had as a trainer. Seamus McGrath, of course, came from the well-known Irish family who had started the Irish Sweep, and he was a director of the Irish Hospitals Trust, the Irish Glass Bottle Company and the Waterford Glass Company.

Le Levanstell, the sire of Levmoss, had also been trained by McGrath, and came from one of the well established families at the McGrath's Brownstown Stud. As a two-year-old he won three times from four starts including the Railway Stakes at the Curragh and the Birdcatcher's Stakes at Naas. At three he failed to win in seven outings but did manage second place in the Jersey Stakes at Royal Ascot. It was as a four-year-old that he revealed his true merit. He won the Woodbine Stakes (7f) at Leopardstown, the Sussex Stakes (8f) at Goodwood and the Queen Elizabeth II Stakes (8f) at Ascot in which race he defeated Petit Etoile. When Le Levanstell retired he had won six races from twenty-four starts worth £11,361. Apart from Levmoss and Sweet Mimosa, probably the best of Le Levanstell's progeny were Le Moss, a full brother to Levmoss whose achievements are related later; Myastrid, who was third to Sleeping Partner and Frontier Goddess in the 1960 Oaks; and Allangrange, the winner of the 1970 Irish St Leger.

Levmoss's dam, Feemoss by Ballymoss, also came from a well established Brownstown Stud family. On the racecourse she was useful but not top class; she won a fourteen-furlong maiden at Gowran Park as a three-year-old, a twelve furlong handicap at the Curragh and the Blandford Stakes by a short head from Wily Trout. In the Irish Guinness Oaks in 1963 she could only manage sixth place to Hibernia III, and she filled the same position behind Christmas Island in the Irish St Leger. As a broodmare, however, Feemoss was a resounding success. Levmoss was her first living foal and her second, Santamoss by Santa Claus, won four races including the Vauxhall Trial Stakes. Her next foal Sweet Mimosa, a full sister to Levmoss, gained four victories including the 1970 Prix de Diane and the Trigo Stakes, in addition to finishing third to Santa Tina and Parmelia in the Irish Guinness Oaks. Both Feemoss's next two foals, Pale Jasmine by Continuation and King Mousse by Kauai King, were winners. Levmoss's full brother, Le Moss, however, established Feemoss as one of the best broodmares in the last twenty seasons. As a three-year-old Le Moss won four races including the Queen's Vase, the March Stakes at Goodwood and the Tennent Trophy at Ayr, besides finishing second to Crow in the St Leger. As a four-year-old he proved himself the best stayer in England winning the Ascot Gold Cup, the Goodwood Cup and the Doncaster Cup. As a five-year-old he again achieved this rare treble, and at Ascot and Goodwood it was the redoubtable Ardross who had to play second fiddle to him.

Feemoss's dam, Feevagh by Solar Slipper, was also a useful racemare. She was owned by Joseph McGrath but trained in England by Willie Stephenson who had won the Derby for him in 1951 with Arctic Prince. Feevagh gained three victories from seven starts as a three-year-old, the most important of which was when she beat the Ribblesdale Stakes winner, Sweet One, a head in the Yorkshire Oaks. She also won the Westhanger Plate (10f) at Folkestone and the Fairfield Stakes (8f) at York, but was outclassed behind Festoon in the 1954 1000 Guineas and behind Sun Cap in the Oaks. Feevagh also was a notable success as a broodmare. She was the dam of eight winners, the best of which were Feemoss and Laurence O by Saint Crespin, whose five victories included the Queen Alexandra Stakes at Ascot, and who was runner-up in the Doncaster Cup and placed behind Connaught in the Great Voltigeur Stakes and behind Gioila Mear in the Irish St Leger. In all, her produce gained sixteen victories and her winning filly Aliceva by Alcide was herself the dam of three extremely useful racehorses: Nikoli (Irish 2000 Guineas), Captain James (Waterford Crystal Mile) and Sutton Place (Coronation Stakes).

Levmoss's third dam, Astrid Wood by Bois Roussel, had two victories from nine starts both as a three-year-old in the Balgriffin Plate at Baldoyle and the Greystones Plate (8f) at Phoenix Park. She was, however, out of her depth in the Irish 1000 Guineas and Oaks though she managed third place as a juvenile in the Beresford Plate (8f) at the Curragh. Astrid Wood was the dam of three winners but more importantly the great grandam of Twig, whose five victories included the Prix de la Côte Normande; Twig Moss, who won the Prix

Noailles and was runner-up to Youth in the Prix du Jockey Club; and finally Tip Moss, who won the Grand Prix d'Evry and Grand Prix de la Ville de Toulouse. Astrid Wood was also the great grandam of Fleet Host whose six victories included the Hollywood Derby and San Luis Rey Handicap.

Astrid Wood's dam, Astrid by Fair Trial, was a very fast two-year-old who won four of her seven races over five furlongs in Ireland. She was owned originally by G.P. Gilpin, then by R.J. Hilliard, and was bought by Joseph McGrath after her third victory in May 1940 in the Dublin Plate at the Curragh. She had originally cost only twenty guineas as a yearling. She proved to be one of McGrath's foundation mares, for amongst her produce was Panastrid, who won the Irish 1000 Guineas amongst her five victories and was herself the dam of the Irish Derby winner, Panaslipper.

It was not really surprising that Astrid cost so little as a yearling as her dam Ethereal only managed two victories from twenty-one outings in a 5-furlong selling plate at Beverley as a juvenile and in the Severn Selling Handicap Plate (6f) at Worcester the next season. In addition Astrid was her only winning produce.

Levmoss stood at the McGrath's Brownstown Stud until 1976 when he was exported to France where he unfortunately died the following year. It is, therefore, impossible to estimate what might have been his true value as a sire after such a short period at stud. His best offspring were Moonlight Night (Musidora Stakes), M'Loshan (Irish St Leger and Grosser Preis von Baden), Shafaraz (Prix du Cadran), Duchamp (Premio Emanuele Filiberto), Moss Trooper (Prix Kergorlay), Shantallah (Chester Vase), Ravel (Dee Stakes), Nuthatch (Nijinsky Stakes) and Small World (Park Stakes). He has already made his mark as a broodmare sire, for Le Melody, the dam of Ardross, was sired by the 1969 Prix de l'Arc de Triomphe winner.

PRIX DE L'ARC DE TRIOMPHE 2400 metres
at Longchamp on Sunday 5 October 1969. For three-year-olds and upwards. Going: Good.
1,000,000 francs to the winner; 400,000 fr. to the 2nd; 200,000 fr. to the 3rd; 100,000 fr. to the 4th. 100,000 fr. to the breeder of the winner; 40,000 fr. to that of the 2nd; 20,000 fr. to that of the 3rd. Value to the winner: 1,181,175 francs.

LEVMOSS	b.c.	by Le Levanstell – Feemoss by Ballymoss	4.60k	Seamus McGrath	W. Williamson	1
PARK TOP	b.f.	by Kalydon – Nellie Park by Arctic Prince	5.58½k	Duke of Devonshire	L. Piggott	2
GRANDIER	b.c.	by Tapioca – Girga by Fine Top	5.60k	Mme Pierre Ribes	M. Philipperon	3
CANDY CANE	b.c.	by Crepello – Candy Gift by Princely Gift	4.60k	Lady Sassoon	G. Thiboeuf	4
PRINCE REGENT	br.c.	by Right Royal – Noduleuse by Nosca	3.55½k	Comtesse de la Valdène	G. Lewis	5
CHAPARRAL	b.c.	by Val de Loir – Niccolina by Niccolo Dell'Arca	3.55½k	Mme Georges Weisweiller	F. Head	6
REMAND	ch.c.	by Alcide – Admonish by Palestine	4.60k	J.J. Astor	J. Mercer	7
BELBURY	b.c.	by Exbury – Belle Sicambre by Sicambre	3.55½k	Mme Léon Volterra	H. Samani	8
BLAKENEY	b.c.	by Hethersett – Windmill Girl – by Hornbeam	3.55½k	Arthur Budgett	E. Johnson	9
ZBOR	b.c.	by Balto – Zaria by Agregat	4.60k	Haras Voshod	A. Zekachev	10
FIASCO		by Pall Mall – Too Friendly by Chulmleigh	7.60k	Mme H. Hausmann	A. Klimscha Jnr.	
COPSALE		by King's Bench – Chrysoprase by Arctic Star	6.60k	Marcos Lemnos	J. Lindley	
SPEED SYMBOLI		by Royal Challenger – Sweet In by Rising Light	6.60k	T. Wada	Y. Nohira	
CARMARTHEN		by Devon – Kuwait by Persian Gulf	5.60k	Mme R.B. Strassburger	J.L. Durry	
ROSELIERE		by Misti – Peace Rose by Fastnet	4.58½k	Charles Engelhard	Y. Josse	
RIBOFILIO		by Ribot – Island Creek by Khaled	3.55½k	Charles Engelhard	R. Hutchinson	
SHOEMAKER		by Saint Crespin – Whipcord by High Treason	3.55½k	Peter Goulandris	D. Keith	
GOODLY		by Snob – Alizetta by Alizier	3.55½k	Maurice Lehmann	A. Barclay	
BONCONTE DI MONTEFELTRO		by Charlottesville – Buontalenta by Zuccarello	3.55½k	Alarico Palmieri	B. Agriformi	
DJAKAO		by Tanerko – Diagonale by Ribot	3.55½k	Baron Guy de Rothschild	J. Derforge	
BEAUGENCY		by Val de Loir – Biobelle by Cernobbio	3.55½k	Mme Pierre Wertheimer	J. Taillard	
YELAPA		by Mossborough – Your Point by Nirgal	3.55½k	Daniel Wildenstein	W.B. Pyers	
CREPELLANA		by Crepello – Astana by Arbar	3.54k	Marcel Boussac	R. Poincelet	
SARACA		by Shantung – Hevea by Herbager	3.54k	Arpad Plesch	Y. Saint-Martin	

24 runners.
LEVMOSS trained by S. McGrath in Ireland; bred by S. McGrath in Ireland.
Distances: ¾ length; 3 lengths; 1 length; short head; neck; 2½ lengths; neck; short neck; ½ length. Time: 2 mins 29.00 seconds.
Pari-Mutuel: Win: 52fr. 50. Places: 11fr. 30; 2fr. 70; 12fr. 50.
S.P.: 2 to 1 Prince Regent; 4 to 1 Park Top; 11 to 1 Roselière and Ribofilio (couplé); 12 to 1 Bonconte di Montefeltro and Djakao; 13 to 1 Saraca; 14 to 1 Chaparral; 19 to 1 Belbury; 24 to 1 Crepellana; 27 to 1 Remand; 30 to 1 Carmarthen; 37 to 1 Goodly; 47 to 1 Yelapa; 52 to 1 LEVMOSS; 59 to 1 Beaugency; 60 to 1 Blakeney; 70 to 1 Grandier and Shoemaker; 93 to 1 Candy Cane; 99 to 1 Zbor, Fiasco, Speed Symboli and Copsale.

LEVMOSS bay colt 1965	Le Levanstell b. 1957	Le Levandou b. 1944	Djebel b. 1937	Tourbillon b. 1928	Ksar
					Durban
				Loïka ch. 1926	Gay Crusader
					Coeur A Coeur
			Lavande br. 1936	Rustom Pasha b. 1927	Son-in-Law
					Cos
				Livadia b. 1927	Epinard
					Lady Kroon
		Stella's Sister ch. 1950	Ballyogan ch. 1939	Fair Trial ch. 1925	Fairway
					Lady Juror
				Serial gr. 1932	Solario
					Book Talk
			My Aid ch. 1933	Knight of the Garter b. 1921	Son in Law
					Castelline
				Flying Aid b. 1918	Flying Orb
					Aideen
	Fee Moss b. 1960	Ballymoss ch. 1954	Moss Borough ch. 1947	Nearco br. 1935	Pharos
					Nogara
				All Moonshine ch. 1941	Bobsleigh
					Selene
			Indian Call ch. 1936	Singapore b. 1927	Gainsborough
					Tetrabbazia
				Flittermere ch. 1926	Buchan
					Keysoe
		Feevagh b. 1951	Solar Slipper b. 1945	Windsor Slipper b. 1939	Windsor Lad
					Carpet Slipper
				Solar Flower br. 1935	Solario
					Serena
			Astrid Wood b. 1944	Bois Roussel br. 1935	Vatout
					Plucky Liège
				Astrid b. 1938	Fair Trial
					Ethereal

Sassafras 1970

Sassafras

The Field

That there were but three horses, Grandier, Blakeney and Beaugency, who had run in the previous season's Prix de l'Arc de Triomphe in 1970 and that the field was the smallest since Exbury won in 1963 can be explained by one name: Nijinsky. Charles Engelhard's champion by Northern Dancer had such a reputation that many possible rivals had been frightened off.

On his racecourse début Nijinsky won the Erne Maiden Stakes at the Curragh. He easily accounted for Decies by 5 lengths in the Railway Stakes, and had 3 lengths to spare over Everyday in the Anglesey Stakes. His most difficult race as a juvenile was when a much improved Decies ran him to three parts of a length in the Beresford Stakes. Decies beat Great Heron, who was to run second to Approval in the Observer Gold Cup, in the National Stakes, so Nijinsky's victory over him bore close inspection. Finally Nijinsky, ridden by Piggott for the first time, had a facile 3-length victory over Recalled, who was still a maiden, in the Dewhurst Stakes at Newmarket.

On his seasonal début in the Gladness Stakes at the Curragh, Nijinsky trounced the very useful four-year-old, Deep Run, by 4 lengths with the 1969 Coventry Stakes winner, Prince Tenderfoot, only third. In the paddock before the 2000 Guineas where he was once more ridden by Piggott, he totally outclassed his rivals and he went down to the start with all the majesty of a Russian prince. Piggott covered him up in the early stages of the race but as soon as he gave him his head, the Northern Dancer colt effortlessly cut down his opponents. On the line he had Yellow God, the Gimcrack Stakes winner, toiling 2½ lengths in his wake with the Prix Djebel victor, Roi Soleil, the same distance away third. It was a display that could not be faulted.

Inevitably Nijinsky started favourite for the Epsom Derby but at the generous odds of 11 to 8 against. Like Sir Ivor in 1968 there had been doubts and questions about his ability to stay a mile and a half. His young sire, Northern Dancer, had been the champion of North America in 1964 when he won the Kentucky Derby, the Preakness Stakes and the Queen's Plate over ten furlongs at Woodbine, but he had only managed third place in the Belmont Stakes. Nijinsky's dam, Flaming Page by Menow, also won the Queen's Plate and Canadian Oaks but had never won over a mile and a half. The question mark about Nijinsky's stamina was one reason why he was permitted to start odds against at Epsom; the other was the presence in the field of the Sea Bird colt Gyr. He was trained in France by Etienne Pollet who, after Gyr had finished a highly promising fourth to Breton in the Grand Critérium, decided to postpone his retirement for one year in order to train him as a three-year-old. Gyr stood over 17 hands and was not a particularly elegant colt. On his reappearance in the Prix Daru in April he was severely hampered in the straight and still had over 3 lengths to make up on the leader Armos with only 150 metres to go. Yet Gyr produced such a devastating burst of speed that he caught Armos on the line and beat him a nose. On 10 May he defeated Grad and Larko 2 lengths and a neck in the Prix Hocquart in workmanlike rather than brilliant fashion.

Nijinsky and Approval were the pick of the paddock at Epsom; Gyr was very excited and tense and banged his head on the stalls when they opened. Piggott kept Nijinsky in fifth or sixth place and he was still on the bridle as the field swung round Tattenham Corner, which Gyr negotiated with unexpected ease for such a big colt. A little over two furlongs out Gyr went into a clear lead and for a few seconds looked like the winner. Piggott, however, was merely biding his time; when he asked Charles Engelhard's colt to challenge the response was immediate and electrifying. He closed the gap between him and Gyr in a matter of strides and after one slap of the whip, Nijinsky swept by to win by 2½ lengths with the Prix Lupin winner, Stintino, 3 lengths away third. It was the first time Charles Engelhard had seen Nijinsky run. The South African magnate had bought him for $84,000 from E.P. Taylor at the Woodbine Yearling Sale in Toronto. Amongst his guests at Epsom that afternoon was the widow of the immortal Russian dancer after whom his champion had been named.

The opposition for Nijinsky in the Irish Sweeps Derby at the Curragh where he was ridden by Liam Ward, was far from formidable. Meadowville, a well-beaten fifth at Epsom, seemed the only possible danger. Although he sweated up in the preliminaries the Northern Dancer colt easily accounted for Meadowville and Master Guy by 3 lengths and the same.

Nijinsky now took on the older generation in the King George VI and the Queen Elizabeth Stakes. Not one other three-year-old dared to oppose him. On paper the field appeared quite strong; his opponents included Karabas (Washington D.C. International and Hardwicke Stakes), Caliban (Coronation Cup), Blakeney (Epsom Derby), Hogarth (Derby Italiano 1968, Premio Presidente della Repubblica 1970) and Crepellana (Prix de Diane). In reality this was a false impression. Karabas was nowhere as good a horse as the previous season; Caliban had been a very lucky winner of the Coronation Cup; Blakeney had been a disappointment since his Derby victory; and Crepellana was not as good a filly as she had been in 1969 for she had only finished seventh to Grandier in the Prix Ganay and had been beaten 2½ lengths by the Prix du Jockey Club third Caro in the Prix d'Ispahan. It was a great shame that Park Top had injured a tendon in the Coronation Cup and could not run. With the exception of Nijinsky this was not up to the usual class of field one associates with England's most important weight for age race. Nijinsky, however, made mincemeat of all his rivals. Coming from behind he cut them down in a matter of a few strides approaching the furlong pole. Piggott was able to look round and ease him fifty yards from the winning post. His nearest pursuer, Blakeney, was a very flattering 2 lengths behind with Crepellana 4 lengths back third. Nijinsky could not have done more than win so effortlessly. Nonetheless the adulation, praise and superlatives that echoed around Ascot that afternoon might have been tempered with a little more caution.

Misfortune now struck Nijinsky, for in August he suffered a severe attack of ringworm and lost all his hair from one flank. Inevitably this held up his preparation for the St Leger and it was only decided to run him at Doncaster at the beginning of September. No colt had won the Triple Crown since Bahram in 1935 and the opposition Nijinsky was to face did not appear very strong. In addition it was difficult for O'Brien to find a preparatory race where Nijinsky would not have to concede large quantities of weight to his opponents before he ran in the Prix de l'Arc de Triomphe.

The Northern Dancer colt started at 7 to 2 on, and understandably Piggott nursed him to a length victory over Meadowville with Politico half a length away third. In reality it was a hollow victory; Meadowville had been a most disappointing eighth to Schönbrunn when even money favourite in the Grand Prix de Deauville, while Politico had only managed fifth place to Roll of Honour in the Grand Prix de Paris. Both could be reckoned to be a stone behind Nijinsky at his best. The St Leger showed that he had made a recovery from the ringworm but in fact told us nothing about his ability to stay either the distance of the St Leger or even a mile and a half, as the opposition at Doncaster was too inferior to give him a proper test.

On account of the gradients on the course at Epsom the Derby can be won by a high-class colt, who does not really stay a mile and a half, but is in reality a ten or eleven furlong horse. Sir Ivor was a notable example. After the St Leger all doubts about Nijinsky's stamina had evaporated. He was hailed by some to be as great as Ribot or Sea Bird; by others, more enthusiastic and uncritical, as the horse of the century. His victory at Ascot had silenced all but a few sceptics regarding his stamina. Nevertheless there is no doubt that the mile and a half course at Ascot is less stiff and demanding than that at Longchamp. The first five furlongs on the Berkshire course are all downhill until Swinley Bottom is reached. A horse with suspect stamina can be held up at the rear of the field and not have to begin to race in earnest until meeting the long but gradual climb to the straight and the none too steep ascent to the winning post. At Longchamp, however, the first five furlongs are all uphill and a runner, whose staying ability is in question, cannot be conserved in the same fashion if he is to keep in touch with the leaders. On the descent to the straight it is difficult to give such a horse a breather, especially in the Prix de l'Arc de Triomphe, since all the runners are jostling to get into a challenging position once the straight is reached. The straight, like Ascot, is a gradual climb to the winning post. It is probable that Longchamp's mile and a half course is the stiffest and most searching test of stamina of any major racecourse in Europe. Therefore the 1970 Prix de l'Arc de Triomphe, for which he started 5 to 2 on favourite, was likely to provide Nijinsky with the greatest test he had ever faced. Although ostensibly on paper and in the great majority of people's minds, the outcome was a mere formality, there remains a suspicion that some people's judgment at that time was mesmerised by the brilliant and explosive bursts of acceleration that Nijinsky had shown at Newmarket, Epsom and Ascot.

The connections of the Italian champion, Ortis and his pacemaker Lar, were in no way deterred by Nijinsky's reputation. Dr Carlo Vittadini's colt was regarded in Italy as a potential successor to Ribot. He needed the race when beaten a head by Alcamo in the Premio Calvirate in the early spring, but he reversed the form in the Premio Emanuele Filiberto at San Siro on 9 April and beat his conqueror 4 lengths. Ortis next took the Derby Italiano in Rome on 7 May by 3½ lengths from Alcamo. It was, however, in the Gran Premio d'Italia at San Siro on 7 June that he put up his most impressive performance. Taking the lead 2½ furlongs out, Ortis drew away from the Irish 2000 Guineas winner, Decies, to win unextended by 5 lengths. Dr Vittadini's colt was rested until the Premio Besana (10f) at San Siro on 18 September when he ran out an 8-length winner over the filly Marietta Robusti who was 5 lengths clear of Alcamo. The latter's victory in the St Leger Italiano on 27 September naturally increased the confidence in Ortis, which had already been raised when Lar ran a respectable third to A Chara and Tapalque in the Prix du Prince d'Orange on 20 September. Lar, who had won several races at Varese between 7 furlongs and 10 furlongs, ran in that trial in order to give Ortis's jockey A. Di Nardo some experience of riding at Longchamp. Ortis was to start third favourite on 4 October at 35 to 4.

Winston Guest's Gyr was slightly preferred to the Italian challenger by the French *turfistes* at 15 to 2. He also was accompanied to the post by a pacemaker, the five-year-old Golden Eagle. Gyr ran once between the Derby and the Prix de l'Arc de Triomphe in the Grand Prix de Saint-Cloud on 5 July. Before the race he sweated profusely and then was slowly away. He pulled hard throughout, but took the measure of Grandier inside the final furlong and won by 2 lengths with Hallez a short head away third. He was certainly a colt of ability but, like so many of Sea Bird's colts, was nervous and temperamental. It was felt that Longchamp would suit him more than Epsom and it was hoped he would give Nijinsky a harder fight on 4 October.

Gerald Oldham's Stintino also challenged the Derby winner again. Before coming to Epsom he was unbeaten, having won both his outings as a juvenile, including the Critérium de Maisons-Laffitte from the future Prix Greffulhe and Prix Lagrange winner Magic Hope. He had begun his three-year-old campaign with a facile win in the Prix de Guiche and then, coming from last to first in the straight in the Prix Lupin, had beaten the Prix Noailles winner, Dragoon, and Sassafras 2 lengths and a neck with Caro, the victor of the Poule d'Essai des Poulains, 3 lengths away fourth. Unfortunately Stintino had developed the very bad trait of losing ground at the start. Slowly away at Longchamp in the Prix d'Ispahan at the end of June, he gave himself an impossible task at the weights. He trailed the field by 20 lengths after 200 metres and did well to finish third, 2½ lengths and 3 lengths behind Caro and Crepellana, with such useful horses as Lorenzaccio and Prince Regent behind him. Stintino gave the useful English colt, Golden Monad, 7 lbs in the Prix de Chantilly on 11 September and beat him and Armos a length and 2 lengths. Nevertheless the best that Stintino's connections could hope for on 4 October was one of the minor berths.

The other French three-year-old colt in the line-up against Nijinsky, Arpad Plesch's Sassafras, had raced entirely in France and had done little wrong in 1970. He had not been considered one of the best juveniles in François Mathet's yard. In four outings as a two-year-old he had won once, the Prix de la Besbré (7f) by 10 lengths at Vichy, but had only been runner-up in the Grand Critérium de Bordeaux and fourth in the Prix de Chênes. A slightly unlucky runner-up to Magic Hope in the Prix Lagrange at Saint-Cloud in March, Sassafras then accounted for moderate opposition in the Prix Androcles at Maisons-Laffitte in April. He had comfortably won the Prix la Force before finishing third to Stintino in the Prix Lupin. In the Prix du Jockey Club Sassafras launched his challenge inside the final 200 metres and beat Roll of Honour and Caro three-quarters of a length and half a length. This performance was made to look even more meritorious when Roll of Honour won the Grand Prix de Paris, while Caro took the Prix d'Ispahan and then put up an excellent display when running Highest Hopes, who was receiving 12 lbs, to 1½ lengths in the Prix Eugène Adam at Saint Cloud on 19 July. Sassafras did not run between the Prix du Jockey Club and the Prix Royal Oak for which he started 5 to 2 on favourite, but almost certainly was in need of the race. He was crossed by Hallez, the former pacemaker for Dragoon, who had broken a leg at Chantilly, and Saint-Martin had to snatch him up. By the time he got going again Hallez was gone beyond recall and passed the post 4 lengths ahead of Arpad Plesch's Prix du Jockey Club winner. The Stewards relegated Hallez to second place. There was a body of opinion who, on account of the distance of Hallez's victory, felt the interference did not affect the result, but that was purely a subjective point of view. It took Sassafras some time to get going again and it is impossible to say what the result would have been if no interference had taken place. The merit of Sassafras's performances in 1970 was quite definitely undervalued by the controversy surrounding Hallez's disqualification in the Prix Royal Oak, and he started at the long odds of 19 to 1 on 4 October. Nevertheless Sassafras was undoubtedly a genuine and consistent colt, who got a mile and a half well, and in all probability would be concerned in the finish of the Prix de l'Arc de Triomphe.

Achille Fould's Miss Dan was the solitary three-year-old filly in the field against Nijinsky. On 20 June she had won the Prix Fille de l'Air at Saint-Cloud by 5 lengths, and had run fourth 2½ lengths behind Santa Tina in the Prix de Pomone on 23 August at Deauville. Santa Tina had won the Irish Guinness Oaks. Miss Dan showed how quickly she was improving when she ran Highest Hopes to 1½ lengths in the Prix Vermeille with such good fillies as the Epsom Oaks heroine, Lupe, and Sweet Mimosa trailing behind her. Highest Hopes was aimed at the Champion Stakes to avoid tackling Nijinsky at Longchamp. She had been runner-up to Sweet Mimosa in the Prix de Diane but by September could be considered the best filly on either side of the Channel. Although she started at 38 to 1 in the Prix de l'Arc de Triomphe, Miss Dan's rapid improvement in the late summer made her the most attractive outsider on Sunday 4 October.

Both Prince Regent, a bitter disappointment as a four-year-old, and Park Top, whose tendon injury to her forelegs kept her off the racecourse after her unlucky defeat by Caliban in the Coronation Cup until her victory over Yellow River and Highline in the Cumberland Lodge Stakes at Ascot, were absentees from the field of the 1970 Prix de l'Arc de Triomphe. Grandier, third to Levmoss and Park Top in 1969, appeared the best of the older rivals of Nijinsky. In need of the race on his seasonal reappearance behind Yelapa in the Prix d'Harcourt in April, he emphatically reversed the form in the Prix Ganay on 3 May when he had 2½ lengths and three-quarters of a length to spare over Yelapa and Prince Regent. On his only other outing before 4 October Grandier had finished runner-up to Gyr at Saint Cloud. Starting at 32 to 1 Mme Pierre Ribes's horse was just preferred to Pierre Wertheimer's Beaugency, who started a point longer. He was far from disgraced when beaten a neck and a short head by Shoemaker and Yelapa in the Prix Jean de Chaudenay at Saint-Cloud on 18 May. He landed the Gran Premio di Milano at San Siro on 21 June by 2 lengths and half a length from Bacuco and Hogarth, but his limitations were shown up when only fourth, a little over six lengths behind the Deutsches Derby winner Alpenkönig in the Grosser Preis von Nordrhein-Westfalen at Düsseldorf on 19 July. Nevertheless he put up a game display to give Lorenzaccio a kilo and run him to a half a length in the Prix Foy at Longchamp on 6 September, but on balance he would have to improve several pounds to be concerned in the finish of the Prix de l'Arc de Triomphe.

The 1969 hero, Blakeney, again took on Charles Engelhard's champion on 4 October. He started at the almost insulting odds – for a Derby winner – of 43 to 1, but that was because his reputation had become so tarnished since his greatest victory. In 1970 Arthur Budgett's home-bred colt had only finished fifth in the Jockey Club Stakes at Newmarket in April and was not at all impressive when accounting for the handicapper, Hey Ho, in the Ormonde Stakes at Chester. In the Ascot Gold Cup he appeared to have the race at his mercy when he drew level with Precipice Wood two furlongs from home, but the latter ran on with greater fortitude and beat Blakeney three-quarters of a length; on his running in the King George VI and the Queen Elizabeth Stakes it seemed he would be very fortunate to reach the frame.

The five-year-old provincial mare, La Bijute, who started at the exceptionally ungenerous odds of 45 to 1, had no place in a field of this class and quality. Trained in Brittany by her owner François Carnus where she normally raced, La Bijute was to be partnered by a young jockey called Guérinet, who had never ridden on any Parisian racecourse. At Deauville she had finished second in a minor handicap but had carried the feather weight of 7 st 6 lbs; that was one of her few ventures away from the provincial tracks in Brittany.

The field for the 1970 Prix de l'Arc de Triomphe was completed by three complete outsiders. Lady Brooke's A Chara had won a substandard Prix du Prince d'Orange and the Prix Guillaume le Conquérant at Deauville. Soyeux, tenth behind Vaguely Noble in 1968, was but a shadow of his former self and his victory over the English handicapper New Member in the Prix de Reux at Deauville was his first for two seasons. Finally, Hubert de Chaudenay's four-year-old Quinquet, who had won a minor event at Saint-Cloud over twelve and a half furlongs, had only managed fourth place to Shoemaker in the Prix Gontaut-Biron and sixth place to Schönbrunn in the Grand Prix de Deauville. None of these three could be seriously considered on 4 October.

The Final Preparations

Arthur Budgett had stated that he did not wish to risk Blakeney at Longchamp if the going was firm. Fortunately the fine weather in Paris broke in the last week in September, and Budgett announced Blakeney a definite runner on Thursday 1 October. Gyr did a very good ten and a half furlong gallop in company with Prince Regent and Golden Eagle on 22 September, after which Pollet stated that Winston Guest's colt was 'a much better horse than he was in the spring, easier to handle' and that 'his only problem is nervousness'. Ortis, a big powerful colt, impressed all who saw him work at Maisons-Laffitte and Enrico Camici was adamant that he was as good as Ribot. On 29 September Stintino did a particularly good spin over nine and a half furlongs while Sassafras, with Yves Saint-Martin up, was accompanied by five stable-companions and covered ten furlongs in workmanlike fashion. Later his trainer, François Mathet, remarked that 'races are very tough these days and it is difficult to see how a three-year-old, who has not been rested, can remain at the top of his form from the 2000 Guineas to the Prix de l'Arc de Triomphe'. These doubts about Nijinsky's inevitable triumph were very much in the minority and were in sharp contrast to the ebullient statement made by Liam Ward, who had partnered Charles Engelhard's colt to six of his victories: 'I'll walk to Tipperary in my birthday suit if Nijinsky ever gets beaten.'

Président Pompidou, who was to go to the Soviet Union later the following week, was not at Longchamp on Sunday 4 October. But an enormous crowd, even by the normal standards of the Prix de l'Arc de Triomphe, gathered at Longchamp to see Nijinsky run. William Rees-Davis, the M.P. for Thanet, had chartered two aircraft to bring members of the Racegoers' Society over for the weekend. There were also a great many visitors from across the Atlantic attracted by the colt's ever-increasing reputation. James Goldsmith was seen at Longchamp with his sixteen-year-old daughter, Isabel. Her mother had been the Bolivian-born tin heiress, Isabel Patino. She had eloped with Goldsmith when she was 18 and had married him in Kelso, Scotland. Sadly she had died in May 1954 only a few hours after the young Isabel's birth. James Goldsmith was to be seen proudly introducing his daughter to his friends the Earl of Suffolk and Sir Max and Lady Jane Rayne, who were also at Longchamp that afternoon. Omar Sharif and Kevin McClory were noted discussing how much they should invest on Nijinsky as if the possibility of defeat was as remote as either of them taking part in an Apollo mission to the moon.

The Race

The afternoon began on a high note for both Lester Piggott and the cross-Channel visitors. Piggott guided Balidar to a neck victory over Huntercombe in the Prix de l'Abbaye de Longchamp. In the paddock Gyr's temperament let him down: he was awash with sweat and in an extremely nervous state. When, after the parade, he turned to go down the course to the start, he set off with such prodigious and headstrong bounds that the crowd gasped. There were innumerable photographers around Nijinsky in the preliminaries and he also was sweating. In fact he did not look as well as he had done before either the Epsom Derby or the King George VI and the Queen Elizabeth Stakes. The photographers were, in fact, treating the Triple Crown winner as they might have done Vaslav Nijinsky had he returned from the dead, and this mobbing of the colt can only have had a detrimental effect on his chance in the race. The scenes in the paddock were more reminiscent of a Hollywood première than a racecourse, but it was Nijinsky who was the star. Blakeney, on the other hand, was totally undisturbed by the hiatus and looked fit and well.

When the field was dispatched, Lar and Golden Eagle, the pacemakers for Ortis and Gyr, immediately took up the running at a good pace, tracked by Sassafras, Ortis and Blakeney. Gyr in the mid-division was pulling violently for his head and refused to settle. Piggott slotted Nijinsky into about tenth place while Stintino, who as normal had dwelt at the start, was amongst the backmarkers. After 300 metres Golden Eagle just headed Lar, but the order behind remained virtually the same with Gyr still endeavouring to pull himself into the chasing group. Approaching the top of the hill Golden Eagle and Lar were followed by Miss Dan, Sassafras, Ortis, Blakeney and Gyr. The Triple Crown winner was still lying tenth, approximately 15 lengths behind the leaders. The order did not alter much coming down the hill, but Piggott slowly began to close the gap between Nijinsky and the leaders, although he was still 10 lengths in arrears as Golden Eagle, Lar and Ortis, who had improved a couple of places, began the turn into the straight. The two pacemakers, their jobs done, rapidly dropped out of contention at this stage and Miss Dan came through to take it up from Ortis, who to the consternation of the Italians was clearly coming under pressure. They were followed by Sassafras, Gyr, Blakeney and Beaugency who had improved steadily on the descent. Nijinsky was still only tenth 10 lengths behind. The race was on in earnest and the pace increased. With 400 metres to go Sassafras moved up easily to dispute the lead on the outside of Miss Dan who was racing on the rails, and he soon mastered the filly and went into the lead. Piggott, meanwhile, had been unable to find an opening on the inside

Sassafras (rails) inflicting the first-ever defeat on Nijinsky (9) by a head with Miss Dan (15) and Gyr (10) 2 lengths and 1½ lengths away third and fourth

and so had switched Nijinsky to the outside to unleash his challenge, but he still had at least 8 lengths to make up on Sassafras. The Triple Crown winner's smooth but powerful acceleration quickly took him past all his rivals except the Prix du Jockey Club winner, but 200 metres out he had reached his quarters. A hundred metres from the winning post Nijinsky caught Sassafras and smoothly went a neck to the good. Everyone at Longchamp then expected him to surge past Arpad Plesch's colt. For a moment the result of the 1970 Prix de l'Arc de Triomphe looked as if it were to be what all had anticipated. But in a split second the entire picture altered dramatically. Nijinsky began to veer to the left under pressure and to lose momentum. Yves Saint-Martin, realising his opponent was in difficulty, called on his mount for a renewed effort. Arpad Plesch's genuine and courageous colt fought back tenaciously, got his head in front again 25 yards from the post where he was a head in front of Charles Engelhard's champion, with Miss Dan 2 lengths away third and Gyr 1½ lengths behind her. Blakeney ran on to be a similar distance away fifth, a short head in front of Beaugency, who was followed by Soyeux, Quinquet, Grandier, A Chara, with the disappointing Ortis and Stintino in rear. The inquest began immediately, because for some the unthinkable had occurred: Nijinsky had been beaten for the first time in his life.

Immediately after the race Piggott was of

the opinion that 'Nijinsky did not have it today. He was not the same horse. Three furlongs from the post I knew that I would have a hard time.' Yves Saint-Martin related that he 'was ahead and looking back for Nijinsky. When Piggott ranged up alongside me I thought it was all over and I would be second. I saw Nijinsky was in difficulties and noticed he was veering left. I called on Sassafras for one final effort and we made it.' Vincent O'Brien was convinced that the severe attack of ringworm had had a much graver effect on the Triple Crown winner than was realised at the time. After he won the St Leger, Nijinsky had lost 29 lbs in weight on his return to Ireland which made his trainer feel that he was not as fit as he should have been at Doncaster. Although he quickly regained the weight, O'Brien felt that the Prix de l'Arc de Triomphe came too soon after the St Leger.

Piggott inevitably was blamed by many people for giving Nijinsky far too much to do in the straight; but there was no apparent change of tactics at Longchamp compared with Epsom and Ascot. He had always ridden the Northern Dancer colt in that manner. There had never been any criticism of the English champion's tactics when Nijinsky had won. Some argued that Nijinsky had run below his best form at Longchamp but that seems unlikely since he finished a length further in front of Gyr at Longchamp than he had done at Epsom and he had increased his Ascot margin over Blakeney by 3 lengths. The Derby winner had shown his astonishing burst of acceleration to head Sassafras, the same overdrive that had impressed so many in the eleven previous victories. Then nevertheless he had been unable to go through with his effort. Sassafras was, in fact, the first three-year-old colt whose temperament and stamina were never questioned. While both Gyr and Stintino were potentially brilliant, the former's excitability and the latter's quirk for losing ground at the start meant that they could not be guaranteed to display their undoubted ability on all occasions. Although Nijinsky had beaten Karabas into fourth place at Ascot, the 1969 Washington D.C. International Stakes winner was so badly struck into that he was forced to retire to stud immediately afterwards and cannot therefore be used as a yardstick. Meadowville's defeat by Allangrange in the Irish St Leger moreover did little to enhance the value of Nijinsky's success at Doncaster.

There were probably three reasons for the Triple Crown winner's failure to beat Sassafras. First the attack of ringworm and his interrupted training schedule must have been contributory causes. Secondly the fact that he had not had a genuine rest for nearly six months was another factor in his disfavour. Thirdly and almost certainly the major reason was that he did not truly stay the very stiff and demanding mile and a half course at Longchamp. Like Sir Ivor, Nijinsky was a brilliant ten and eleven furlong colt with a burst of explosive acceleration, even superior to that of the 1968 Derby hero. When held up over a mile and a half, this astonishing burst of speed had swamped his previous opponents, but on 4 October he met in Sassafras a really high-class genuine stayer and for the first time tasted defeat. Lester Piggott had some compensation in the race immediately after the Prix de l'Arc de Triomphe for he steered Gold Rod to victory in the Prix du Moulin de Longchamp.

All the controversy and astonishment at Nijinsky's defeat took the attention away from the victory of Arpad Plesch's Sassafras. It was very unfortunate that the inquest detracted from his greatest success for it had been most meritorious and deservedly gained. Sassafras had, in fact, missed setting a new record for the race by less than a second and that was in conditions not ideal for a very fast time. Surprisingly, considering he was unquestionably the best French jockey riding for several years, it was the first time Yves Saint-Martin had ridden the winner of the Prix de l'Arc de Triomphe. In 1966 François Mathet had said of Saint-Martin: 'I have often dreamed of being able to communicate with a jockey during a race. With Saint-Martin it is not necessary because he is always in the right spot. He has a great feel for a horse and a nervous one will settle under him immediately. He is cool, has exceptionally quick reflexes and has grown into a strong finisher.' This comment was indeed praise from a trainer whose standards were the highest but he had seen Saint-Martin's career develop from the outset for it was to his stable in Gouvieux on 15 September 1955 that the fourteen-year-old Agenais had arrived in Chantilly to begin his apprenticeship. For François Mathet it was, of course, the third success in Longchamp's great race, for he had been responsible for the dual winner Tantième in 1950 and 1951.

Sassafras's owner Arpad Plesch, a Hungarian financier, and his wife had been enthusiastic supporters of the Turf for several years. They had won the Prix du Jockey Club with Talpaque in 1968, the Poule d'Essai des Poulains with Thymus in 1959, the Prix Vermeille with Saraca in 1969 and the Epsom Derby in 1961 with the total outsider Psidium. This was the first occasion, however, when their colours were carried successfully in the Prix de l'Arc de Triomphe. After the race Madame Plesch exclaimed: 'Never in my wildest dreams did either my husband or I think that Sassafras could topple Nijinsky.' Not merely did they own him but Sassafras was also bred by them in Ireland.

The Rest of the Season

Unfortunately Nijinsky ran again in the Champion Stakes at Newmarket. Prior to the Prix de l'Arc de Triomphe the original intention had been to run him in the United States in the Man o'War Stakes at Belmont Park but the plan was changed. The Champion Stakes comes much too soon after the Longchamp race and its history is littered with the failures of winners and runners who have been tested to their limit in the Prix de l'Arc de Triomphe. Sir Ivor is one of the few exceptions to this general rule. There was an enormous crowd of 20,000 to bid farewell to Nijinsky but all the attention and adulation lavished on him made him upset and nervous in the preliminaries.

Charles St George's Lorenzaccio took the lead from the start. A furlong from home the Triple Crown winner, lying second, looked to have every chance but none of the old sparkle was there. He failed to quicken and was beaten 1½ lengths with Hotfoot only a neck behind third. Piggott reported that Nijinsky had worked himself up into such a state of nerves before the start that he 'never felt we would win'. It was a very sad ending to such a brilliant and illustrious career on the racecourse and in hindsight both Vincent O'Brien and Charles Engelhard's racing manager, David McCall, regretted the decision to run him at Newmarket.

The other runners in the 1970 Prix de l'Arc de Triomphe, who ran later in the autumn, performed reasonably. Ortis, who did not entirely satisfy his trainer in his gallops on his return to Italy, ran Bacuco to a head in the Gran Premio del Jockey Club at San Siro on 18 October. Miss Dan, however, ran a very fine race to be second, a length behind Fort Marcy, in the Washington International D.C. Stakes at Laurel Park with Bacuco, who had won the Premio Roma on 1 November, 5 lengths away third, Lorenzaccio fifth and Beaugency last. Quinquet, however, proved no match for Armos in the Prix du Conseil Municipal and finished out of the frame.

Pedigree and Stud Record

Sassafras never ran again after his victory in the Prix de l'Arc de Triomphe. He was by Sheshoon by Precipitation, the very talented half-brother to Charlottesville. Sheshoon's victories included the Ascot Gold Cup, the Grand Prix de Saint-Cloud, the Grosser Preis von Baden, the Prix de Barbeville and the Prix Gontaut-Biron. He had run ninth to Puissant Chef in the 1961 Prix de l'Arc de Triomphe but had been totally unsuited by the extremely heavy going. In addition to Sassafras he was the sire of Pleben (Grand Prix de Paris and Prix Royal Oak), Samos III (Prix Royal Oak), Mon Fils (2000 Guineas), Stintino (Prix Lupin), Vela (Critérium des Pouliches), Oak Hill (Critérium des Pouliches), Riverside (Prix de Royallieu), Flaming Heart (Prix de Minerve), Green Shoon (Grand Handicap International d'Ostende) and Society (Blue Riband Trial Stakes).

When Sassafras went to stud in Ireland, where he stood at the Ballylinch Stud in Co. Kilkenny, he had run eleven times; he had won six races, had been runner-up twice, third once and fourth once, and he had collected 3,306,406 francs for his owner in prize money. His dam, Ruta by Ratification, was bred by Maurice Hennessy and was bought by Arpad Plesch at the 1961 Deauville sales for a little under five thousand guineas. Plesch sent her to Ireland where she ran twice as a juvenile in 1962 and won a small race over five furlongs worth £495. Ruta proved a useful broodmare and was the dam of six winners, who won fourteen races between them in France. Sassafras stood head and shoulders above the other five, the best of which was Sorbi by Stupendous, who was successful on three occasions and was runner-up in both the Prix Saint-Roman and Prix la Force.

Ruta's dam, Dame d'Atour by Cranach, was unraced and was owned by Baron Guy de Rothschild. She was the dam of three winners, the best of which unquestionably was Roi Dagobert by Sicambre. As a two-year-old he won the Prix de Blaison and was second in the Critérium de Maisons-Laffitte and runner-up to Silver Cloud in the Grand Critérium. He was unquestionably the best three-year-old colt in France the following year winning the Prix Greffulhe, the Prix Noailles and Prix Lupin after which he suffered a setback, but on his only other run as a three-year-old he finished a close fourth to Topyo in the Prix de l'Arc de Triomphe and could be regarded as that year's moral winner. Dame d'Atour was also the dam of Dame des Clefs by Cernobbio, who won twice and ran third in the Prix Fille de l'Air. She was the dam of Dame des Ondes, whose three victories included the Prix Vanteaux, and the grandam of Water Boy, who won four times including the Prix d'Orgemont, the Prix des Yearlings, the Prix Gladiator and was third in the Prix Lupin.

Dame d'Atour's dam was Barley Corn by Hyperion. Barley Corn was owned and bred by Lord Derby and was trained in England by Harry Peacock. She ran twice as a juvenile and ran third in a field of four in the East Lancashire Stakes (5f) at Haydock Park. She had three outings as a three-year-old but only managed second place to Fair Flindra, who had whipped round at the start but still won in the Castle Maiden Plate (10f) at Carlisle. In her final race at Hamilton Park she first refused to leave the paddock and her jockey had to mount her on the course; she then refused to start at all. It was not, therefore, surprising that Lord Derby sent her to the Newmarket December Sales in 1953 where she was bought by Baron Guy de Rothschild for 4,100 guineas. She proved an absolute bargain, for before her death in 1968 she produced six winners. Her first foal was the brilliant but unlucky Shantung by Sicambre, whose three victories included the Prix de la Rochette, but he is best remembered as the unluckiest loser of the Derby since the Second World War, when he finished third to Parthia and Fidalgo; he also ran Harbager to a head in the Grand Prix de Saint Cloud. Barley Corn's second foal, Imberline by Ocarina, won twice including the Prix Edgard de la Charme and was third to Never Too Late II and Paimpoint in the Oaks. Imberline herself became a very useful broodmare and was the dam of Corduroy, who won five races and was runner-up in the Prix Greffulhe and Prix Eugène Adam, of Percale, whose two wins included the Prix de Flore and who was second in the Prix Vermeille, Prix de Royallieu and Prix de Pomone, and of Cigaline, who won two races including the Prix Pénélope. Imberline was also the grandam of Poil de Chameau (Prix Hocquart) and Paysanne, who won three races including the Prix Vermeille when she dead-heated with San San. Barley Corn's third foal, Diagonale by Ribot, achieved one victory and ran second in the Prix Perth. Diagonale was the dam of Djakao (Grand Prix de Deauville) and Cheryl (Prix de l'Opéra). Barley Corn's other three winning progeny, Faille by Alizier, Crepon by Crepello and Calcot by Tanerko, won nine races between them.

Barley Corn's dam, Schiaparelli by Schiavoni, was a useful if not outstanding racemare. In fifteen outings at two and three she won five races, the Zetland Plate (12f) at Stockton, the Three Tuns Plate (12f) at Thirsk, the Spinner's Plate (12f) at Manchester, the Park Plate (12f) at Haydock Park, and the Muswell Plate (13f) at Alexandra Park. But when she tackled a higher class in the Gold Vase at Ascot she was unplaced. As a broodmare, however, she was an enormous success. Her first foal Herringbone by King Salmon won two of her five starts as a juvenile. In 1943 Herringbone beat Ribbon a neck in the 1000 Guineas, finished fourth in the Oaks to Why Hurry and then beat Ribbon and the Derby winner Straight Deal a short head and three-quarters of a length in the St Leger. Herringbone was the great grandam of King's Lake by Nijinsky (Irish 2000 Guineas, Sussex Stakes and Joseph McGrath Memorial Stakes) and Cloonara by Sir Ivor (Phoenix Stakes) and champion two-year-old filly in Ireland in 1976.

Schiaparelli's Blue Peter colt Masthead won four races including the Newmarket St Leger, and was twice runner-up to Migoli. But he did not compare with Swallow Tail by Bois Roussel, whose six victories included the Chester Vase and the King Edward VII Stakes, and who came third to Nimbus and Amour Drake in the very Nimbus and Amour Drake in the very controversial 1949 Derby. Some considered Amour Drake very unlucky, but Swallow Tail, only a head and the same behind Nimbus, certainly was hampered by the winner inside the final furlong. Schiaparelli's best other winner was North Cape by Borealis and she produced four others in addition who won eight races between them.

Schiaparelli's dam, Aileen by Nimbus, was French bred and had been bought as a ten-year-old by Lord Derby from the Duc Decazes in

1928. She had won eleven flat races and four hurdle races in France in 1921 and 1922 and had bred four winners of several races under both rules in France before her export to England.

Aileen's dam was Yveline who won five races including the Prix de Diane and produced six winners the best of which was Ivanoe, third in the 1928 Prix du Jockey Club. Photine, her dam, was a daughter of Primavista, the dam of Priori, the victor of the 1925 Prix de l'Arc de Triomphe, the Prix Royal Oak and the Prix du Cadran.

Thus it can be seen that Sassafras was extremely stoutly bred on his maternal side and indeed came from an exceptional Anglo-French family. At stud Sassafras was rather neglected by English and Irish breeders and many of his best winners have run abroad. Nevertheless he has got Henry le Balafré (Prix Royal Oak), Dom Alaric (Grand Prix de Deauville twice, Niagara Handicap), Galway Bay (Coventry Stakes, V.R.C. Craven 'A' Stakes), Floressa (Prix de Royaumont), Nasaari (Prix Greffulhe), Glenaris (Hollywood Oaks), Marmolada (Oaks d'Italia, Grand Premio Citta di Torino, Premio Lydia Tesio, Premio Federico Tesio), Sourbancies (Premio Parioli), Moulouki (Prix Jean de Chaudenay), Lotar (St Leger Italiano, Premio Emanuele Filiberto, Premio Principe Amadeo), Ho Han Wai (Prix Corrida), Dogberry (Old Newton Cup), Hokuei One (a Group winner in Japan), Sassabunde (2nd Irish Guinness Oaks), Gap of Dunloe (3rd Prix du Jockey Club), Salora Lady (2nd Park Hill Stakes) and the useful stayer Military Band. He was exported to the U.S.A. in 1980 and is now standing at the Pillar Stud in Lexington Kentucky. It might have been better to have stood him in France where his true merit was better known than to have sent him to Ireland, for possibly in France he would have got better quality mares.

There can be no doubt however, that his record at stud, reasonable as it has been, in any way compares with the brilliant success of Nijinsky as a stallion. He may well prove to be even more influential than his own sire Northern Dancer. Nevertheless on Sunday 4 October Sassafras proved the better colt over a mile and a half at Longchamp and the distinction he earned that afternoon was to become the first horse ever to defeat Nijinsky, a feat which should not be underrated.

PRIX DE L'ARC DE TRIOMPHE 2400 metres
at Longchamp on Sunday 4 October 1970. For three-year-olds and upwards. Going: Good.
1,200,000 francs to the winner; 480,000 fr. to the 2nd; 240,000 fr. to the 3rd; 120,000 fr. to the 4th. 120,000 fr. to the breeder of the winner; 48,000 fr. to that of the 2nd; 24,000 fr. to that of the 3rd. Value to the winner: 1,379,800 francs.

SASSAFRAS	b.c.	by Sheshoon – Ruta by Ratification	3.55½k	Arpad Plesch	Y. Saint-Martin	1
NIJINSKY	b.c.	by Northern Dancer – Flaming Page by Bull Page	3.55½k	Charles Engelhard	L. Piggott	2
MISS DAN	ch.f.	by Dan Cupid – Miraloma by Clarion	3.54k	Achille Fould	J. Taillard	3
GYR	ch.c.	by Sea Bird – Feria by Toulouse Lautrec	3.55½k	Winston F.C. Guest	W. Williamson	4
BLAKENEY	b.c.	by Hethersett – Windmill Girl by Hornbeam	4.60k	Arthur Budgett	A. Barclay	5
BEAUGENCY	b.c.	by Val de Loir – Biobelle by Cernobbio	4.60k	Pierre Wertheimer	F. Head	6
SOYEUX	ch.c.	by Fils de Roi – Soie Sauvage by Sériphos	5.60k	R. Gribe	S. Leonardos	7
QUINQUET	br.c.	by Net – Wild Queen by Wild Risk	4.60k	Hubert de Chaudenay	Y. Josse	8
GRANDIER	b.c.	by Tapioca – Girga by Fine Top	6.60k	Mme Pierre Ribes	M. Philipperon	9
A CHARA	b.c.	by Worden – Cannon Ball by By Thunder	4.60k	Lady Brooke	A. Lequeux	10
GOLDEN EAGLE		by Right Royal – Aquilla II by Nasrullah or Princequillo	5.60k	Mme A. Manning	R. Jallu	
LA BIJUTE		by Tarquin – Poppéa by Fast Fox	5.58½k	François Carnus	J.D. Guérinet	
STINTINO		by Sheshoon – Cynara by Grey Sovereign	3.55½k	Gerald Oldham	G. Thiboeuf	
ORTIS		by Tissot – Orientale by Nagami	3.55½k	Carlo Vittadini	A. Di Nardo	
LAR		by Exar – Licence by Tropique	3.55½k	Carlo Vittadini	C. Castalvi	

15 runners.
SASSAFRAS trained by F. Mathet; bred by Arpad Plesch in Ireland.
Distances: Head; 2 lengths; 1½ lengths; 1½ lengths; short head; 2 lengths; ¾ length; 4 lengths. Time: 2 mins 29.70 seconds.
Pari-Mutuel: Win: 19fr. 90. Places: 3fr.; 1fr. 50; 4fr. 20.
S.P.: 4 to 10 Nijinsky; 15 to 2 Gyr and Golden Eagle (couplé); 35 to 4 Ortis and Lar (couplé); 19 to 1 SASSAFRAS; 25 to 1 Stintino; 32 to 1 Grandier; 33 to 1 Beaugency; 38 to 1 Miss Dan; 43 to 1 Blakeney; 45 to 1 La Bijute; 88 to 1 A Chara; 99 to 1 Soyeux and Quinquet.

SASSAFRAS bay colt 1967	Sheshoon ch. 1956	Precipitation ch. 1933	Hurry On ch. 1913	Marcovil ch. 1903	Marco
					Lady Villikins
				Tout Suite ch. 1904	Sain Foin
					Star
			Double Life ch. 1926	Bachelor's Double ch. 1906	Tredennis
					Lady Bawn
				Saint Joan b. 1918	Willbrook
					Flo Desmond
		Noorani ch. 1950	Nearco br. 1935	Pharos b. 1920	Phalaris
					Scapa Flow
				Nogara b. 1928	Havresac II
					Catnip
			Empire Glory b. 1933	Singapore b. 1927	Gainsborough
					Tetrabbazia
				Skyglory ch. 1922	Sky-Rocket
					Simone
	Ruta b. 1960	Ratification b. 1953	Court Martial ch. 1942	Fair Trial ch. 1925	Fairway
					Lady Juror
				Instantaneous b. 1931	Hurry On
					Picture
			Solesa b. 1940	Solario b. 1922	Gainsborough
					Sun Worship
				Mesa b. 1932	Kircubbin
					Mackwiller
		Dame d'Atour b. 1955	Cranach b. 1938	Coronach ch. 1923	Hurry On
					Wet Kiss
				Reine Isaure b. 1931	Blandford
					Oriane
			Barley Corn b. 1950	Hyperion ch. 1930	Gainsborough
					Selene
				Schiaparelli b. 1935	Schiavoni
					Aileen

Mill Reef 1971

Mill Reef

The Field

There were four foreign challengers for the 1971 Prix de l'Arc de Triomphe, three of which were trained in England: the four-year-old Ortis, Mill Reef, who was to be the second consecutive foreign runner to start at odds-on, and the unbeaten three-year-old Royalty. There was also the first challenger from the U.S.A. since Tom Rolfe in 1965, the five-year-old One for All.

The American Paul Mellon's Mill Reef had a record almost comparable with Nijinsky when he came to contest the Prix de l'Arc de Triomphe. He appeared to have the best chance for many years of becoming the first English-trained winner of the race since Migoli in 1948. A beautifully proportioned medium-sized bay colt, he had been the second best juvenile in England in 1970. After an effortless 5-length victory in the Salisbury Stakes (5f) in May, he beat a moderate Coventry Stakes field by 8 lengths at Royal Ascot. When he beat Green God 10 lengths in very heavy going in the Gimcrack Stakes at York, it was the most impressive display by any two-year-old in England that season. He then defeated Hecla a length in the Imperial Produce Stakes (6f) at Kempton and ended the season by thrashing Wenceslas and Lombardo by 4 lengths and half a length in the Dewhurst Stakes at Newmarket. His solitary defeat came at the hands of My

Swallow by a short head in the Prix Robert Papin at Maisons-Laffitte on 20 July. Some considered him a little unlucky as he had a less favourable draw than his conqueror. My Swallow, however, more than franked the form when he defeated Bonami in the Grand Critérium at Longchamp.

Mill Reef was by Never Bend by Nasrullah out of the Kentucky Oaks winner Lalun by Djeddah. He was the first colt by Never Bend ever to race in England and, like Sir Ivor and Nijinsky, there was a slight doubt about his stamina over twelve furlongs prior to the Epsom Derby. As a juvenile Never Bend was the best colt of his generation and was placed top of the Experimental Free Handicap. He won the mile Champagne Stakes by an astonishing 8 lengths. Never Bend was placed in both the Kentucky Derby and the Preakness Stakes, but ran disappointingly on the sole occasion he ran beyond ten furlongs.

Mill Reef reappeared in the Greenham Stakes at Newbury in April and trounced the opposition by 4 lengths. Paul Mellon's colt was just preferred to My Swallow in the market for the Two Thousand Guineas, with the unbeaten Middle Park Stakes winner, Brigadier Gerard, who had not had an outing in 1971, the third favourite. My Swallow set a scorching pace from the start closely tracked by Mill Reef. Two furlongs out as Mill Reef was gradually gaining the upper hand over My Swallow, Brigadier Gerard came with a wet sail and convincingly beat him 3 lengths with My Swallow three-quarters of a length back third. There was no question that over a mile Brigadier Gerard was the better colt.

Mill Reef started 100 to 30 favourite for the Epsom Derby – Brigadier Gerard was an absentee – and his most formidable rivals appeared to be Mme Pierre Wertheimer's Prix de Guiche and Prix Hocquart winner, Bourbon, and the Chester Vase and Observer Gold Cup victor, Linden Tree. Mill Reef looked extremely well in the paddock and moved beautifully past the stands to the start. Bourbon, however, was sweating profusely and was most fractious. In front of the stands his antics caused his bridle to snap in two places. When a new one was fitted, he threw his jockey Freddie Head and then only consented to go to the start when a mounted policeman escorted him. Bourbon's trainer Alec Head said he would have withdrawn his charge had it not been the Derby.

Geoff Lewis settled Mill Reef into about sixth place 5 lengths behind the leaders. As Linden Tree led Homeric, the Lingfield Derby Trial winner, into the straight Mill Reef was beautifully poised in fourth place a little behind Lombardo. Homeric came under pressure two furlongs out, where Lewis pulled Mill Reef to the outside to make his challenge. Paul Mellon's colt accelerated smoothly past Lombardo and drew level with Linden Tree. When asked for an extra effort to overtake the latter, Mill Reef quickened impressively and had two lengths advantage at the finishing post. Emile Littler's French-trained Irish Ball, who was both unbalanced and hanging to the left early in the straight, finished the fastest of the remainder and caught Lombardo near the line to take third place, 2½ lengths behind Linden Tree.

Mill Reef next took on the older generation in the Eclipse Stakes at Sandown at the beginning of July. The field included the best four-year-old in France, Caro, who was unbeaten in 1971 and who had improved enormously since the previous year. Two furlongs out the race rested between Caro and Mill Reef, but Paul Mellon's colt unleashed a scintillating burst of acceleration a furlong and a half from the winning post and won unextended by 4 lengths, with Welsh Pageant 2½ lengths away third. After this race Caro's trainer, Albert Klimscha, exclaimed, 'It is no disgrace to be beaten by a good horse, and Mill Reef is a great horse, better than Nijinsky.'

Not surprisingly Mill Reef started at 13 to 8 on for the King George VI and the Queen Elizabeth Stakes. When he launched his challenge two furlongs out the race was over in a couple of strides and it became a procession. Mill Reef beat Ortis and Acclimatization 6 lengths and 3 lengths with Stintino and Irish Ball a further short head and 6 lengths back fourth and fifth. Paul Mellon's champion was now rested until the Prix de l'Arc de Triomphe, for which, like Sea Bird, he was not given a preparatory race. Nevertheless he started a worthy favourite on Sunday 3 October.

Dr Carlo Vittadini's Ortis had been sent from Italy at the end of 1971 to be trained in England by Peter Walwyn. On his seasonal reappearance he found the concession of 5 lbs to the useful Pembroke Castle too much for him in the Coronation Stakes over ten furlongs at Sandown and went down by a neck. The distance also was probably too sharp for him, so it was surprising that he was sent to Leopardstown in

May for the Players Wills Stakes over the same distance. He finished runner-up 4 lengths behind the subsequent Derby fourth Lombardo. Ortis came back to his best in the Hardwicke Stakes over twelve furlongs and run in very heavy going at Royal Ascot. He trounced Pembroke Castle and Laurence O 8 lengths and 1½ lengths; many felt it was the most impressive performance of the meeting. He had clearly proved the best of Mill Reef's rivals at Ascot in July and had demolished moderate opponents in the Valdoe Stakes at Goodwood in September. Nonetheless the best that he could hope for was one of the minor places at Longchamp and it cannot have helped his chance that the going was firm on 3 October.

It was difficult to gauge the merit of Lady Beaverbrook's Royalty, an unbeaten three-year-old. He was a half-brother to the Gimcrack Stakes and Prix Robert Papin winner, Double Jump, and had won his first four races – maiden plates and handicaps – at Newmarket, Lingfield and Kempton Park. In the higher class Warren Stakes at Goodwood he put up a game performance to beat the Irish raider Power Ruler and Frascati three-quarters of a length and 4 lengths. He easily accounted for mediocre opposition in the North Yorkshire Stakes at Redcar on 7 August. It was a bold policy to take on the best in Europe with Royalty but in reality his starting price of 38 to 1 seemed a little ungenerous.

The challenger from the United States, One for All, for whom Willie Carson was engaged to ride, had run forty-one times and won $307,655 in prize money. His successes had included the Sunset Handicap at Hollywood Park and immediately prior to coming to Paris he had won two races in Canada, one of which was the valuable Niagara Handicap (11f) at Fort Erie on 29 August. His trainer, Horatio Luro, had been given permission to gallop him right-handed round the turf course at Fort Erie, a most unusual sight in North America where all the racetracks are left-handed. One for All was not in the same class as previous American challengers like Tom Rolfe and Carry Back, and his starting price of 39 to 1 was probably an overestimate of his real chance on 3 October.

Alec Head's stable was represented by both Bourbon and the filly Pistol Packer in the 1971 Prix de l'Arc de Triomphe, but it was felt that the latter was the greatest French threat to Mill Reef. Madame Alec Head's filly had won the Prix de la Fourrière (8½f) by 3 lengths at Longchamp the previous October and had been runner-up on her only other outing as a juvenile. On her reappearance she had run third in the Prix Imprudence but then had won her following five races. On 24 April she beat Sega Ville three-quarters of a length in the Prix Chloë (9f) at Longchamp. She next had only a neck to spare over Brainstorm in the Prix Saint-Alary on the same course on 23 May. As a result Alec Weisweiller's Cambrizzia, who had won a maiden race over eight and a half furlongs at Longchamp in April and the Prix Finlande by an impressive 5 lengths at Maisons-Laffitte in May, was made favourite for the Prix de Diane with Pistol Packer the second market choice. Pistol Packer took the lead 400 metres out and held the very strong late challenge of Cambrizzia by a short head. The runner-up had met with interference in the straight, and so close was the result that the judge deliberated for twenty minutes before announcing his verdict. Cambrizzia had a clear advantage two strides after the post and some felt a dead-heat would have been a fairer result, but on the evidence of the photograph it appeared Pistol Packer had managed to hold on to her advantage, albeit a very narrow one.

Pistol Packer was rested until the Prix de la Nonette on 5 September when she was obliged to concede 5 lbs to Cambrizzia, who had beaten Her Majesty the Queen's useful filly Albany in the Prix de Minerve (10f) at Chantilly on 24 July. At the weights it was expected Alec Weisweiller's filly would turn the tables on Pistol Packer, but she again proved superior, on this occasion by an emphatic 2½ lengths. Pistol Packer underlined her class and her chance in the Prix de l'Arc de Triomphe when she again accounted for Cambrizzia by a length in the Prix Vermeille a fortnight later, with Pink Pearl and the Oaks d'Italia heroine, Tingitana, well beaten third and fourth. Pistol Packer had thus proved herself to be one of the best fillies in France for many years and looked sure to pose a live threat to Mill Reef. Freddie Head chose to ride her in preference to Bourbon who was the mount of Yves Saint-Martin; the two stable companions were coupled in the betting at 17 to 4. It was surprising, therefore, to find the consistent Cambrizzia who, with the exception of her racecourse début at Maisons-Laffitte in April, had never been out of the first two, virtually ignored in the market at 39 to 1. Although

Cambrizzia had the disadvantage of always having to be held up for a very late run, all logic suggested that this genuine filly would not be far from Pistol Packer at the winning post.

Bourbon had slightly redeemed his reputation in France after his mulishness at Epsom. He had only managed seventh place behind Ramsin in the Grand Prix de Saint-Cloud, but was not disgraced when a short head, a nose and three-quarters of a length fourth behind Crucible, Hallez and Hagen in the Prix Maurice de Nieuil at Saint-Cloud on 14 July. Bourbon's task in the Prix Royal Oak was made much easier by a series of misfortunes that struck the very best French three-year-old colts. The best juvenile the previous year, Bonami, had broken a leg in the Prix Greffulhe. The Prix du Jockey Club and Grand Prix de Paris hero, Rheffic, had broken down in training and had gone to stud. Tarbes and Leading Man, the two principals in the Prix Lupin had also gone wrong; all were of course absentees from the Prix de l'Arc de Triomphe. Bourbon took the lead a furlong from home in the Prix Royal Oak and ran out a clever length winner from Oarsman with the Irish colt Parnell half a length away third. A fortnight later the latter was to beat Wenceslas 4 lengths in the Irish St Leger. On 3 October, therefore, Bourbon was probably one of the two best three-year-old colts in training in France, the other being Irish Ball.

Before Epsom, Irish Ball had won the Prix Daru by 1½ lengths from Acclimatization with Sharapour 1½ lengths back fourth. Piggott had ridden him into third place behind Tarbes in the Prix Lupin but decided to ride the Noel Murless-trained The Parson in preference to Irish Ball at Epsom. The refusal of Linden Tree to race at all in the Irish Sweeps Derby removed Irish Ball's principal rival, and taking the lead two furlongs out Emile Littler's colt had no difficulty in holding Lombardo by 3 lengths. He had clearly run below his best behind Mill Reef at Ascot and could not be entirely left out of the calculations for the minor places on 3 October, although he was allowed to start at 28 to 1.

Michael Sobell's Oarsman had been trained in England by Dick Hern until midsummer. In three runs he had won once, beating Tammuz in the Pattingham Maiden Plate at Wolverhampton. He showed improved form when he took the Prix de Menneval by a short neck from Bobari at Deauville after he had been sent to be trained by John Cunnington Jr. His proximity to Bourbon in the Prix Royal Oak was something of a revelation and it tended to cast doubts on the value of the form.

Baron Guy de Rothschild's Arlequino had run only three times in his life when he took on the best at Longchamp on 3 October. It was there he had made a winning début in a small race in April. He had then run Bourbon to a neck in the Prix Hocquart but unfortunately pulled a muscle in that event. He was not seen out again until the Prix de Chantilly on 12 September, when he beat the useful English-trained three-year-old Rugged by half a length. Arlequino, however, was the second string of Geoffrey Watson's stable in the race; its main hope was Baron Thierry de Zuylen de Nyevelt's Ramsin. Although Ramsin won the Prix Maurice de Nieuil as a three-year-old, he was several pounds below the best of his generation, but by the time of the Prix de l'Arc de Triomphe he had improved so much that he was reckoned to be the best staying four-year-old in France. In the Prix de Barbeville in early April Ramsin defeated Hallez a comfortable 3 lengths. Giving Hallez two kilos, Ramsin finished a short head in front of his rival in the Prix Jean Prat, but was relegated – some thought unluckily – when Hallez's jockey Freddie Head objected. In the Prix du Cadran Ramsin had no difficulty in beating the very useful English stayer Rock Roi 3 lengths with Hallez 6 lengths away third. In his final outing before the Prix de l'Arc de Triomphe, Ramsin gamely accounted for Hokkaido and Tarbes three-quarters of a length and the same in the Grand Prix de Saint-Cloud on 4 July, with Acclimatization and Hallez a neck and 5 lengths away fourth and fifth and Bourbon and Armos in rear. Despite this victory, there were several who felt that Ramsin would find the distance of the Prix de l'Arc de Triomphe on the sharp side for him, especially as his best form had been shown on soft going. The third colt from George Watson's stable was Ramsin's pacemaker, Ossian, and the trio started at 7 to 1.

Mme Georges Weisweiller's Hallez, the mount of Lester Piggott and the last runner Willie Head was to saddle in the Prix de l'Arc de Triomphe, came next in the betting at 21 to 1. He had had a slightly disappointing season in 1971. Clearly held by Ramsin, Hallez had been beaten a short head by Crucible in the Prix Maurice de Nieuil, and although he won the Prix de Prince d'Orange on 19 September by 2 lengths, the opposition was not as strong as usual.

The four-year-old who probably had the best chance of posing a threat to Mill Reef was the Eclipse Stakes runner-up, Comtesse Batthyany's Caro. He had given Miss Dan 8 lbs and beaten her a length in the Prix d'Harcourt on 12 April. In the Prix Ganay three weeks later he defeated Amadou and Stintino three-quarters of a length and 2 lengths with Miss Dan another 5½ lengths away seventh. In the Prix Dollar Caro gave Mister Sic Top 7 lbs and a 5-length thrashing. He was rested after the Eclipse Stakes until the Prix Foy over 11 furlongs at Longchamp on 5 September. Caro probably needed the race but was not disgraced when he narrowly failed to give the good English colt Prominent 7 lbs. He again had Mister Sic Top 3 lengths behind him in fourth place. The principal question about Caro was whether he would stay the final furlong at Longchamp on 3 October.

Lucien Deshayes's Mister Sic Top started the complete outsider at 95 to 1, because there was absolutely no reason why he should reverse his placings with Caro. His best performance in 1971 had been to win the Prix d'Ispahan by 5 lengths from Lithiot on 27 June. Mme Edouard Lyon's Armos, the winner of the Prix de Conseil Municipal in 1970, had performed inconsistently as a four-year-old. His best display had been to beat Amadou and Hokkaido a head and a length in the Prix Jean de Chaudenay at Saint-Cloud on 31 May, with Miss Dan a little over 3 lengths back fifth. He never reproduced that form, however, and was beaten 6 lengths by Miss Dan in the Prix Kergorlay and was only seventh to her in the Grand Prix de Deauville.

Miss Dan was probably not such a good performer in 1971 as she had been the previous season. Her Prix Kergorlay victory at Deauville showed she was on the way back and when she beat the Grand Prix de Paris third, Valdrague, and Crucible a length and a short head in the Grand Prix de Deauville, the hopes of her connections were given a boost. Nevertheless on balance Miss Dan's chance seemed about the same as that of Hallez.

The field for the 1971 Prix de l'Arc de Triomphe was completed by the Aga Khan's three-year-old Sharapour, who represented François Mathet's stable in the absence of the stable star Rheffic. Sharapour had won his only two outings as a juvenile including the Prix Saint-Roman. He had been, however, a little disappointing as a three-year-old. He had failed when favourite for the Prix Daru in April and had only finished seventh to Zug in the Poule d'Essai des Poulains. On 2 July he beat Toulon by three-quarter of a length over ten furlongs at Saint-Cloud and in September, receiving 2 lbs from Lithiot, had beaten him 3 lengths over an extended nine furlongs at Longchamp. He certainly appeared about a stone behind the best of his generation as a three-year-old and his price of 31 to 1 seemed most ungenerous.

At the beginning of September Enrico Camici had intended to send Weimar, whom he had trained to win the Gran Premio d'Italia and Gran Premio di Milano, to Paris. Although he started at 100 to 1 on for the Premio Besana and won by 4 lengths, the manner of his victory was insufficiently decisive for Camici, who re-routed his charge to the St Leger Italiano which he easily took. Weimar gained his twelfth victory when he beat Parnell in the Gran Premio del Jockey Club a fortnight after the Prix de l'Arc de Triomphe. It was a pity he did not run at Longchamp for he was clearly a useful colt. Besides One for All, two other American horses were still left in the race on 14 September: Bold Reason, the Hollywood Park Derby winner; and the extremely useful five-year-old mare, Princess Pout. They were scratched shortly afterwards, but it is interesting that Princess Pout's first foal, a colt by Hoist the Flag named Alleged, was to distinguish himself but six years later.

The Final Preparations

Mill Reef's young trainer, Ian Balding, decided to make special arrangements to transport Paul Mellon's Derby winner to Paris to minimise any stress he might suffer on the journey. He obtained permission from the U.S. Embassy to fly the colt from the Greenham Airbase which was leased to the U.S. Air Force. It was only ten minutes by horsebox from his yard at Kingsclere. The base commander, Colonel Dwight D. Blanton, ordered his men to give Balding every assistance, and he provided the necessary lifting equipment. Only two hours after Mill Reef had left Newbury, he was cleared through Le Bourget. This smooth and quick operation can only have helped his cause on 3 October. Balding had been delighted by his charge's final gallop on Newbury racecourse on 19 September in company with his stable-companions Aldie, National Park and Bright

Beam. Afterwards his jockey, Geoff Lewis, remarked, 'He's magnificent, better than ever'.

For Lewis it was only his third mount in Longchamp's most prestigious race. He had earlier finished unplaced on Soderini and Prince Regent. Before the race Lewis prepared a colour chart of all the silks of his rivals and learned them in order that he knew exactly whom he had to beat turning for home. The plan of the race was that Mill Reef should try to hit the front 300 metres from the winning post and make the best of his way home. Charlie Smirke, who had ridden the last English trained winner, Migoli, in 1948, felt before the race that Mill Reef was 'a good thing'. He added, 'He's a different class to Migoli, who had run second to Pearl Diver in the Derby the previous year. Provided Mill Reef doesn't get into trouble, there should be no danger.' In fact not many *turfistes* seriously thought the Never Bend colt would be beaten, but a few who did put their faith in Pistol Packer who, with her Prix Vermeille victory, had shattered the European record earnings for a filly. She had then won £163,095 whereas the previous record had stood at £126,419.

There was an enormous and very cosmopolitan crowd at Longchamp on Sunday 3 October. Président Pompidou was the first French president to attend the Prix de l'Arc de Triomphe for over a decade. Amongst the English spectators taking a very keen interest in all the arrangements was Sir Philip Oppenheimer, whose company, De Beers, was to sponsor the King George VI and the Queen Elizabeth Stakes at Ascot the following July. Lord Cornwallis, the 79-year-old Lord Lieutenant of Kent, had travelled over for the weekend by Air France who had given him V.I.P. treatment. The airline's computer paid scant respect to the titled and had listed the peer as plain 'My Lord'. The computer, however, made no mistake about the woman accompanying him, Mrs Betty Kenward, better known for countless years as that assiduous scribe of the social scene, 'Jennifer'.

The Race

Lewis started the afternoon on a high note by winning the Prix de l'Abbaye de Longchamp on Sweet Revenge by ¾ length from Swing Easy, the mount of Lester Piggott. In the paddock Mill Reef, the very attractive filly Pistol Packer, and Ortis, were the pick of the field. Paul Mellon's champion strode out magnificently on his way to the start. The ground was firm and the presence of Ramsin's pacemaker, Ossian, ensured that the race would be run at a fast pace.

One for All was first out of the stalls, but as soon as the field settled down Heurteur had sent his mount Ossian into the lead and set a cracking pace to Ramsin, Sharapour, Hallez, Ortis and Mill Reef. They were tracked by One for All and Irish Ball. At this stage Pistol Packer and Cambrizzia were in the rear just behind Oarsman and Bourbon. Ossian kept up his strong gallop to the top of the hill with the order behind him virtually unchanged. Ossian continued in the lead down the hill closely followed by Ramsin, Hallez, Ortis, Sharapour, Mill Reef and Miss Dan. During the descent both Caro and, from further behind, Pistol Packer began to improve their positions on the outside, but Cambrizzia still seemed almost rooted in last place. As the field began to swing into the straight, Ossian weakened quickly and Sharapour and Ortis took up the running, closely followed by Hallez, Ramsin, Mill Reef and Miss Dan. Once in the straight the first to come under pressure were Sharapour and Ramsin, who found the gallop too strong, and Lester Piggott pushed Hallez into a narrow lead over Ortis. For a moment it looked as if Mill Reef, on the inside, would be hampered by the fading Sharapour, but Lewis sensibly pulled his mount off the rails. Four hundred metres out Ortis began to hang under pressure and Lewis was able to launch his challenge in the gap that had opened up between him and Hallez. Miss Dan, about 2 lengths behind, had no more to give whilst Pistol Packer on the outside, Caro and Arlequino were all moving into challenging position and Cambrizzia, who had been last entering the straight almost 10 lengths behind the leaders, could at least be seen making progress through beaten horses. Nonetheless as soon as Mill Reef had quickly got the measure of Hallez, he rapidly put 2 lengths between himself and his nearest pursuers. Two hundred metres from the winning post Paul Mellon's champion was going so well within himself that none seemed likely to catch him, not even Pistol Packer, whose challenge on the outside took her past the one-paced Hallez. Mill Reef was always holding her throughout the final 150 metres and he had 3 lengths to spare over Pistol Packer. So fast was Cambrizzia travelling in the final furlong that

Mill Reef (13) has 3 lengths and 1½ lengths to spare over Pistol Packer (17) and Cambrizzia (18) with Caro (almost hidden on rails) and Hallez fourth and fifth

she was only a neck behind her great rival in third place. She had made up more ground than any other runner in the straight and was three-quarters of a length in front of Caro, who was the first of the older generation to finish, but he had been unable to sustain his challenge in the final furlong. Half a length behind in fifth place came Hallez who had held Royalty, who had run on gamely in the final 300 metres, by a neck. Then in a group together came Bourbon, Arlequino and One for All, who had run creditably considering he had always been travelling on the wrong leg. Irish Ball, who began his effort far too late, finished tenth.

On his return to the unsaddling enclosure Mill Reef was greeted by tumultuous applause. Geoff Lewis stated that he had enjoyed a good run all the way and that he knew he had had the race won turning into the straight. Mill Reef had, in fact, set a new twelve-furlong course record of 2 minutes 28·30 seconds. Freddie Head's immediate reaction was that 'Mill Reef was the best horse I've ever seen', but felt that Pistol Packer was still the best filly he had ever ridden. He also thought that had his mount not been drawn in the very worst position – eighteenth of eighteen runners – and had he not been forced to challenge on the wide outside, she might have given Mill Reef a harder struggle. The French racing press were generously unanimous in their praise of the champion the following morning: 'Mill Reef has joined the ranks of the greatest horses', 'Mill Reef accomplished a feat worthy of Sea Bird', and 'Mill Reef is a great horse' are examples of the praise lavished on him in *Paris Jour*, *Le Figaro* and *L'Aurore*. In the excitement after the race Paul Mellon could only gasp 'Wonderful' before he was presented to Président Pompidou. His wife said that it had been 'an unbelievable feeling' to see her husband's horse win.

For Geoff Lewis it was his first victory in the Prix de l'Arc de Triomphe. He had worked as a page-boy in a London hotel before he was apprenticed to Ron Smyth at Epsom. The 1971 season had been an exceptional one for him since not merely had he won the Derby for the first time as well but also the Oaks on Altesse Royale and the Coronation Cup on Lupe. His only previous classic victory had been on Right Tack in the 1969 2000 Guineas.

It was also a first success in Longchamp's most important race for Mill Reef's trainer, the 32-year-old Ian Balding. He was the younger brother of Toby Balding, who had trained the Grand National winner Highland Wedding and whose father was the famous polo player Gerald Balding. Ian Balding was also a very talented sportsman. While at Cambridge he won a Rugby blue playing full-back and he was a very useful amateur steeplechase jockey in 1963 he had won the National Hunt Chase at Cheltenham.

Paul Mellon, for whom Mill Reef had won his first Derby and Prix de l'Arc de Triomphe, had been an enthusiastic supporter of English racing under both rules since his undergraduate days in England before the Second World War. He had had horses in training with the Hon. Aubrey Hastings, Ivor Anthony and the former's son, Captain Peter Hastings-Bass. When the latter tragically died in his mid-forties, Ian Balding, who was his assistant and later married his daughter, took over the Kingsclere stable. Paul Mellon had bred Mill Reef himself at his stud, Rokeby Farms in Kentucky. He had seventy-nine horses in training in the U.S.A. and England at the beginning of 1971. At his stud the broodmares included Key Bridge, the dam of Fort Marcy, First Feather, the dam of Run the Gantlet, and All Beautiful, that of Arts and Letters. Mill Reef's victory in the Prix de l'Arc de Triomphe was exceptionally popular on both sides of the Channel. What was even more surprising was that Mill Reef had provided him with his first classic victory in England after a series of near-misses. Mellon had already reserved the London nightclub, Annabel's, the Wednesday after Mill Reef's Longchamp triumph to celebrate his Derby victory. All the women guests were presented with a specially designed Givenchy scarf in his black and gold colours.

Mill Reef did not race again in 1971 but was kept in training as a four-year-old. He put up a scintillating display on his seasonal début in the Prix Ganay at Longchamp on 30 April; he thrashed the very useful Amadou by a remarkable 10 lengths, though some considered that an underestimate on the part of the judge. In the Coronation Cup he did not look as well in himself as usual in the paddock and only managed to scrape home by a neck from Homeric. He had easily beaten Homeric into fifth place in the Derby the previous year and although Homeric had finished a close runner-up to Athens Wood in the St Leger, clearly something was wrong with Mill Reef. This lack-lustre display was explained a few days later when the Prix de l'Arc de Triomphe winner went down with the virus. The entire racing world in England had been looking forward to his proposed meeting with his only conqueror as a three-year-old, the unbeaten Brigadier Gerard, in the Eclipse Stakes at Sandown, but unfortunately this meeting was never to take place. It had also been hoped that Pistol Packer would join issue again with Mill Reef but unfortunately she was very badly struck into when beating Amadou 3 lengths in the Prix d'Harcourt at Longchamp in April and she never recovered her form. Brigadier Gerard won both the Eclipse Stakes and the King George VI and the Queen Elizabeth Stakes in Mill Reef's absence.

By August 1972 Paul Mellon's champion was sufficiently recovered for Balding to begin to get him ready for another attempt at the Prix de l'Arc de Triomphe, but tragedy struck again. In late August Mill Reef fractured his left foreleg in four places during a half-speed gallop at Kingsclere. Fortunately the Never Bend's colt's temperament was so equable that the veterinary surgeon was able to save him for stud, despite the fact that his leg was encased in plaster for several weeks.

The Rest of the Season

None of the first six in the 1971 Prix de l'Arc de Triomphe ran again that autumn. Miss Dan was a respectable third a neck and 2 lengths behind Lombard and Cortez in the Preis von Europa at Cologne on 17 October. Lombard had won the Deutsches St Leger the previous season and the Grosser Preis von Nordrhein-Westfalen earlier in July, whilst Cortez had accounted for a reasonable field in the Grosser Preis von Baden. Irish Ball and Ortis both accepted invitations to the Washington D.C. International Stakes at Laurel Park in November. Irish Ball was not disgraced when runner-up to Run the Gantlet, though Ortis could only manage a distant sixth place.

Pedigree and Stud Record

Mill Reef was syndicated for stud at the then record of £2 million. It was an enormously generous gesture of Paul Mellon not to re-import him into the United States but to stand him at the National Stud in England. His sire, Never Bend, ran ten times as a two-year-old winning seven races, including the Futurity (6½f), the Cowdin Stakes (7f) and the Champagne Stakes (8f), and $402,969, a very large sum in 1962. He won six races as a three-year-old and was placed five times from thirteen starts. His victories included the Flamingo Stakes (9f) and the Yankee Handicap (9f). He was runner-up to Châteaugay in the Kentucky Derby and to Kelso in the Woodward Stakes (10f) and was third to Candy Spots and Châteaugay in the Preakness Stakes. It is fair to say that he was a better juvenile than a three-year-old and that Mill Reef and Riverman were the two best colts he got at stud.

Mill Reef's dam, Milan Mill by Princequillo, only ran once unsuccessfully. Princequillo, however, had more than proved his worth as a stallion with such notable offspring as Round Table, Hill Prince and Prince Simon. In the paddocks Milan Mill proved an excellent broodmare getting four winners from her first five foals. The first, Milan Meadow by Never Bend, won three times in the U.S.A. whilst her second was Mill Reef. The third, Millicent by Cornish Prince, was unraced but is herself the dam of winners including Marston's Mill who won four races in the U.S.A. Her fourth foal, Mille Fleurs by Jacinto, was a winner as a three-year-old in England whilst the fifth, a full sister to Mill Reef, Memory Lane, won two races including the Princess Elizabeth Stakes and was third in the Mill Reef Stakes.

Milan Mill's dam, Virginia Water by Count Fleet, the 1943 Triple Crown winner in the U.S.A. and sire of Kelso, was unraced. She, however, was another excellent broodmare. She was the dam of five winners, the best of which was Berkeley Springs whose three victories included the Cheveley Park Stakes, the Strensall Stakes and the Rosemary Stakes, and she was runner-up to Glad Rags in the 1966 1000 Guineas and to Valoris in the Oaks. Goose Creek, another of Virginia Water's foals, won six races in England and the U.S.A.; he also ran second in the Royal Lodge Stakes at Ascot and the Rous Memorial Stakes. He is now a sire in the United States, as is another of her produce, Fountain Hill, who won five times in his native land.

Virginia Water's dam, Red Ray by Hyperion, was also unraced. She had been owned by Lord Portal who died in May 1949. On the orders of his executors all his bloodstock was sold at Newmarket in July 1949. On the advice of Clive Graham, with whom he worked closely for many years, Jim Ryan bought Red Ray for Paul Mellon's stud for 12,000 guineas. She only produced two foals before her premature death, Virginia Water and Claret, the winner of two races in the U.S.A.

The reason why Clive Graham had advised Jim Ryan to buy Red Ray for Paul Mellon had been because her dam was Infra Red by Ethnarch and her grandam Black Ray by Black Jester. Infra Red had been bred by the Chicago millionaire Marshall Field at the Tally-Ho Stud in Ireland and won two races, the Great Surrey Foals Stakes as a juvenile and the Princess Elizabeth Stakes the following season, both at Epsom. In addition she had run fourth in the 1939 1000 Guineas to Galatea II, Aurora and Olein. Infra Red bred only one foal at the Tally-Ho stud for Marshall Field, the sprinter Magic Red who won three races and subsequently became a successful sire. In 1942 Marshall Field decided to dispose of all his bloodstock as he found it difficult to reconcile all the suffering the war was causing in Europe and his having a stud and horses in training in the British Isles. Infra Red then was bought on the advice of Captain Cecil Boyd-Rochfort for his friend Lord Portal for 5000 guineas at Newmarket on 13 July 1942.

She was in foal to Nearco and the resulting filly, Riding Rays, won twice and ran second in the Nassau Stakes and was third in the Cheveley Park Stakes. She herself became the grandam of Dubasoff, Regal Light, a leading sire in Australia, and Sovereign Gleam. Infra Red was also the dam of Red Briar, who won three races and ran third in the Ebbisham Stakes. Red Briar was a very successful broodmare; she was the dam of Pipe of Peace, the leading two-year-old in England in 1956, who won the Middle Park Stakes, the Greenham Stakes and the Gordon Stakes, and was third to Crepello in both the 2000 Guineas and the Derby in 1957. Red Briar was also the dam of Bigberry, the grandam of Ring Mistress (Milady Handicap). Infra Red's last offspring was Excelsa who won six races for Captain Boyd-Rochfort in Ireland between twelve and twenty furlongs.

Excelsa was also a highly successful broodmare, for she was the dam of Exar (Gran Premio d'Italia, Gran Premio di Milano, Premio Ambrosiano, Doncaster Cup, Goodwood Cup and runner-up to Sheshoon in the Ascot Gold Cup.) She was also the grandam of Wollow, the champion English two-year-old in 1975 and winner of the 2000 Guineas, Eclipse Stakes, Dewhurst Stakes, Champagne Stakes, Benson and Hedges Gold Cup and Greenham Stakes, and of Red Rumour, one of the leading stallions in Australia in the last few years.

Infra Red's dam was that remarkable broodmare, Black Ray by Black Jester, who won the St Leger. Captain Boyd-Rochfort had bought her in foal to Sunstar together with a filly foal at foot by Thunderer from her breeder Jim Joel for £1000 for his patron Marshall Field. Black Ray was then seven, she had the remarkable record of producing nineteen consecutive foals. The solitary year she was barren was the one she died at the age of twenty-three in July 1942. The best two of her colts were Jacopo by Sansovino and Foray by Tetrametra. Both were brilliant two-year-olds who were rated top of the Free Handicap in 1930 and 1936 respectively, though Jacopo had to share his position with Portlaw. Besides Infra Red, Black Ray's best filly foal was her full sister Eclair who won seven races and was one of the best fillies of her generation. Eclair also was a most influential broodmare getting Khaled, the winner of the Coventry Stakes, Middle Park Stakes, the St James's Palace Stakes, and runner-up to Happy Knight in the 1946 2000 Guineas. Khaled became the sire of Swaps. Another of Eclair's progeny was Lady Electra, a very good racemare owned by Phil Bull of Timeform and the grandam of Eudaemon.

Black Ray's dam, Lady Brilliant, was by Sunbridge out of the Oaks winner, Our Lassie, who was a half sister to the St Leger winner, Your Majesty. Black Ray was also a full sister to Black Gem, the third dam of Imprudence who won the Oaks, the 1000 Guineas and the Poule d'Essai des Pouliches. It can thus be seen that Mill Reef was bred in the purple and it was certainly to be expected that he would be a success at stud and so it has proved in his career to date.

Mill Reef was the sire of Acamas who won the Prix du Jockey Club and Prix Lupin in 1978 and was runner-up in the King George VI and the Queen Elizabeth Diamond Stakes, whilst in the same year his son Shirley Heights won the Epsom Derby, the Irish Sweeps Derby and the Mecca Dante Stakes at three and the Royal Lodge Stakes as a juvenile. Glint of Gold won the Gran Critérium and Derby Italiano in 1981, the Grand Prix de Paris and the Grand Prix de Saint-Cloud, the Preis von Europa in Germany, the Great Voltigeur Stakes and the John Porter Stakes, and was runner-up in the Epsom Derby and the St Leger in England. Mill Reef's filly, Fairy Footsteps, won the 1000 Guineas. Other notable winners by Paul Mellon's champion include Milford (Princess of Wales's Stakes, Derby Trial Stakes, White Rose Stakes); Main Reef (July Stakes and Cumberland Lodge Stakes); Idle Waters (Park Hill Stakes); Millionaire (Queen's Vase); Water Mill (2nd St Leger and 3rd Grand Prix de Paris); English Harbour (Predominate Stakes); Jenne Loup (Prix de Reux) and Special Leave (Hyperion Stakes). There is no absolutely question that Mill Reef is going to be as influential a stallion as he was brilliant on the racecourse. Dionne (Prix de l'Opéra, Twickenham Stakes), Pas de Seul (Hungerford Stakes, Prix Eclipse), Diamond Shoal (3rd St Leger, John Porter Stakes, Grand Prix d'Evry), Simply Great (Mecca-Dante Stakes), and Jolly Bay (Pretty Polly Stakes) are the most notable of Mill Reef's recent winners.

PRIX DE L'ARC DE TRIOMPHE 2400 metres
at Longchamp on Sunday 3 October 1971. For three-year-olds and upwards. Going: Firm.
1,200,000 francs to the winner; 480,000 fr. to the 2nd; 240,000 fr. to the 3rd; 120,000 fr. to the 4th. 168,000 fr. to the breeder of the winner; 48,000 fr. to that of the second; 24,000 fr. to that of the 3rd. Value to the winner: 1,399,150 francs.

MILL REEF	b.c.	by Never Bend – Milan Mill by Princequillo	3.55½k	Paul Mellon	G. Lewis	1
PISTOL PACKER	ch.f.	by Gun Bow – George's Girl by Ossian II	3.54k	Mme Alec Head	F. Head	2
CAMBRIZZIA	br.f.	by Cambremont – Aliziane by Alizier	3.54k	Alec Weisweiller	A. Barclay	3
CARO	gr.c.	by Fortino – Chambord by Chamossaire	4.60k	Comtesse Margit Batthyany	J. Lindley	4
HALLEZ	b.c.	by Le Fabuleux – Hagueneau by Krakatao	4.60k	Mme Georges Weisweiller	L. Piggot	5
ROYALTY	br.c.	by Relko – Fair Bid by My Babu	3.55½k	Lady Beaverbrook	J. Mercer	6
BOURBON	b.c.	by Le Fabuleux – Biobelle by Cernobbio	3.55½k	Mme Pierre Wertheimer	Y. Saint-Martin	7
ARLEQUINO	br.c.	by Exbury – Isoline by Klairon	3.55½k	Baron Guy de Rothschild	J. Massard	8
ONE FOR ALL	b.c.	by Northern Dancer – Quill by Princequillo	5.60k	John A. Bell III	W. Carson	9
IRISH BALL	b.c.	by Baldric II – Irish Lass by Sayajirao	3.55½k	Emile Littler	W.B. Pyers	10
MISTER SIC TOP		by Tiffauges – Madame Daurade by Sicambre	4.60k	Lucien Deshayes	J.C. Desaint	
ARMOS		by Mossborough – Ardelle by Supreme Court	4.60k	Mme Edouard Lyon	G. Thiboeuf	
ORTIS		by Tissot – Orientale by Nagami	4.60k	Carlo Vittadini	D. Keith	
RAMSIN		by Le Haar – Maritchia by La Varende	4.60k	Baron Th. de Zuylen de Nyevelt	H. Samani	
OSSIAN		by Charlottesville – Dreida by La Varende	4.60k	Baron Th. de Zuylen de Nyevelt	L. Heurteur	
MISS DAN		by Dan Cupid – Miraloma by Clarion	4.58½k	Achille Fould	J. Taillard	
SHARAPOUR		by Tanerko – Paola II by Palestine	3.55½k	H.H. Aga Khan	W. Williamson	
OARSMAN		by Aureole – Ship Yard by Doutelle	3.55½k	Michael Sobell	M. Philipperon	

18 runners.
MILL REEF trained by I. Balding at Kingsclere, England; bred by Paul Mellon in the U.S.A.
Distances: 3 lengths; 1½ lengths; neck; ¾ length; ½ length; neck; short neck; ½ length; short head. Time: 2 mins 28.30 seconds.
Pari-Mutuel: Win: 1fr. 70. Places: 1fr. 30; 1fr. 90; 4fr. 30.
S.P.: 7 to 10 MILL REEF; 17 to 4 Pistol Packer and Bourbon (couplé); 7 to 1 Arlequino, Ramsin and Ossian (couplé); 21 to 1 Hallez; 24 to 1 Caro; 27 to 1 Miss Dan; 28 to 1 Irish Ball; 31 to 1 Sharapour; 38 to 1 Royalty; 39 to 1 Cambrizzia and One for All; 51 to 1 Ortis; 71 to 1 Oarsman; 74 to 1 Armos; 95 to 1 Mister Sic Top.

MILL REEF bay colt 1968	Never Bend b. 1960	Nasrullah b. 1940	Nearco br. 1935	Pharos b. 1920	Phalaris
					Scapa Flow
				Nogara b. 1928	Havresac II
					Catnip
			Mumtaz Begum b. 1932	Blenheim br. 1927	Blandford
					Malva
				Mumtaz Mahal gr. 1921	The Tetrarch
					Lady Josephine
		Lalun b. 1952	Djeddah ch. 1945	Djebel b. 1937	Tourbillon
					Loika
				Djezima ch. 1933	Astérus
					Heidilann
			Be Faithful br. 1942	Bimelech b. 1937	Black Toney
					La Troienne
				Blood Root b. 1932	Blue Larkspur
					Knockaney Bridge
	Milan Mill b. 1962	Princequillo b. 1940	Prince Rose b. 1928	Rose Prince b. 1919	Prince Palatine
					Eglantine
				Indolence b. 1920	Gay Crusader
					Barrier
			Cosquilla b. 1933	Papyrus br. 1920	Tracery
					Miss Matty
				Quick Thought b. 1918	White Eagle
					Mindful
		Virginia Water gr. 1953	Count Fleet br. 1940	Reigh Count ch. 1925	Sun Reigh
					Contessina
				Quickly br. 1930	Haste
					Stephanie
			Red Ray gr. 1947	Hyperion ch. 1930	Gainsborough
					Selenè
				Infra Red gr. 1936	Ethnarch
					Black Ray

San San 1972

San San

The Field

The absence of both Mill Reef and Pistol Packer from the 1972 Prix de l'Arc de Triomphe was immensely regrettable. After her injury in the Prix d'Harcourt, when W. Pyers's mount ran into her as she was pulling up, Pistol Packer reappeared in the Prix Gontaut-Biron at Deauville but narrowly failed to overhaul Boreen by a nose. In the Prix Foy at Longchamp on 10 September she started at 5 to 1 on but was beaten 2 lengths and 8 lengths by Snow Castle and Crazy Rhythm. It was clear she was nowhere near her best and she was retired to stud.

There were no less than six foreign-trained runners in the 1972 Prix de l'Arc de Triomphe: two came from Ireland, Roberto and Boucher, three from England, Steel Pulse, Parnell and Erimo Hawk. My Mourne was the first ever challenger from Spain. In addition Homeric, Snow Castle and Mejiro Musashi were trained in France in October 1972, but they had all raced for most of their careers in their native lands, namely England, Argentina and Japan.

Vincent O'Brien trained both Roberto and Boucher. John Galbreath's Roberto won his first three races as a juvenile in Ireland including the Anglesey Stakes by 6 lengths and the National Stakes by 5 lengths. In the Grand Critérium at Longchamp, however, Roberto was only able to finish fourth to Hard to Beat, Steel Pulse and Prodice. He easily won the Vauxhall Trial Stakes at Phoenix Park on 1 April on his seasonal reappearance. His chief opponent in the 2000 Guineas was High Top, who had slammed the 1000 Guineas heroine, Waterloo, by 7 lengths at Thirsk only two weeks earlier and was to start favourite. High Top led from the start, and Roberto, held in rear by Williamson, looked to have a lot to do at the Bushes. He made such rapid progress, however, that he almost caught the favourite coming out of the Dip and eventually only went down by half a length with Sun Prince 6 lengths away third and Steel Pulse fourth. Bill Williamson was again supposed to ride Roberto in the Derby but was unfortunately injured in a fall at Kempton Park ten days earlier. Although he declared himself fit, Roberto's owner decided to give the mount to Lester Piggott who had ridden him in the Grand Critérium. He, however, promised Williamson his percentage if the colt won at Epsom. A large section of the racing public were very critical of this substitution, but in view of the value of a Derby winner as a stallion, it seemed only logical if there were the slightest doubt about Williamson's fitness. It also seemed just to those people whose money was staked on the ante-post favourite.

That Roberto won the Derby by a short head from Rheingold was entirely due to Piggott's supreme genius at Epsom regardless of the controversy. After the two colts had engaged in a titanic struggle throughout the final furlong, Piggott literally lifted the Hail to Reason colt over the line. The lightweight jockey Ernie Johnson on Rheingold admittedly had some difficulty in keeping him balanced and on an even keel; had the jockeys been switched, it is almost certain the result would have been reversed.

In the Irish Sweeps Derby Roberto, still feeling the effects of his very hard race in the Derby, made no show behind Steel Pulse, who had been only eighth at Epsom. Roberto reappeared in the Benson and Hedges Gold Cup at York where Piggott elected to ride Rheingold, who in the meantime had won the Grand Prix de Saint-Cloud. The Epsom Derby winner was partnered by the crack Panamanian jockey, Braulio Baeza, whom John Galbreath had specially flown over from the U.S.A. Both colts were opposed by the unbeaten Brigadier Gerard, the 3 to 1 on favourite. Roberto led for a furlong and was then steadied by Baeza. He took up the running again with 6 furlongs to go and the Panamanian jockey thereafter set such a scorching pace that all his rivals, except Brigadier Gerard a length behind him, were clearly beaten. When Mercer went for his whip a little over a furlong out, Brigadier Gerard found nothing and Roberto strode away to win unchallenged by 3 lengths in a new course record. The runner-up duly won the Queen Elizabeth II Stakes and the Champion Stakes in the autumn so he had not gone over the top. Roberto thus came to Longchamp on Sunday 8 October with the distinction of being the only colt ever to defeat Mrs John Hislop's champion.

Surprisingly, in the Prix Niel (11f) at Longchamp in September Baeza did not repeat the same forcing tactics. Piggott dictated the pace on Hard to Beat the whole way and easily held the Irish colt's challenge by a length. It was the second time in two meetings Hard to Beat had proved superior. Although Roberto started joint second favourite for the Prix de l'Arc de Triomphe at 9 to 2, he was essentially a ten-furlong horse and might well run out of stamina in the last furlong at Longchamp.

The other Vincent O'Brien colt, Boucher, had initially only been left in the race if the going became soft and Roberto was withdrawn. At the last moment, however, it was decided to run him anyway. As a juvenile he had won the Beresford Stakes and run a creditable fourth to High Top and Steel Pulse in the Observer Gold Cup. After winning the Nijinsky Stakes at Leopardstown in April he finished last in the Chester Vase, but was found to be suffering very badly from the virus on his return to Ireland. In August he took a ten-furlong handicap at Leopardstown and then won the Desmond Stakes by 2½ lengths from the Irish 1000 Guineas winner Pidget. Although Boucher dwelt at the start of the St Leger, he still beat the Derby fourth, Our Mirage, and the Oaks winner Ginevra. He had yet, however, to meet colts of the calibre of Hard to Beat and Roberto, and his price of 17 to 1 probably reflected his chance at Longchamp.

Rory More O'Ferrall's Parnell, who had been trained in Ireland by J. Quirke the previous year, joined Bernard van Cutsem's Newmarket stable in 1972. Under top weight he slaughtered moderate opposition in the mile and a half Rouge Dragon handicap at Ascot in April. Parnell then accounted for Arlequino by three-quarters of a length in the Prix Jean Prat (15½f) at Longchamp on 22 April. He was outstayed by Rock Roi by a length in the Prix du Cadran on 21

May, but then gave possibly the finest performance of his career in the King George VI and the Queen Elizabeth Stakes. Sent into the lead by Carson from the outset, Parnell was only caught by Brigadier Gerard two furlongs from home. Although he struggled gamely Rory More O'Ferrall's four-year-old was beaten 1½ lengths by Brigadier Gerard, but he decisively accounted for Riverman in third place and both Steel Pulse and the Derby Italiano winner Gay Lussac were further behind. He won a minor race at Windsor by 20 lengths in September, but unfortunately ten days before the Prix de l'Arc de Triomphe, Parnell was found lame in his box with bruised ligaments on his rear hind leg. Although his work was only interrupted for three days, such a setback in his training schedule can only have had a detrimental effect on his chance in Paris. Nevertheless he was the best of the English contingent, and at 19 to 2 was the shortest priced runner of the older generation.

The four-year-old Erimo Hawk, trained by Geoffrey Barling at Newmarket, was an out and out stayer who would definitely need a longer distance than that of the Prix de l'Arc de Triomphe. After victories in the Selfridge's Queen's Prize (16f) and the Paradise Stakes at Ascot, Erimo Hawk ran Rock Roi to a head in the Ascot Gold Cup. Since he had been continually hampered by the winner throughout the final furlong, the Stewards reversed the placings. He added the Goodwood Cup to his victories but was a disappointing fourth in the Doncaster Cup. His price of 99 to 1 was a fair estimate of his chance on Sunday 8 October.

The other English raider, Steel Pulse, was trained by Sam Armstrong in 1971 and won four races including the Critérium de Maisons-Laffitte when he beat Riverman and Mata Hari. He looked an outstanding classic prospect after he had finished runner-up to Hard to Beat in the Grand Critérium and to High Top in the Observer Gold Cup. After finishing second in the Craven Stakes and fourth in the 2000 Guineas, Steel Pulse was moved to Scobie Breasley's Epsom stables. When eighth to Roberto at Epsom it was reported that he had been hampered by Lyphard, but in the Prince of Wales's Stakes at Royal Ascot he proved no match for Brigadier Gerard and was beaten 5 lengths. Steel Pulse, however, then won the Irish Sweeps Derby by a length and 6 lengths from Scottish Rifle and Ballymore, the Irish 2000 Guineas winner, who patently did not stay a mile and a half. Steel Pulse was 7 lengths behind Parnell, the runner-up to Brigadier Gerard, when fourth in the King George VI and the Queen Elizabeth Stakes. He was then a very disappointing last to Boucher in the St Leger. Despite his victory at the Curragh, Steel Pulse was probably not such a good three-year-old as he had been as a juvenile and his price of 37 to 1 at Longchamp seemed a little ungenerous.

The Spanish challenger, My Mourne, who had cost 6,000 guineas as a foal at the Newmarket Sales, had won five races in Spain including the Gran Premio de Madrid (2500 metres) in June and the Copa d'Or (2400 metres) in August. It was very difficult to regard his chance as anything but remote in the Prix de l'Arc de Triomphe.

It was also almost impossible to gauge the chance of the ex-Japanese five-year-old, Mejiro Musashi, who in the autumn of 1972 was trained by Freddie Palmer. He had won five of his fourteen starts in Japan. As a three-year-old Mejiro Musashi was a little below top class, but in 1971 he won, as a four-year-old, the Spring Tennon-Shon, the Takarazuka Kinen and the Hakodate Kinen, and was the second leading money-winner for the season.

Mme R. A. Franklyn's four-year-old mare Snow Castle had won two races in Argentina over eleven furlongs earlier in 1972. As a three-year-old, however, she was a little below the best of her generation and was unplaced to Urugaya in the Gran Premio Carlos Pelligrini (3000 metres) in November. She won the Prix Foy from Crazy Rhythm and Pistol Packer, who was way below her best, in September when she was trained by Roger Poincelet; but that form was inevitably suspect. It was reported that she had gone well on the gallops but it was hard to envisage her playing a prominent part in the finish on 8 October.

Although Sir Michael Sobell's Homeric started at 39 to 1, he could not be totally ruled out of contention. Trained by Dick Hern in England in the spring and early summer, Homeric was 2 lengths runner-up to Selhurst in the Aston Park Stakes at Newbury and then in the Coronation Cup at Epsom ran Mill Reef, admittedly sickening for the virus, to a neck. He was again beaten 2 lengths by Selhurst and Frascati in the Hardwicke Stakes and immediately afterwards joined the Chantilly stable of John Cunnington Jr. On 14 July he beat Santa Luigi and Crucible a neck and 1½ lengths in the Prix Maurice de Nieuil at Saint-Cloud and then took the Prix Kergorlay by a neck and three-quarters of a length from Arlequino and Falkland. He was a disappointing seventh to Novius when warm favourite for the Grand Prix de Deauville, but the race was run at a very slow pace initially and then developed into a sprint. When it was clear that Homeric was not going to take a hand in the finish, his jockey did not give him a hard race. Homeric could be regarded as a possible outsider for the 1972 Prix de l'Arc de Triomphe.

There was absolutely no doubt that the three-year-old colt Hard to Beat was the best of the French defenders. Successful in three of his four outings as a two-year-old, his victory in the Grand Critérium established him as the best colt in France, if not the best on both sides of the Channel. He had been bought by Stephen Sokolow for only 920 guineas as a yearling, which made him a real bargain, probably because he was by the slightly unfashionable stallion Hardicanute. On his seasonal début in the Prix de Fontainbleau Hard to Beat comfortably defeated Tamiran and he started 8 to 10 favourite for the Poule d'Essai des Poulains on 23 April. He was not well away, and Piggott was unable to employ the front-running tactics advised by his trainer Dick Carver. Riverman led throughout and beat Gift Card three-quarters of a length with the favourite 5½ lengths away fifth. It is also possible that a mile was then too short for him, but Riverman proved to be an excellent miler and there were virtually no horses who could give him several lengths start and beat him over eight furlongs. Over the longer distance of the Prix Lupin on 14 May Mr Sokolow's colt, again partnered by Piggott, beat Vintner and Gift Card an easy length and 1½ lengths with Lyphard three-quarters of a length back fourth and Sancy seventh. Lyphard, reputedly the best three-year-old in Alec Head's yard, had started favourite at 7 to 10, but excuses were made for his defeat. It had been intended to run Hard to Beat at Epsom, but when Lyphard was announced a definite runner in the Derby, it was decided to run the Hardicanute colt in the Prix du Jockey Club. The Japanese textile magnate, Junzo Kashiyama, was so impressed by his display in the Prix Lupin that he agreed to buy Hard to Beat from Sokolow for a million dollars providing he won at Chantilly. In the Prix du Jockey Club on 4 June Hard to Beat took the lead shortly before the straight and thereafter was never in danger. He was 3 lengths and 1½ lengths ahead of Sancy and Flair Path at the winning post with Pleben seventh; he had also lowered the record time for the race, set by Auriban in 1952 and Val de Loir in 1962, by 1·30 seconds.

Carver was not entirely happy with the Prix du Jockey Club winner during his work before the Grand Prix de Saint-Cloud, feeling he might have gone over the top, but he still started favourite at 1 to 2. His fears were justified for, when he was challenged by Rheingold 200 metres out, he found nothing and was even passed and beaten half-length by Arlequino who was 3 lengths behind the Epsom Derby runner-up. Hard to Beat was rested until the Prix Niel on 10 September when he showed himself back to his best, for he accounted for Roberto by a comfortable length, the pair finishing 10 lengths ahead of Toujours Prêt. As a result of this victory the Hardicanute colt started 5 to 4 favourite on Sunday 8 October.

George Watson's stable had two runners in the 1972 Prix de l'Arc de Triomphe, Pleben and Rescousse, both owned by Baron de Rédé. The Sheshoon colt Pleben, who cost 135,000 francs as a yearling at Deauville, was unraced as a two-year-old. In the Prix de Grenelle, a twelve-furlong race confined to débutants, he so easily disposed of the opposition that his next outing was in the Prix du Jockey Club, when he ran on strongly in the later stages to finish seventh. In the Grand Prix de Paris he was opposed by both Talleyrand and Novius who had finished in front of him at Chantilly. Pleben, however, decisively beat Sukawa and Talleyrand 2 lengths and 2½ lengths with Novius only sixth. In his only other race before the Prix de l'Arc de Triomphe, the Prix Royal Oak on 17 September, he again ran out an impressive winner by 1½ lengths and the same from Novius and Lassalle. The doubt about the Sheshoon colt was whether he had the speed to come back to a mile and a half.

There was no such question about the speed of Baron de Rédé's filly Rescousse. As a juvenile she had won once over 5½ furlongs and had been a close seventh to Dark Baby in the Critérium des Pouliches. After running third in the Prix La Camargo on her seasonal reappearance, Rescousse won the Prix Pénélope at Saint-Cloud from the useful La Troublerie. She only managed third place behind Dame des Ondes in the Prix Vanteaux, as a result of which she was the least fancied of the three runners from her stable in the Prix de Diane, the other two being Paysanne and Caprea. In the Prix de Diane Rescousse was ridden by Yves Saint-Martin; she led throughout to win unchallenged by 3 lengths and half a length from Paysanne and Prodice, who had finished respectively third and first in the Prix Saint-Alary. Baron de Rédé's filly was rested until the Prix de la Nonette in September. Giving San San 7 lbs, Rescousse beat her 3 lengths with Paysanne a short neck back third; such good fillies as Pidget, La Troublerie, who had taken the Prix de la Côte Normande at Deauville, and Licata, the Prix de Malleret winner, were never able to put in a challenge. Rscousse started a very warm favourite for the Prix Vermeille on 24 September but was beaten into fifth place by the dead-heaters San San and Paysanne. It was the first time she had finished out of the frame as a three-year-old; some felt she might have found the distance too far for her, but perhaps she just had an off-day. Nevertheless in Rescousse and Pleben Baron de Rédé had two serious threats to Hard to Beat and Roberto on 8 October and they started joint second favourites at 9 to 2, the same price as the Epsom Derby winner.

Comtesse Margit Batthyany's San San had had eight outings as a two-year-old but had only won once, the Prix de la Châtaigneraie (1650 metres), a nursery handicap at Saint-Cloud on 30 November. She had, however, finished runner-up on three occasions, once behind La Troublerie in the Prix de Newmarket (1500 metres) at Longchamp on 4 October. San San won the Prix des Tulipes, a mile handicap for fillies at Maisons-Laffitte, on her first appearance in April 1972, but only managed fourth place, a little over a length behind La Troublerie and Licata in the Prix Chloë on 29 April, and second place to Valmarena in the Prix Finlande on 31 May. A muscular injury kept her off the track till the Prix Marly-le-Roi at Saint-Cloud on 30 July in which she finished fifth. On 13 August, however, San San won the Prix de Psyché (2000 metres) at Deauville from Liva. She gave a very much improved display in the Prix de la Calonne nine days later: giving weight to Marisela, the runner-up to Waterloo in the 1000 Guineas, San San ran her to a short neck. After her defeat by Rescousse in the Prix de la Nonette, Comtesse Batthyany's filly started at 20 to 1 for the Prix Vermeille. Jean Cruguet, who had ridden her in all her races in 1972, was given orders to hold San San up for as long as possible. Entering the straight she had so much ground to make up that it seemed he had overdone the waiting, but once he let her go, San San finished so strongly that she flew past all her rivals with the exception of Paysanne. They passed the post together and the judge declared a dead-heat; had San San not stumbled slightly twenty metres out she probably would have won outright. The pair were 2 lengths and half a length ahead of Decigale and the Epsom Oaks heroine Ginevra, with Rescousse fifth and Regal Exception in the rear. Although she had been clearly improving with every race since the Prix de Psyché, San San was permitted to start at the generous odds of 37 to 2. Certainly on her running in the Prix Vermeille she had a better chance than that on Sunday 8 October.

The other three-year-old filly in the 1972 Prix de l'Arc de Triomphe was Robin Scully's Regal Exception, who had won a maiden race at Saint-Cloud in March, but had made no show in either the Prix Vanteaux or in the Prix d'Ièna. Nevertheless she was sent to Epsom to contest the Oaks, for which she started at 100 to 1. Her trainer John Fellows failed to declare blinkers for her overnight, which had been his intention. Regal Exception still nearly brought off a major surprise for she led the field into the straight and appeared likely to hold on until Ginevra challenged and overtook her a furlong out. Regal Exception, however, still finished second ahead of the Irish 1000 Guineas runner-up, Arkadina. In the Irish Guinness Oaks Regal Exception, fitted with blinkers, comfortably beat Arkadina 3 lengths with Pidget third. Her disappointing display in the Prix Vermeille was put down to the fact that Robin Scully's filly was found to be amiss after the race. Her starting price of 94 to 1 was almost an insult to her ability and her chance in the Prix de l'Arc de Triomphe was certainly better than that.

There were two other three-year-old colts in the field on 8 October. Sancy had won his first three races in the spring, the Handicap Optional, the Prix Greffulhe and the Prix Noailles. Although runner-up at Chantilly behind Hard to Beat in June, he was very decisively beaten. He disappointed when only tenth in the Prix Royal Oak and it was difficult to enthuse about his chance. The other colt, Toujours Prêt had won three races over ten furlongs in 1972 but since he had been 10 lengths behind Hard to Beat and Roberto in the Prix Niel, his chance seemed as remote as his odds of 99 to 1 were long.

The field was completed by Sharapour and Mister Sic Top, both of whom had run without distinction behind Mill Reef in 1971, and Raphaël Hakim's Card King. The Aga Khan's Sharapour had shown consistently improved form throughout the spring. After he had won a mile handicap at Saint-Cloud in April and gained an 8-length victory at Longchamp over ten furlongs, Sharapour beat Mazus by half a length in the Prix Dollar. He was far from disgraced when running Riverman to 2 lengths in the Prix d'Ispahan on 25 June, with Ballyhot and Mister Sic Top 3 lengths and a short head away third and fourth. Sharapour was rested until the Prix du Prince d'Orange, in which, running on under pressure, he courageously resisted the challenges of Mister Sic Top and the very useful English colt Moulton. Sharapour, however, hampered the latter in the final furlong and the stewards relegated him to third place, awarding the race to Mister Sic Top. Although he had clearly improved as a four-year-old, Sharapour still had never won at more than ten furlongs and would probably find the extra two furlongs of the Prix de l'Arc de Triomphe too far for him.

The same applied with even more certainty to Mister Sic Top. He had won the Prix Exbury from his elder full brother Monsieur D at Saint-Cloud in March, but had always given the impression throughout his career that he was best at a mile to a mile and a quarter.

Card King, who had been bought out of a claiming race at Chantilly, had run Ex Libris to three-quarters of a length in the Prix du Conseil Municipal in October 1971. He had, however, failed to win in seven outings as a four-year-old, although he did finish runner-up to Beau Charmeur in the Gran Premio di Milano and was third to Ticeron and Weimar in the Gran Premio d'Italia. He was, however, only fifth to Novius in the Grand Prix de Deauville and last of seven to Rheingold in the Grand Prix de Saint-Cloud, and not surprisingly was allowed to start at 84 to 1.

Marcel Depalmas, who had partnered Rescousse in the Prix Vermeille, elected to ride Pleben on 8 October, so Yves Saint-Martin took the mount on the filly he had ridden to victory in the Prix de Diane. Unfortunately San San's regular jockey Jean Cruguet was injured in a fall a few days before the 1972 Prix de l'Arc de Triomphe. On Wednesday 4 October her trainer Angel Penna engaged the current champion jockey in France, Freddie Head, to ride Comtesse Batthyany's filly. The same morning Lester Piggott partnered his fourteenth mount in Longchamp's most prestigious race in his final gallop at Chantilly. So well did Hard to Beat go that Piggott afterwards announced: 'This must be my best chance of winning the Prix de l'Arc de Triomphe.' Despite his interrupted training schedule, Parnell also had a satisfactory final workout at Newmarket.

The Race

Sunday 8 October was a beautifully warm sunny afternoon at Longchamp where again the stands were packed with a very cosmopolitan crowd. In the Prix de l'Abbaye de Longchamp Piggott, partnering the Irish trained favourite Home Guard, was beaten into second place by Williamson on David Robinson's Deep Diver. In the Critérium des Pouliches, Piggott rode the unbeaten Irish filly Fiery Diplomat, but she was defeated by 2½ lengths and a nose by Allez France and Kerlande. Daniel Wildenstein's filly thrilled the crowd with a scintillating performance; coming from last to first in the straight, she cut down her rivals with an amazing burst of acceleration to win virtually unextended. She was certainly a star in the making.

In the paddock Hard to Beat looked fit and hard-trained and was his usual composed self. Roberto, however, was sweating and appeared rather wound up. Steel Pulse was sweating even more profusely and was extremely nervous. Parnell on the other hand looked superb and he and the favourite were the pick of the field.

As soon as the stalls opened, Baeza shot Roberto into the lead and set a blistering pace to Regal Exception, Snow Castle, Boucher, Parnell, Sharapour and Hard to Beat. Pleben, Erimo Hawk, Mejiro Musashi, Rescousse, My Mourne and San San were the backmarkers. As the runners emerged from behind the Petit Bois Roberto was still taking the field along at breakneck speed from Regal Exception and Snow Castle who were followed by Parnell, Sharapour, Boucher and Hard to Beat who in turn were tracked by Homeric and Toujours Prêt. Mejiro Musashi, Rescousse, San San and the already struggling My Mourne were still at the rear about 12 lengths behind the leaders. Roberto covered the first half of the race five seconds faster than the leaders had done the previous year. Coming down the hill the Epsom Derby winner was still bowling along a length in

San San (18) beating Rescousse (17) and Homeric (white cap, partially hidden) with Regal Exception (19) fourth and Card King (4) fifth

front of Regal Exception and Snow Castle. Then in a group together came Boucher, Parnell and Sharapour, with Hard to Beat and Homeric, both going easily, hard on their heels. Steel Pulse, already under pressure, Sancy, Pleben and Card King were in the mid-division whilst the last three were still San San, who was beginning a forward move, Rescousse and My Mourne. As Roberto began to swing into the straight, Baeza stole a glance over his shoulder at Regal Exception but he had not shaken off his nearest pursuer, although Snow Castle was clearly beaten and starting to lose her place. Once in the straight Roberto was closely attended by Regal Exception on the rails. They were tracked by Boucher and Homeric together, Sharapour and Hard to Beat, who was in a perfect position if he was good enough, and San San, who had made rapid headway to join the leading group on the outside. Four hundred metres out Roberto gave way to Regal Exception, who was tracked by Homeric and San San, who had continued to close on the leading three on the outside. Meanwhile Boucher and Parnell were rapidly dropping out of contention, Sharapour was beginning to struggle and it became obvious that Hard to Beat was not responding to Piggott's urgings. Three hundred yards from the finish Homeric got the better of Regal Exception and the fading Roberto with San San still going easily on the outside. On the wide outside Rescousse, who had been almost last entering the straight, could be seen making rapid progress past beaten horses, while Card King was threading his way through the centre of the field. With 200 metres to go it seemed that Homeric was going to hold on to his lead, but then Head unleashed San San and in a few strides she had shot past Sir Michael Sobell's four-year-old. Rescousse meanwhile was still 3 lengths behind her in about eighth place, but she finished so fast that she was only 1½ lengths behind Comtesse Batthyany's filly at the winning post, half a length in front of Homeric who 1½ lengths to spare over Regal Exception. She held off Card King by half a length; he had been slightly hampered in the final furlong by the weakening Sharapour, who was sixth another half length behind him. Roberto was 1½ lengths away seventh, with the disappointing Hard to Beat, Parnell and Mister Sic Top the next to finish.

San San had never looked in serious danger of defeat once she struck the front. Hers was a thoroughly deserved victory for she had equalled Mill Reef's course record of 2 minutes 28·30 seconds. Nevertheless Maurice Philipperon, Homeric's jockey, lodged an objection to the winner for interference in the final furlong, but this was quickly overruled by the Stewards. Homeric, however, may have been a slightly unlucky loser in spite of the ease of San San's victory. He hobbled into the unsaddling enclosure exceptionally lame in both his forelegs. Philipperon reported that he had felt Homeric's legs go when he was still in the lead and going easily 200 metres from home. To have finished third in such a condition was a most courageous and meritorious performance by Homeric.

There were, however, few excuses for the other runners. Piggott commented, 'Hard to Beat was not good enough,' although he felt the going might have been a little on the firm side for him. There was some criticism of Baeza's tearaway tactics on Roberto, but they had been successful at York. When restraining tactics had been employed on the Derby winner in the Prix Niel, he had not been able to catch Hard to Beat, whereas in this race he finished in front of his rival for the first time.

San San's owner, Comtesse Margit Batthyany, who then lived in Lugano and had interests in banking, was a most enthusiastic supporter of European racing. She owned over sixty broodmares and the Gestüt Erlenhof in Germany and the Ballykeane Stud in Ireland. She was so thrilled by San San's triumph that she exclaimed, 'This is the greatest day of my life. I simply can't believe it.' Her racing manager José Mora, however, was not so surprised and remarked, 'I knew that San San had a chance if the race was run at a fast pace.' It was the first victory for a filly since that of La Sorellina in 1953. Comtesse Batthyany had enjoyed great success in 1972; she had also won the Poule d'Essai des Pouliches with Mata Hari. She was no stranger to classic victories, for Caro had won the Poule d'Essai des Poulains, Samos III the Prix Royal Oak, and Pia the Oaks, in the late 1960s and early 1970s. In Germany that excellent colt Orsini, whose fourteen victories included the Deutsches Derby, had also belonged to her. What was perhaps most remarkable about San San's victory in the 1972, however, was not the fortune that smiled on her owner but that three fillies should have finished in the first four. In fact it was the first time since Coronation beat Double Rose that fillies had filled the first two places.

It was also the first time that San San's trainer, the Argentinian Angel Penna, had won the Prix de l'Arc de Triomphe. As a juvenile San San had been trained by Albert Klimscha, but Penna had come to Chantilly in 1972 to train all Comtesse Batthyany's horses in training in France. He had been champion trainer in Argentina and Venezuela, which he had left in 1967 to train in the United States where he was also remarkably successful, training the Hollywood and American Derby winner, Bold Reason, and Czar Alexander. To have won the Prix de l'Arc de Triomphe, the Prix Vermeille, the Poule d'Essai des Pouliches and the Prix Morny with Filiberto was an incredible feat in his first season as a trainer in France.

San San's jockey, Freddie Head, was certainly fortunate to get the mount, which gave him his second success in the Prix de l'Arc de Triomphe. It was, in fact, the only occasion in her career that Head ever rode San San.

The Rest of the Season

Only one of the runners in the 1972 Prix de l'Arc de Triomphe, Card King, ran in the Prix du Conseil Municipal, but he finished in rear behind Monsieur D. Steel Pulse, rather surprisingly, was brought out for the Champion Stakes six days after making no show in Paris. He again ran very disappointingly behind Brigadier Gerard and Riverman. San San, Parnell, Steel Pulse and Mejiro Musashi all accepted invitations to run in the Washington D.C. International Stakes at Laurel Park in November. Parnell did the best, finishing 4 lengths runner-up to Droll Role with Steel Pulse and San San 9 lengths and 1 length behind third and fourth and Mejiro Musashi only seventh. Shortly after half way, however, Boreen and Jumbo Jet fell. San San, whom Cruguet was holding up at the back of the field as usual, was severely impeded and thereafter had no chance of victory. During the winter Comtesse Batthyany sold San San to the Japanese millionaire Tadao Tamashima, and she ran in his colours in 1973.

San San, however, was nowhere near the same filly as a four-year-old. She finished out of the first ten behind Rheingold in the Prix Ganay. She did not run again till the Prix de Hastings (10f) at Deauville in August when she only finished fifth to Dark Baby. In the 1973 Prix de l'Arc de Triomphe San San was a well-beaten twelfth. When she retired to stud in Japan, she had won five races from her twenty-two outings and had finished runner-up six times. In France San San won 2,216,405 francs and in the U.S.A. $5,000.

Pedigree and Stud Record

San San's sire, Bald Eagle by Nasrullah, had been bred by Harry F. Guggenheim in the United States and had been sent to be trained by Captain Cecil Boyd-Rochfort at Newmarket. He won the Duke of Edinburgh Stakes as a juvenile and, after victories in the Craven Stakes and Dante Stakes, he started second favourite for the Epsom Derby but he disappointed badly behind Hard Ridden. He had a reputation in England for being potentially brilliant but inconsistent. He had been well-fancied when failing behind Pall Mall in the 2000 Guineas. After finishing runner-up in the St James's Palace Stakes at Royal Ascot, he returned to the U.S.A. The style of racing there suited him much better, and as a four-year-old and five-year-old he won the Washington D.C. International Stakes twice, the Gallant Fox Handicap in record time, the Saratoga Handicap, the Suburban Handicap, the Aqueduct Handicap, the Metropolitan Handicap in a new course record, the Widener Handicap also in record time, and the Gulf Stream Handicap. At stud, however, he frequently passed on his awkward temperament to his offspring. His best offspring before San San was probably Barely Once (St Louis Derby and Fairmount Derby), though one of his colts, National Seal, won the Norwegian 2000 Guineas, not normally a race associated with the sire of a winner of the Prix de l'Arc de Triomphe. Bald Eagle was imported into France for $50,000 in 1971 and stood at the Haras de Courtomer. San San's victory at Longchamp meant that he finished fourth in the French list of successful stallions in 1972.

San San was bred by Harry F. Guggenheim in the United States. Comtesse Batthyany bought her as a foal for $15,000 at the dispersal sale of his bloodstock. Her dam, Sail Navy by Princequillo, was put into training but in fact never appeared on the racecourse. She had not achieved any notable success in the paddocks before she produced San San. In 1962 her foal died; her next foal Brace Up, a colt by Turn-to, won a ten-furlong maiden race as a three-year-old at Ripon in England in 1967. Midway Island by Turn-to was Sail Navy's second live foal but was unraced. She, however, has produced six foals, five of them winners, including Current Charge by Little Current (Tetrarch Stakes) and Very Special Lady by Buckpasser (Nursery Stakes and Osunitas Stakes). Sail Navy's filly, Havfrue by Bald Eagle, was foaled the year before San San but was unplaced as two-year-old.

Sail Navy's dam, Anchors Aweigh by Devil Diver, had two victories as a three-year-old, one of which was in the Jasmine Stakes. She produced seven foals to race, six of whom were winners. The best of these was Never Bow by Never Bend, whose ten victories included the Widener Handicap, the Brooklyn Handicap, the Stuyvesant Handicap and the Westchester Handicap; he netted a total of $415,046 in prize-money. Her filly, Make Sail by Ambiorix, won seven races including the Kentucky Oaks, the Alabama Stakes and the Top Flight Handicap. Make Sail herself was the dam of six winners, including Captain's Gig who won the Futurity Stakes and the Jim Dandy Stakes. She was also the grandam of Huguenot, the winner of the Beresford Stakes and Prix Daphnis. Three of Anchors Aweigh's colts became sires: Hearts of Lettuce (Louisiana Handicap, 2nd Ohio Derby), Sheet Anchor (3rd Gallant Fox Handicap), and George Patrick (3rd St Louis Derby).

San San's third dam was the chestnut filly, True Bearing by Sir Gallahad. There is, however, some mystery about her sire. Sir Gallahad was a pure dominant bay. According to the laws of genetics a pure dominant bay stallion cannot sire a chestnut foal. There is no question of Sir Gallahad being an impure bay; he sired 567 named foals and only one, True Bearing, was registered as a chestnut and there is no doubt that she was one. Who the real sire of True Bearing was must remain unknown. She was unraced. In addition to Anchors Aweigh she produced two other winners: On Course, whose five victories won him $23,410 and who became a sire; and Sharon Mary, who won two races as a three-year-old and was a successful broodmare. She was the dam of Golden Don (Michigan Mile and One-Eighth Handicap, Manhattan Handicap) and Don Dougold (Hamilton County Handicap).

True Bearing's dam, the chestnut Dead Reckoning by Man o'War, was also a successful broodmare. Her best offspring were the Stakes winner Director J.E. and later a sire and the stakes-placed filly Moonsight the dam of Moon Mad. This family has flourished in the United States since the 1870s when Blairgowrie by Breadalbane, the tenth dam of San San, was imported from England. The line traces back to Gibside Fairy, the grandam of Cotherstone and the third dam of Western Australian.

At stud in Hokkaido in Japan San San produced four foals between 1977 and 1981 but was barren in 1979. Riki Sansan, a colt by Sir Ivor foaled in 1977, gained four victories from thirteen outings. Her foal the following year, Speed Sansan, a filly by Silver Shark, ran sixteen times and won three races. San San foaled a colt by Partholon in 1980 and a filly by Northern Taste in 1981.

PRIX DE L'ARC DE TRIOMPHE 2400 metres
at Longchamp on Sunday 8 October 1972. For three-year-olds and upwards. Going: Good to firm.
1,200,000 francs to the winner; 480,000 fr. to the 2nd; 240,000 fr. to the 3rd; 120,000 fr. to the 4th. 288,000 fr. to the breeder of the winner; 48,000 fr. to that of the 2nd; 24,000 fr. to that of the 3rd. Value to the winner: 1,415,250 francs.

SAN SAN	b.f.	by Bald Eagle – Sail Navy by Princequillo	3.54k	Comtesse Margit Batthyany	F. Head	1
RESCOUSSE	b.f.	by Emerson – Bella Mourne by Mourne	3.54k	Baron de Rédé	Y. Saint-Martin	2
HOMERIC	ch.c.	by Ragusa – Darlene by Dante	4.60k	Sir Michael Sobell	M. Philipperon	3
REGAL EXCEPTION	b.f.	by Ribot – Rajput Princess by Prince Taj	3.54k	Robin Scully	J. Mercer	4
CARD KING	b. or br. c.	by Cardington King – Nantua II by Again II	4.60k	Raphaël Hakim	J. Lindley	5
SHARAPOUR	b.c.	by Tanerko – Paola II by Palestine	4.60k	H.H. Aga Khan	W.B. Pyers	6
ROBERTO	b.c.	by Hail to Reason – Bramalea by Nashua	3.55½k	John W. Galbreath	B. Baeza	7
HARD TO BEAT	b.c.	by Hardicanute – Virtuous by Above Suspicion	3.55½k	Junzo Kashiyama	L. Piggott	8
PARNELL	ch.c.	by St. Paddy – Nella by Nearco	4.60k	Rory More O'Ferrall	W. Carson	9
MISTER SIC TOP	br.c.	by Tiffauges – Madame Daurade by Sicambre	5.60k	Lucien Deshayes	H. Samani	10
PLEBEN	ch.c.	by Sheshoon – Devastating by Honeyway	3.55½k	Baron de Rédé	M. Depalmas	10
MEJIRO MUSASHI		by Wallaby – Kiyoha by Primero	5.60k	Toyokichi Kitano	Y. Nohira	
ERIMO HAWK		by Sea Hawk – Nick of Time by Niccolaus	4.60k	Shin-Ishi Yamamoto	P. Eddery	
SNOW CASTLE		by Snow Cat – Luna Park by Claro	4.58½k	Mme R.A. Franklyn	G. Thiboeuf	
MY MOURNE		by Mourne – Chanter by Chanteur	3.55½k	Antonio Blasco	J.A. Borrego	
SANCY		by Sanctus – Wordys by Worden	3.55½k	Myran Ekrayan	J.C. Desaint	
TOUJOURS PRET		by Val de Loir – Dundee by Chingacgook	3.55½k	Lawrence M. Gelb	J. Taillard	
BOUCHER		by Ribot – Glamour by Nasrullah	3.55½k	Ogden Phipps	L. Pincay	
STEEL PULSE		by Diatome – Rachel by Tudor Minstrel	3.55½k	Ravi N. Tikoo	W. Williamson	

19 runners.
SAN SAN trained by A. Penna; bred by H.F. Guggenheim in the U.S.A.
Distances: 1½ lengths; ½ length; 1½ lengths; ½ length; ½ length; 1½ lengths; 1½ lengths; 1½ lengths; head. Time: 2 mins. 28 seconds.
Pari-Mutuel: Win: 19fr. 60. Places: 4fr. 60; 3fr. 0; 6fr. 80.
S.P.: 5 to 4 Hard to Beat; 9 to 2 Pleben and Rescousse (couplé) and Roberto; 19 to 2 Parnell; 17 to 1 Sharapour and Boucher; 37 to 2 SAN SAN; 37 to 1 Steel Pulse; 38 to 1 Sancy; 39 to 1 Homeric; 41 to 1 Mejiro Musashi; 46 to 1 Snow Castle; 67 to 1 My Mourne; 84 to 1 Card King; 94 to 1 Regal Exception; 99 to 1 Erimo Hawk, Toujours Prêt and Mister Sic Top.

SAN SAN bay filly 1969	Bald Eagle b. 1956	Nasrullah b. 1940	Nearco br. 1935	Pharos b. 1920	Phalaris
					Scapa Flow
				Nogara b. 1928	Havresac II
					Catnip
			Mumtaz Begum b. 1932	Blenheim br. 1927	Blandford
					Malva
				Mumtaz Mahal gr. 1921	The Tetrarch
					Lady Josephine
		Siama b. 1947	Tiger ch. 1935	Bull Dog b. or br. 1927	Teddy
					Plucky Liège
				Starless Moment b. 1928	North Star III
					Breathless Moment
			China Face b. 1940	Display b. 1923	Fair Play
					Cicuta
				Sweepilla b. 1931	Sweep
					Camillas
	Sail Navy b. 1958	Princequillo b. 1940	Prince Rose b. 1928	Rose Prince b. 1919	Prince Palatine
					Eglantine
				Indolence b. 1920	Gay Crusader
					Barrier
			Cosquila b. 1933	Papyrus br. 1920	Tracery
					Miss Matty
				Quick Thought b. 1918	White Eagle
					Mindful
		Anchors Aweigh b. 1949	Devil Diver b. 1939	St. Germans b. 1921	Swynford
					Hamoaze
				Dabchick br. 1931	Royal Minstrel
					Ruddy Duck
			True Bearing ch. 1945	Sir Gallahad b. 1920	Teddy
					Plucky Liège
				Dead Reckoning ch. 1931	Man o' War
					Guesswork

Rheingold 1973

Rheingold

The Field

The 1973 racing season was totally dominated by some outstanding three-year-old fillies in France, England and Ireland. By far the best two, Daniel Wildenstein's Allez France by Sea Bird, and Nelson Bunker Hunt's Dahlia by Vaguely Noble, both renewed rivalry in that year's Prix de l'Arc de Triomphe.

Allez France fully lived up to the enormous promise she had shown in the 1972 Critérium des Pouliches when she produced an astonishing turn of foot to beat Princess Arjumand and Dahlia, the easy winner of the Prix de la Grotte, by 2½ lengths and a neck in the Poule d'Essai des Pouliches on 29 April, after she had looked in a hopeless position on the turn into the straight. It was then tentatively planned to run her in the Epsom Derby so she next ran in the Prix Lupin on 13 May. She finished a disappointing seventh, however, a little over 4 lengths behind Kalamoun, the Poule d'Essai des Poulains winner. The plan to run her in the Derby was, therefore, scrapped. In the Prix de Diane on 10 June Allez France again displayed all her scintillating brilliance decisively accounting for Dahlia and Virunga, the first two in the Prix Saint-Alary, by 2½ lengths and 4 lengths, with Lady Berry 1½ lengths away sixth. Daniel Wildenstein's filly was then rested until the Prix de la Nonette at Longchamp on 2 September. In the paddock she looked very much in need of the race and without ever looking dangerous finished nearly 5 lengths fourth to Gay Style, My Great Aunt and Snobbishness, with only her two pacemakers behind her. She, therefore, was permitted to start at 6 to 4, coupled with her owner's Virunga, in the Prix Vermeille on 23 September, whereas Dahlia started at 9 to 10 favourite. In the paddock Allez France looked in much better condition and she proceeded to beat Hurry Harriet and El Mina by a most emphatic 2 lengths and 2½ lengths. Dahlia was a little over 2 lengths away fifth just ahead of Virunga, but she had slipped on the turn and returned to the unsaddling enclosure lame on her near hind. Thus on the three occasions they had met, Allez France had always proved Dahlia's superior. Her class and her blinding burst of speed inevitably meant that Allez France started favourite at 7 to 4 against on Sunday 7 October.

Dahlia's connections had blamed her defeat in the Prix de Diane, for which there had been twenty-five runners, on the fact that she had had an unlucky run. She was sent to the Curragh for the Irish Guinness Oaks on 21 July where she faced a formidable and then unbeaten opponent in Mysterious, the heroine of the 1000 Guineas and the Oaks. Dahlia, nevertheless, slammed Mysterious and Hurry Harriet by 3 lengths and 4 lengths. But one week later she reappeared against the best four-year-olds in Europe in the King George VI and the Queen Elizabeth Stakes at Ascot. Turning for home she was in last place but produced a remarkable run on the rails to take the lead two furlongs out. She proceeded to draw 6 lengths clear of Rheingold with Our Mirage another 2 lengths away third, Weaver's Hall fourth, and such good horses as Parnell, Hard to Beat, Card King and Roberto, trailing in her wake. The Prix Niel at Longchamp in September was run at a slow pace and Dahlia's blistering finishing speed carried her to a comfortable half a length victory over Tennyson. When Dahlia slipped in the Prix Vermeille, she either struck into herself or caught the heels of another runner. The lameness severely interrupted her training schedule in the next fortnight and it was only four days before the Prix de l'Arc de Triomphe that she was announced a definite runner. In all probability it was an unwise decision to have run her at all as she was unlikely to do herself justice in the race. Her reputation, however, caused Dahlia to start at odds of 8 to 1.

The other French three-year-old filly in the field on Sunday 7 October was Baron Guy de Rothschild's Lady Berry. Her only defeat in six outings as a three-year-old had been in the Prix de Diane which was certainly too short for her. In the spring Lady Berry had won a twelve-furlong event at Longchamp and the Prix des Tuileries. She easily accounted for Amira and El Mina by 2 lengths and the same in the Prix de Pomone at Deauville on 19 August. She avoided meeting Allez France and Dahlia in the Prix Vermeille and went instead for the Prix Royal Oak. The first two favourites, Authi and Balompie, were engaged in a battle royal in the straight when Lady Berry came past them both with a wet sail 200 metres out, and drew away to beat Authi 1½ lengths with Balompie and Récupéré 2 lengths and a short neck away third and fourth. Thus Lady Berry became the first filly to win the Prix Royal Oak for almost fifty years, and her victory merely went to underline the superiority of the 1970 generation of fillies over the colts. The one doubt about Baron Guy de Rothschild's filly was whether she might find a mile and a half a little on the sharp side for her but she was still very well fancied at odds of 35 to 4.

Mysterious, who had easily beaten the Prix Vanteaux and the Prix de Malleret winner, Virunga, declined another meeting with Dahlia and had the Champion Stakes as her autumn objective. The Irish filly Hurry Harriet, however, again took on Dahlia and Allez France although clearly held by both on her running at the Curragh and in the Priz Vermeille. Runner-up to Hail the Pirates in the Desmond Stakes (10f) at the Curragh on 18 August and to the surprise 100 to 1 winner, Miss Thérèse in the Blandford Stakes on 1 September, her solitary success in 1973 before the Prix de l'Arc de Triomphe had been when she beat the Irish 1000 Guineas runner-up, Annerbelle, 1½ lengths in the Pretty Polly Stakes at the Curragh on 30 July. She was certainly overgenerously priced at 96 to 1 but were she to have won on Sunday 7 October, it would have been a surprise.

Despite Allez France's immense reputation there were four other challengers from Ireland besides Hurry Harriet: Miss Thérèse, Bonne Noël, Ragapan and Star Appeal. There were also no less than five from England: Rheingold, Parnell, Buoy, Attica Meli and Firefright. Of these Firefright, a handicap winner at York and a close third under a light weight to Bonne Noël in the Ebor Handicap, and Miss Thérèse, who had only appeared a fair handicapper before her surprise Blandford Stakes victory, had no chance at all in the Prix de l'Arc de Triomphe. Ragapan had run well to be 5½ lengths fifth to Morston in the Epsom Derby, and had beaten Buoy 1½ lengths for second place when 2½ lengths behind Weaver's Hall in the Irish Sweeps Derby, with Star Appeal 5½ lengths away seventh. He was, however, a moderate eighth to Peleid in the St Leger and it was difficult to enthuse about his prospects at Longchamp. Star Appeal, the winner of the Madrid Free Handicap (7f) in the spring, was beaten into third place by Hail the Pirates in the Gallinule Stakes at the Curragh and filled the same position three-quarters of a length and 8 lengths behind Conor Pass and Sunyboy in the Irish St Leger on 22 September. He had earlier won the Preis der Stadt over ten furlongs at Baden Baden but he seemed likely to be outclassed at Longchamp.

Mrs Anne Parker Poe's four-year-old Bonne Noël probably had the best chance of the Irish raiders, if only that of an outsider, in the 1973 Prix de l'Arc de Triomphe. After winning a handicap at Gowran Park, he beat Valuta half a length in the Prix Kergorlay with Pleben and a below par Parnell a distant fourth and fifth. Nevertheless after a stewards' enquiry he was relegated to second place. He gained swift compensation when, under top weight, he defeated Sir Lark and Firefright in the Ebor Handicap at York. Although Bonne Noël was probably better over distances further than a mile and a half, his odds of 29 to 1 were not bad value.

R.D. Hollingsworth's three-year-old Buoy had won the Predominate Stakes at Goodwood in May and the Great Voltigeur Stakes at York in August. He was, however, only third in the Irish Sweeps Derby and 2½ lengths runner-up to Peleid in the St Leger. On both occasions he had been unable to quicken when challenged and had run on at one pace. It seemed sure that he would find several too fast for him in the final 400 metres at Longchamp and for that reason started at the long odds of 61 to 1.

For the second consecutive year Parnell won the Rouge Dragon Handicap at Ascot. He next beat Lassalle 3 lengths in the Prix Jean Prat on 23 April with Pleben and Sang Bleu 2¼ lengths and three-quarters of a length away fourth and fifth and Novius sixth. He was then unfortunately injured and missed both the Prix du Cadran and the Ascot Gold Cup. He was clearly not at his best when well beaten in the Grand Prix de Saint-Cloud, behind Dahlia at Ascot and in the Prix Kergorlay. In September he did easily dispose of Hakodate by 6 lengths in the Campari Cup at Ascot, but by October 1973 it was almost certain that Rory More O'Ferrall's horse required two miles when up against the highest-class opposition.

Louis Freedman's Attica Meli had been the best three-year-old filly in England in 1972, and her five victories had included the Yorkshire Oaks, the Park Hill Stakes and the Princess Royal Stakes. In the spring and summer of 1973 Attica Meli ran well below her best. Roberto beat her an easy 5 lengths in the Coronation Cup whilst Rheingold trounced her 6 lengths in the Hardwicke Stakes. On her fifth outing, in the Geoffrey Freer Stakes at Newbury on 18 August, Attica Meli, for the first time in season, looked very well in the paddock. She drew 6 lengths clear of Ballyhot who had beaten her 2 lengths for third place behind Our Mirage in the Princess of Wales's Stakes on 3 July. Attica Meli followed up this victory with an effortless one by 4 lengths and the same from Bretton Woods and Proverb in the Doncaster Cup. She started at odds of 22 to 1 on 7 October, and it was impossible to rule Louis Freedman's four-year-old mare out of contention.

Nevertheless the most likely English trained challenger to take the 1973 Prix de l'Arc de Triomphe appeared to be Henry Zeisel's four-year-old Rheingold. He had never been beaten in France. Although arguably the best three-year-old in England and France over a mile and a half in 1972, he was retired for the season after his disappointing third in the Benson and Hedges Gold Cup and had not run in the Prix de l'Arc de Triomphe. Yves Saint-Martin had ridden him in the Grand Prix de Saint-Cloud, and during the winter it was arranged that he should ride him in all his races whether in France or in England. The brilliant French jockey, therefore, came to Newbury on 14 April and partnered Rheingold to a 5 length and three-quarters of a length victory over Knockroe and Coup de Feu in the John Porter Stakes. Rheingold next produced a great burst of acceleration to beat the soft ground specialist Bog Road 1½ lengths in the Prix Ganay a fortnight later with Mister Sic Top and Citheron dead-heating for third place a short head behind. Card King was seventh, Moulton ninth and San San and Primette out of the first ten. Rheingold's easy victory over Attica Meli in the Hardwicke Stakes at Royal Ascot was followed by his second consecutive triumph in the Grand Prix de Saint-Cloud when he beat Direct Flight more comfortably than the official three-quarters of a length margin suggests. The pair were 8 lengths and a nose ahead of Roybet and Card King whilst Parnell, Rescousse and Pleben were even further adrift. Henry Zeisel's four-year-old was a short-priced favourite for the King George VI and the Queen Elizabeth Stakes but he played up before entering the stalls, and then was bumped on the turn into the straight. Saint-Martin in fact brought Rheingold very wide on the outside rounding the turn and was not seen to his best advantage. However he had no answer to Dahlia's amazing turn of foot but it is certain that he might have finished much closer to her given more fortune in running.

Yves Saint-Martin flew over to York to partner Rheingold in the Benson and Hedges Gold Cup on 21 August. Piggott was booked to ride Roberto, but on account of the soft going, he was withdrawn on the morning of the race. Piggott had a tentative arrangement to ride Moulton for Harry Wragg if that happened, but one of the joint owners of Rheingold approached him and offered him the mount in place of Saint-Martin, as Piggott had already been booked for Rheingold for the Prix de l'Arc de Triomphe for which Saint-Martin was unavailable. Piggott accepted; Lewis kept the mount on Moulton and a very irritated Saint-Martin was left in the stands without a ride. Rheingold started favourite at 4 to 6, but with three furlongs to go was already under pressure and he was beaten 2½ lengths and 4 lengths by Moulton and the Eclipse Stakes winner Scottish Rifle. Immediately afterwards Piggott reported that he had run 'like a dead horse'. On his return to his stables at Lambourn Rheingold was found to be suffering from a blood disorder. The problem facing his trainer Barry Hills was whether he could be nursed back to health and fitness before 7 October. On his best form Rheingold was the best staying four-year-old in Europe and he was to start at odds of 77 to 10 in the Prix de l'Arc de Triomphe.

There were, in fact, sixteen older horses who took on Allez France on 7 October. Mme François Dupré's Direct Flight, the runner-up to Rheingold at Saint-Cloud, had run the useful English five-year-old Colum to 2 lengths in the Prix Flying Dutchman (10½f) at Evry on 23 May with Primette 9 lengths away sixth. On 11 June he beat Kalidar and Amadou 4 lengths and a short head in the Prix Jean de Chaudenay at Saint-Cloud with Our Mirage three-quarters of a length away fourth. In the Prix Foy at Longchamp in September Direct Flight gave Mister Sic Top a kilo and beat him a comfortable neck, with Sang Bleu a short head away third and Unicornus 3½ lengths back sixth.

Mister Sic Top, running in the Prix de l'Arc de Triomphe for the third consecutive season, had won the Prix Exbury by a length from Blue Chip and had run third to Rheingold in the Prix Ganay in the spring. He had, however, been unplaced to Gift Card in the Prix Dollar and to La Troublerie in the Prix d'Ispahan. Even as a six-year-old ten furlongs was his best distance and he was unlikely to be concerned in the finish on 7 October.

Raphaël Hakim's five-year-old Card King had run a little inconsistently in the spring and summer. He reappeared in the Prix d'Harcourt on 8 April and ran Toujours Prêt to 3 lengths, with Mister Sic Top a head behind third and Hard to Beat last of ten. Card King then ran unplaced in the Prix Ganay and Prix Dollar, but he ran well when 2-length runner-up to the German four-year-old Prince Ippi, who had won the Preis von Europa the previous autumn, in the Gran Premio d'Italia at San Siro on 10 June. Fourth in the Grand Prix de Saint-Cloud and not disgraced when sixth to Dahlia at Ascot, Card King was rather surprisingly beaten a neck by Primette in the Prix de Reux (13½f) at Deauville. He totally reversed that form, however, in the Grand Prix de Deauville when he gained his first victory of the season by half a length and 1½ lengths from Balompie and Authi with Primette another 5 lengths back sixth. On his best form Card King might be in contention for one of the minor places and his odds of 75 to 1 seemed really quite generous.

The four-year-old mare Primette, despite her success in the Prix de Reux, had been slammed in both the Prix Flying Dutchman and in the Prix Ganay. In April she had won a 10½ furlong handicap at Saint-Cloud, but looked totally out of her depth in the Prix de l'Arc de Triomphe. Primette was, in fact, a stable-companion of San San, who was trained in 1973 by John Fellows for her new owner Tadao Tamashima. San San had made no show in the Prix Ganay and on her only other run before 7 October had been beaten a little over a length into fifth place by Dark Baby in the Prix Hastings (10f) at Deauville on 16 August. So little chance was given her of repeating her victory of the previous season that San San started at 74 to 1 on 7 October.

Hard to Beat also only had two races in 1973 prior to the Prix de l'Arc de Triomphe. When last in the Prix d'Harcourt in April, his forelegs had been bandaged and he did not look particularly well in the paddock. In the King George VI and the Queen Elizabeth Stakes Hard to Beat looked in much better condition. He was lying third entering the straight but dropped out with 300 yards to go to finish only ninth. In September, however, Hard to Beat began to show his old sparkle in his work at Chantilly. In one gallop with his stable-companion Lassalle and the Prix Maurice de Nieuil and Prix Kergorlay winner, Valuta, he so comprehensively defeated them that Valuta's trainer John Cunnington Jr decided to withdraw his charge from the field of the Prix de l'Arc de Triomphe. Certainly should Hard to Beat be anywhere near the form that won him the Prix du Jockey Club the previous season, he would pose a threat to all his opponents and, despite his lack of races and training problems, his odds of 21 to 1 were a reflection of Carver's confidence in him.

His stable-companion Lassalle had won the Prix du Cadran by 3 lengths from Pleben and the Ascot Gold Cup by 5 lengths from Celtic Cone, but those victories were achieved in the absence of Parnell through injury. The fields for both races had probably been the most moderate for many years. It seemed most unlikely that Lassalle would follow in the footsteps of Levmoss and he started justifiably at odds of 99 to 1.

The other three four-year-olds in the field for the Prix de l'Arc de Triomphe, Sang Bleu, Novius and Unicornus, all appeared out of their depth. In 1972 Sang Bleu had been runner-up to Sancy in the Prix Noailles and to Ticeron in the Gran Premio del Jockey Club. In April 1973 he was second to Filandre in the Prix de Barbeville but only fifth in the Prix Jean Prat and in the Gran Premio di Milano behind Garvin in June. His only victory in 1973 had come in La Coupe (13f) at Longchamp in May when he beat Irish Royal. J.R. Mullion's Novius had been a useful three-year-old the previous season when he had won the Grand Prix de Deauville and been runner-up to Pleben in the Prix Royal Oak. He was sent to be trained by Paddy Prendergast in Ireland in 1973, but had been a disappointing seventh to Parnell in the Prix Jean Prat and was beaten half a length and 1½ lengths by Ormindo and Mezzanine in the Ormonde Stakes at Chester on 10 May, his last outing before the Prix de l'Arc de Triomphe. He had, however, returned to Freddie Palmer's stable in the summer. R.A. Franklyn's Unicornus had won six races, including two Group Three events in his native Argentina. On his first appearance in France he ran a close third to Dark Baby in the Prix Hastings at Deauville, but he was beaten more than 4 lengths when sixth to Direct Flight in the Prix Foy. All these three started at the long odds of 99 to 1.

There were only four French three-year-old colts in the field on 7 October, Comtesse Batthyany's Balompie and El Famoso, Walter Burmann's Tennyson, and Mme Pierre Wertheimer's Authi. Tennyson, in fact, was to start second favourite at 23 to 4. Unraced as a juvenile Tennyson had won a small 10½ furlong event at Maisons-Laffitte in the spring. In the Prix du Jockey Club at Chantilly on 3 June many considered him to have been an unlucky loser. Roi Lear had taken the lead from Eddystone early in the straight with Gunter third on the rails. Four hundred metres out Bill Pyers on Gunter switched his mount to the outside to make his challenge; in doing so Gunter cut right across Tennyson. Alfred Gibert, his jockey, had to snatch up Walter Burmann's colt to avoid running into Gunter. Tennyson lost 4 lengths and at the same time gave a severe bump to the favourite Margouillat on his outside. The latter had thereafter no chance at all; Gibert remarkably was able to get Tennyson going again and so well was he finishing that he was only a neck and half a length behind Roi Lear and Gunter at the winning post. Gibert's objection to Gunter was swiftly upheld and Tennyson was awarded second place. His misfortune at Chantilly was emphasised in the Grand Prix de Paris three weeks later. Authi, who had taken the lead before the straight, was 3 lengths clear of Tennyson 300 metres out, but the Prix du Jockey Club runner-up came with a sustained and steady run to beat Authi a head. The pair finished 5 lengths ahead of Raagavor with Margouillat eighth and Gunter tenth. Although easily beaten by Dahlia in the Prix Niel, Tennyson was unsuited by the slow pace. The principal doubt about him was whether he would be better suited by a distance slightly longer than a mile and a half.

Authi, a half-brother to Hard to Beat,

almost certainly required a longer distance than twelve furlongs. He had been runner-up in the Prix Royal Oak as well as the Grand Prix de Paris and also third in the Grand Prix de Deauville. On each occasion he had been beaten for speed when challenged inside the distance. Admittedly in the spring he won over twelve furlongs at Longchamp but he really needed a stiff test of stamina and his odds of 22 to 1 seemed a little ungenerous.

The same requirement applied to Balompie. He had won an eleven-furlong maiden race at Evry and the Prix des Suresnes in April. He had only managed tenth place to Morston in the Epsom Derby and had failed to cope with the acceleration of either Lady Berry in the Prix Royal Oak or that of Card King in the Grand Prix de Deauville. He was coupled in the betting at 17 to 1 with Comtesse Batthyany's other runner, El Famoso, who only had his racecourse début in June. On his second appearance El Famoso won a ten-furlong maiden event at Saint-Cloud and then, when only receiving 6 lbs, beat the subsequent St Leger winner, Peleid, a neck in the Prix de la Ville de Trouville (10f) at Deauville, where he also won the Prix Michel Houyvet (12½f). Kalpour, who was only third in the Prix Kergorlay and sixth in the Prix Royal Oak, exposed El Famoso's limitations when he beat him a neck at Longchamp in September.

Unfortunately Roi Lear was injured in a gallop in September and had to be retired to stud. The previous month Morston, the Epsom Derby hero, sprained a tendon and also went to stud. Another absentee was Roberto, whose autumn objective was to have been the Champion Stakes, but he too pulled a ligament and returned to the U.S.A. to start his career as a stallion. There were two possible challengers from the U.S.A.: Cougar II, who detested soft going and was eventually scratched ten days before the race; and London Company for whom Pat Eddery was booked. Arrangements had been made to fly the colt from the U.S.A. to Paris on Wednesday 3 October but he met with a setback two days earlier and was also withdrawn. The decision as to whether to run Kalamoun, the winner of the Prix Lupin, the Poule d'Essai des Poulains and the Prix Jacques le Marois, in the Prix de l'Arc de Triomphe or in the Prix du Moulin de Longchamp was influenced by two factors. The first was that the Aga Khan's jockey Henri Samani had received an eight-day suspension which precluded him from riding on Sunday 7 October. Secondly if the going was on the soft side, the doubt about Kalamoun's stamina was made more significant. Eventually it was decided to run Kalamoun in the Prix du Moulin de Longchamp with Freddie Head riding. The Aga Khan was disappointed not to run him in Longchamp's most prestigious race because, since his father's victory with Saint Crespin in 1959, he had only had a colt good enough to run on three occasions, and when Jour et Nuit ran his then stable jockey Yves Saint-Martin was also unable to ride due to a suspension.

The Final Preparations

On Rheingold's return from his disappointing third in the Benson and Hedges Gold Cup, his trainer Barry Hills called in the legendary ex-trainer Syd Mercer who had cured the Fabergé II colt of a wart a fortnight earlier. Mercer had an extraordinary ability to cure horses of illnesses with remedies he would divulge to no one. He concluded that Rheingold's spleen was not working properly and that this was causing his blood disorder. He prescribed some of his herbal powders which were put in Rheingold's feed every day. A little over a fortnight later Mercer pronounced his patient cured. Lester Piggott came to Lambourn to ride Rheingold in his final gallop, over ten furlongs with Our Mirage, Natsun and Doubleglow. He drew away from them impressively and Piggott commented afterwards that the Fabergé II colt felt a totally different horse to that which he had ridden at York. Barry Hills thought he would take 'a great deal of beating' in Paris. In a television interview a few days before Sunday 7 October, Piggott said he felt he had a good chance of winning on Rheingold. He did concede it had not been a lucky race for him, for it was to be his fifteenth attempt to win it.

The other invalid, Dahlia, did her final work on Wednesday 3 October. Pyers reported that she had gone 'great, really great'. The final decision to run her was made not by her trainer, Maurice Zilber, but by Nelson Bunker Hunt who had watched the gallop. The previous Monday Allez France's trainer, Albert Klimscha, said that she 'has never been in better condition before a race than she is now, and she's less nervous than she used to be'. Noel Murless also felt Attica Meli would do herself justice provided she did not become upset by the preliminaries and lose her race before the start.

Rheingold arrived in Paris on Friday 5 October. Early on Sunday morning Barry Hills went to see him in his stable and thought he looked listless; he also found a firm dropping, and his confidence was sapped a little. On his return to his hotel he told the syndicate who owned Rheingold about his doubts. These fears proved groundless. When Rheingold came into the paddock at Longchamp he looked in peak condition, hard trained and really well in himself.

One of the syndicate who owned Rheingold, Charles St George, had Abergwaun, trained by Vincent O'Brien and ridden by Lester Piggott, in the first race at Longchamp that afternoon, the Prix de l'Abbaye de Longchamp. She had won the King's Stand Stakes at Ascot and the Prix de Seine-et-Oise at Maisons-Laffitte in September. Abergwaun never looked like catching the English champion sprinter, Sandford Lad, whom Charles St George and Tim Rogers had bought shortly before the race and who was unbeaten in 1973; Abergwaun finished 2 lengths behind him with another English sprinter, Supreme Gift, third. Charles St George, however, was not at Longchamp that afternoon as he was confined to bed in London with influenza. Nelson Bunker Hunt's Hippodamia was a most emphatic 6 lengths and a head winner from Comtesse de Loir and La Tulipe in the Critérium des Pouliches, with Louis Freedman's Polygamy another head away fourth.

The Race

In the paddock, like Rheingold, Allez France looked a picture, her coat had a brilliant sheen and on looks these two stood out. Hard to Beat also appeared in better condition than he had been at Ascot. The American jockey, Angel Cordero Jr, had been engaged to ride the Irish filly Hurry Harriet. Unfortunately she took an enormous hold going down to the start and refused to be pulled up until she was behind the Petit Bois. She then deposited Cordero on the ground and bolted; the entire incident caused a twenty-minute delay. This wait at the start certainly did not help the cause of Attica Meli and caused more than a little worry amongst Rheingold's connections who remembered how he had played up at Ascot.

Cordero and Hurry Harriet were finally reunited, but this escapade put paid to any chance she might have had. Eventually when the field got away Direct Flight took up the running from Authi, Unicornus, Miss Thérèse, Rheingold, Hard to Beat, Allez France and Tennyson, whilst Dahlia and San San were at the rear of the field. Direct Flight and Authi continued to dispute the lead to the top of the hill and the order remained virtually the same except that Gibert had restrained Tennyson, who was now in the middle of the field. Beginning the descent Authi took a narrow lead over Direct Flight on the rails and Unicornus, with Rheingold fourth and Allez France tracking him fifth, just ahead of Hard to Beat. Both Dahlia and Lady Berry were making progress from the rear to move closer to the leaders on the outside, but Miss Thérèse was already finding the gallop too hot and Tennyson, slightly hampered, dropped back to the rear of the field. In the last 200 metres before the straight Allez France lost a couple of lengths and positions as the pace increased but Balompie began to close on the leaders. On the turn into the straight Authi still led Direct Flight but Piggott on Rheingold, who looked full of running, moved up to challenge on their outside ahead of Balompie, Hard to Beat, who was tracking the two leaders and, going very easily, Allez France. Then came the weakening Unicornus, Dahlia and Lady Berry on the outside and Buoy and Card King who had improved their positions coming down the hill. Piggott was seen to check Rheingold at this stage so as not to take the lead too soon, but both Authi and Direct Flight were coming under severe pressure and Allez France was moving into a challenging position. Dahlia, eighth on the outside, was quickly under pressure and it was clear she was done with, while Tennyson had a hopeless task and certainly was not going to take any hand in the finish. About 500 metres out Piggott gave Rheingold his head and he surged past the two leaders and went 2 lengths clear. Thiboeuf on Hard to Beat had elected to track Direct Flight and Authi, and as Authi dropped out of contention he had to be checked. Switched to the inside his path was still blocked by Direct Flight, so it was left to Allez France to set off in pursuit of Rheingold. She had found a beautiful clear run through on the rails and 200 metres out had closed the gap to 2 lengths. Behind her Buoy, Balompie and Hard to Beat, still trying to

Rheingold (12) defeating Allez France (27) easily by 2½ lengths

find a gap, were being caught by the fast-finishing Card King, Lady Berry and Sang Bleu. When Piggott realised Daniel Wildenstein's filly was steadily catching him, he began to ride out Rheingold with hands and heels. In the final 100 metres Allez France began to tire and hang a little to the left and Rheingold once again increased his advantage over her. At the winning post he was the comfortable 2½-length winner. The pair were 4 lengths in front of Hard to Beat who, when Thiboeuf found an opening, ran on very strongly, finishing 2 lengths ahead of Card King with Lady Berry and Sang Bleu a neck and a short head behind fifth and sixth. They were followed by Balompie, Buoy, Lassalle and Tennyson, both of whom had run on in the final stages. There was absolutely no question that Rheingold's victory was both impressive and well-deserved.

Rheingold and Piggott were greeted by his syndicate of owners headed by Henry Zeisel and Tim Sasse on their return to the unsaddling enclosure to much applause from the Parisian crowd, for the Fabergé II colt was still unbeaten in France. In spite of the ease of his victory the soft ground meant that the time was 7·5 seconds slower than those of San San and Mill Reef in the two previous years. Syd Mercer, who had cured Rheingold of his blood disorder, was not in Paris but watched the race on television in his home near Woodstock in Oxfordshire. He reckoned that there was 'not a horse in the world' that could have beaten Rheingold that afternoon. On dismounting Lester Piggott dryly observed: 'I knew after I had gone a furlong that, barring accidents, I would win.' Thus after finishing second three times and third once, the English champion jockey had won his first Prix de l'Arc de Triomphe on his fifteenth mount. Nevertheless he was in action again for his brother-in-law Robert Armstrong on Sparkler in the Prix du Moulin de Longchamp. Taking the lead early in the straight Sparkler had 2½ lengths to spare over Kalamoun, on whom Freddie Head had possibly exaggerated the waiting tactics. The afternoon of Sunday 7 October was certainly Piggott's at Longchamp.

Rheingold was the first-ever thoroughbred owned by Henry Zeisel, who was born in Vienna. His father had been conductor of the Vienna Opera House orchestra whilst his mother, who was Russian, had danced with Pavlova. Zeisel, a child prodigy, had become the lead violinist in the Vienna Opera House Orchestra at seventeen. The first opera in which he played the lead violin was Wagner's *Rheingold* which was why he named his first yearling by that name. He also owned a night-club in London's Oxford Street called 'The Rheingold'. Shortly before the 1972 Epsom Derby he sold several shares in the colt after his victory in the Dante Stakes at York to a syndicate composed of Charles St George, Peter Richards, Tim Sasse, Sir William Pigott-Brown, Antony Shead and John Weller-Poley, for £100,000. After Rheingold's victory at Longchamp Zeisel said that he would 'much prefer to win the Prix de l'Arc de Triomphe than the Derby, for at Longchamp you beat the best horses in the world'.

For the young English trainer, Barry Hills, only in his fifth season as a trainer, it was a first victory in the Prix de l'Arc de Triomphe. In the 1960s he had been travelling head-lad and then head lad to John Oxley at Newmarket. He had won so much money when Frankincense won the Lincolnshire Handicap that he was able to buy Keith Piggott's stables at South Bank in Lambourn and set up on his own. Certainly he had achieved phenomenal success in his first five seasons training. When Hills went to see Rheingold after the race, he noticed that he had sprained his near fore. An invitation to the Washington D.C International Stakes had, therefore, to be refused and Rheingold was retired to stud.

There were few excuses for any of the horses behind Rheingold at Longchamp with the exception of Dahlia who finished a moderate sixteenth. Bill Pyers felt that lack of work following her setback in the Prix Vermeille was the probable reason for her poor performance and was inclined to think that she should never have run. Saint-Martin thought that Allez France might have been unsuited by the unusually slow gallop in the early stages of the race. He had ridden Rheingold in all his other victories in France and expressed himself a little surprised at the ease of his victory. With a clearer run Hard to Beat might have finished closer to the first two but would never have threatened the winner.

The Rest of the Season

Both Allez France and Hurry Harriet, who had never been seen with a chance at Longchamp and was not subjected to a hard race, ran in the Champion Stakes at Newmarket. Albert Klimscha reported that Daniel Wildenstein's filly was very nervous during the flight from France and viewed the very heavy ground at Newmarket with some trepidation. Saint-Martin sent Allez France into the lead a little after the Bushes but coming out of the Dip she was caught by Hurry Harriet and beaten three-quarters of a length with Sharp Edge, the Irish 2000 Guineas winner 4 lengths away third. There were some, including her owner, who considered that the French jockey had allowed Allez France to hit the front much too soon. Two days earlier Parnell, who also had not been in contention at all at Longchamp, easily defeated Reload by 2½ lengths in the Jockey Club Cup over two miles at Newmarket. On 28 October Parnell finished a short neck and 4 lengths in front of Lassalle and Forceful in the Prix Gladiateur, but was relegated to third place after a stewards' enquiry. Primette ran third to Récupéré and Sarajevo in the Prix du Conseil Municipal on 21 October, but Novius and El Famoso were unplaced. The same afternoon Sang Bleu beat Orsa Maggiore, the Oaks d'Italia winner, 3 lengths in the Gran Premio del Jockey Club at San Siro with Balompie fourth. On 1 November Sang Bleu gained a further success in Italy when he won the Premio Roma by 3 lengths from Veio. Dahlia showed her performance at Longchamp was way below her best when she beat Big Spruce and Scottish Rifle 3½ lengths and a length in the Washington D.C. International Stakes at Laurel Park on 10 November with Card King fourth and Hurry Harriet last. The latter's saddle slipped shortly after the start, so her jockey spent the race precariously balanced on her withers. The results of all these races, however, only went to underline the merit of Rheingold's victory on 7 October.

Pedigree and Stud Record

The 1973 Prix de l'Arc de Triomphe winner had been bought as a yearling for only 3000 guineas. He was by Fabergé II, a colt by Princely Gift, who in 1970 stood at the Old Fairyhouse Stud in Co. Meath for a fee of £148. As a two-year-old Fabergé II won the Errol Stakes and the Virginia Water Stakes at Ascot. He was only beaten a neck and three-quarters of a length by Talahasse and News Item in the Champagne Stakes at Doncaster and he was also third to Showdown and Smooth Jet in the Middle Park Stakes at Newmarket. His best performance as a three-year-old was to finish runner-up 2 lengths behind Baldric II in the 1964 2000 Guineas in front of Balustrade, Showdown and Derring-Do. That year he also won the Severals Stakes over a mile at Newmarket. Fabergé II certainly did not stay more than eight furlongs and was some way below the best of his generation. He had also sired the 1973 Gimcrack and Champagne Stakes winner, Giacometti, who was to win the Champion Stakes and to be placed in the Derby, St Leger and 2000 Guineas in 1974; the 1972 Derby Italiano winner, Guy Lussac and McKenna's Gold, who won the top weight for age race in Venezuela the Premio Clasico Simon Bolivar the same year. He was not a sire one would normally have expected to produce a winner of the Prix de l'Arc de Triomphe and he was exported to Japan.

Rheingold's dam, Athene by Supreme Court, was totally undistinguished. She was bought as a yearling for 140 guineas by P.J. Poston, the Newmarket trainer of mostly platers. She only ran as a juvenile when she achieved third place in the Chelsea Plate at Hurst Park in April 1962, the only time she reached the frame in eight outings. In September that year she was given as the prize for the raffle at the Saints and Sinners Charity Meeting. Athene was sold as a three-year-old for 300 guineas at the Ascot January Sales and then at the Newmarket December Sales for 700 guineas in foal to Pandofell. That foal named Pando Belle won two small races over 16 furlongs in Ireland in 1969. Before Rheingold was foaled in 1969 she produced Ritsa by Nice Guy, a winner in Greece, and Hope of Clivarn by Whistling Wind, who won in South Africa. Athene has got three other winners since the birth of Rheingold: Harry Churchills by Quisling; Holernzaye, a filly by Sallust; and a full sister to him, Invitation, whose four victories were achieved in minor events at Folkestone, Nottingham and Wolverhampton. At the age of twenty Athene produced a filly by Lypheor, Rhein Silver, who ran once as two-year-old in a seven-furlong maiden stakes at Newmarket in October 1982.

Athene's dam, Necelia by Nearco, was unraced but she produced four winning progeny of seven races: Trade Gap by Set Fair, Debrac by Combat and Internos by Auralia were only moderate; Nectar by Honeyway, however, was trained by Sam Hall, and won a ten-furlong maiden race at Redcar and three reasonable handicaps at Manchester, Ripon and Redcar at distances from eleven to sixteen furlongs.

Necelia's dam, Cecily by Cecil, was owned by Enid Lady Chesterfield. Because of the Second World War she only had four outings as a three-year-old and never reached a place. Her first winning foal, Xerxes by Nearco, won the Stockil Stakes (10f) at Doncaster in May 1949, whilst his full-brother Cyrus, trained by W. Elsey for Lady Joicey, won three of his six outings as a juvenile over five furlongs and six furlongs at Stockton and Newcastle and was considered good enough to be sent to run in the Windsor Castle Stakes at Royal Ascot in which he was unplaced to Tabriz. Cecily's only other winning foal was Pipregan by Anamnestes who won three minor races over eleven and twelve furlongs in 1958.

Cecily's dam, Matanilla by Friar Marcus, was also owned and bred by Enid Lady Chesterfield. She was unraced and produced but one winner, Banco by Loaningdale, who won a mile handicap plate at Manchester in 1939.

Matanilla was out of Reef by Rochester; *she* was owned by Lord Durham. On the racecourse Reef only managed two places from six starts as a three-year-old at Haydock and Stockton. She was, however, a quite useful broodmare producing five individual winners of fourteen races between 1925 and 1935. Unquestionably the best of these was Rattlin the Reefer by Chaucer, who ran in the colours of Lord Woolavington. As a juvenile he won the Richmond Stakes at Goodwood and ran Brienz to half a length in the Dewhurst Stakes. In 1929 he was sixth to Mr Jinks in the 2000 Guineas, eighth to Trigo in the Derby, and won the Jersey Stakes from Ellenborough at Royal Ascot.

Reef's dam, Barrier by Greyleg, also produced five winning foals of twelve individual races; Barrulet by Tracery won five races worth £2,449 in 1920 and 1921, but more importantly Barrier's two fillies by Grey Crusader, Indolence and Hellspont, who both won once, had a considerable influence on the Stud Book, especially the former. Indolence was the dam of Prince Rose by Rose Prince. Not merely was he third to Pearl Cap and Amfortas in the 1931 Prix de l'Arc de Triomphe, but he also was the sire of Princequillo, Prince Bio and Prince Chevalier. Hellspont's daughter, Bouquet by Buchan, was the dam of Airborne by Precipitation, the 1946 Epsom Derby winner. Barrier traced back to Blink Bonny's dam, Queen Mary, one of the greatest mares in the Stud Book. Nevertheless Rheingold's immediate maternal line was not at all impressive and no one would suggest he was bred in the purple.

When Rheingold retired to the Coolmore Stud in Co. Tipperary he had won four races in France worth 3,192,000 francs and five races in England and £74,439. He was not an enormous success at stud if judged entirely by his progeny in England and Ireland. Up to October 1982 they had won ninety-eight races worth £366,059. His best winners were Rhein Bridge (Lancashire Oaks), Meistersinger (Morland Brewery Trophy Handicap), Funny Spring (Irish Sweeps Autumn Handicap), Golden River and Karadar. Rheingold was exported to Japan in July 1979. Nevertheless that is but one side of the picture, for he was not at all unsuccessful as a stallion abroad. Between the time he went to stud until October 1982 he had sired 232 winners of races worth £745,822 outside the British Isles. Amongst Rheingold's better winners were Noir et Or (Prix Conseil de Paris, Prix Saint-Roman, Grand Prix Prince Rose, Grand Prix d'Evry and Prix d'Hedouville), Mot d'Or (Prix Hocquart, 3rd Grand Prix de Paris); Armistice Day (Grand Prix de Marseille, Prix Gontaut-Biron and Prix Exbury); Bater (Premio Principe Amadeo, Premio Merano); Ghazwan (W.L. McKnight Handicap); Rivellino (Grand Prix de Bruxelles); Rheinsteel (Norwegian Derby, Norwegian 2000 Guineas and Dansk Eclipse Stakes); My Franky (2nd Derby Italiano); and Godilko, whose four victories in France and the U.S.A. were worth more than £75,000. In Japan Rheingold got 44 foals in 1981 and 47 foals in 1982. No one would pretend that Rheingold was a resounding success at stud in Ireland but neither could he be described as a failure. Certainly he was a better performer on the racecourse and he is unique in the history of the Prix de l'Arc de Triomphe in that he beat three winners of the race on 7 October 1973; San San (1972), Allez France (1974) and Star Appeal (1975) and he accomplished this feat with impressive ease.

PRIX DE L'ARC DE TRIOMPHE 2400 metres
at Longchamp on Sunday 7 October 1973. For three-year-olds and upwards. Going: Soft.
1,200,000 fr. to the winner; 480,000 fr. to the 2nd; 240,000 fr. to the 3rd; 120,000 fr. to the 4th. 288,000 fr. to the breeder of the winner; 96,000 fr. to that of the 2nd; 48,000 fr. to that of the 3rd. Value to the winner: 1,497,400 francs.

RHEINGOLD	br.c.	by Fabergé II – Athene by Supreme Court	4.60k	Henry Zeisel	L. Piggott	1
ALLEZ FRANCE	b.f.	by Sea Bird – Priceless Gem by Hail to Reason	3.54k	Daniel Wildenstein	Y. Saint-Martin	2
HARD-TO BEAT	b.c.	by Hardicanute – Virtuous by Above Suspicion	4.60k	Junzo Kashiyama	G. Thiboeuf	3
CARD KING	b. or. br.c.	by Cardington King – Nantua II by Again II	5.60k	Raphaël Hakim	G. Dettori	4
LADY BERRY	ch.f.	by Violon d'Ingres – Moss Rose by Mossborough	3.54k	Baron Guy de Rothschild	M. Depalmas	5
SANG BLEU	b.c.	by Right Royal – Royal Girl by Norseman	4.60k	Mme Cino Del Duca	J. Taillard	6
BALOMPIE	ch.c.	by Pardao – Banning by Crafty Admiral	3.55½k	Comtesse Margit Batthyany	J. Cruguet	7
BUOY	ch.c.	by Aureole – Ripeck by Ribot	3.55½k	R.D. Hollingsworth	J. Mercer	8
LASSALLE	b.c.	by Bon Mot – Windy Cliff by Hard Sauce	4.60k	Zenya Yoshida	J. Lindley	9
TENNYSON	b.c.	by Val de Loir – Tidra by Prince Taj	3.55½k	Walter Burmann	A. Gibert	10
MISTER SIC TOP		by Tiffauges – Madame Daurade by Sicambre	6.60k	Lucien Deshayes	R. Litt	
PARNELL		by St. Paddy – Nella by Nearco	5.60k	Rory More O'Ferrall	W. Carson	
DIRECT FLIGHT		by Baldric II – Felicia by Tantième	4.60k	Mme François Dupré	M. Philipperon	
UNICORNUS		by Carapalida – Starlight by Timor	4.60k	R.A. Franklyn	G. Doleuze	
NOVIUS		by Sicambre – Novia II by Acropolis	4.60k	J.R. Mullion	J.C. Desaint	
BONNE NOEL		by Santa Claus – Camilla Edge by Alcide	4.60k	Mme Anne Parker-Poe	W. Williamson	
FIREFRIGHT		by Firestreak – Moana by Zucchero	4.60k	G.J. van der Ploeg	P. Cook	
ATTICA MELI		by Primera – Come on Honey by Never Say Die	4.58½k	Louis Freedman	G. Lewis	
MISS THERESE		by Raise You Ten – No Slipper by His Slipper	4.58½k	P.J. Lally	E. Hide	
PRIMETTE		by Ribot – Primary by Petition	4.58½k	Robin Scully	G. Rivases	
SAN SAN		by Bald Eagle – Sail Navy by Princequillo	4.58½k	Tadao Tamashima	A. Barclay	
EL FAMOSO		by Ragusa – Effervescene by Charlottesville	3.55½k	Comtesse Margit Batthyany	E. Belmonte	
RAGAPAN		by Ragusa – Panaview by Panaslipper	3.55½k	F.O. Sullivan	A. Murray	
AUTHI		by Aureole – Virtuous by Above Suspicion	3.55½k	Mme Pierre Wertheimer	F. Head	
STAR APPEAL		by Appiani II – Sterna by Neckar	3.55½k	Waldemar Zeitelhack	P. Alfi	
DAHLIA		by Vaguely Noble – Charming Alibi by Honey's Alibi	3.54k	Nelson Bunker Hunt	W.B. Pyers	
HURRY HARRIET		by Yrrah Junior – Somnambula by Chanteur	3.54k	Malcolm Thorp	A. Cordero Jr	

27 runners.
RHEINGOLD trained by B. Hills at Lambourn, England; bred by Dr James Russell in Ireland.
Distances: 2½ lengths; 4 lengths; 2 lengths; neck; short head; ¾ length; 2 lengths; head; neck. Time: 2 mins 35.80 seconds.
Pari-Mutuel: Win: 8fr. 70. Places: 3fr. 00; 1fr. 90; 4fr. 90.
S.P.: 7 to 4 Allez France; 23 to 4 Tennyson; 77 to 10 RHEINGOLD; 8 to 1 Dahlia; 35 to 4 Lady Berry; 17 to 1 Balompie and El Famoso (couplé); 21 to 1 Hard to Beat; 22 to 1 Direct Flight, Attica Meli and Authi; 29 to 1 Bonne Noël; 45 to 1 Parnell; 61 to 1 Buoy; 64 to 1 Miss Thérèse; 74 to 1 San San; 75 to 1 Card King; 96 to 1 Hurry Harriet; 99 to 1 Lassalle, Sang Bleu, Ragapan, Star Appeal, Novius, Primette, Firefright, Mister Sic Top and Unicornus.

RHEINGOLD bay Colt 1969	Fabergé II b. 1961	Princely Gift b. 1951	Nasrullah b. 1940	Nearco br. 1935	Pharos
					Nogara
				Mumtaz Begum b. 1932	Blenheim
					Mumtaz Mahal
			Blue Gem b. 1943	Blue Peter ch. 1936	Fairway
					Fancy Free
				Sparkle gr. 1928	Blandford
					Gleam
		Spring Offensive b. 1943	Legend of France br. 1935	Dark Legend b. 1914	Dark Ronald
					Golden Legend
				Francille ch. 1927	Battersea
					Farine Lactée
			Batika b. 1934	Blenheim br. 1927	Blandford
					Malva
				Brise Bise b. 1922	Buchan
					Panne
	Athene b. 1960	Supreme Court br. 1948	(Persian Gulf) or Precipitation ch. 1933	Hurry On ch. 1913	Marcovil
					Tout Suite
				Double Life ch. 1926	Bachelor's Double
					Saint Joan
			Forecourt br. 1943	Fair Trial ch. 1925	Fairway
					Lady Juror
				Overture br. 1937	Dastur
					Overmantle
		Necelia b. or br. 1949	Nearco br. 1935	Pharos b. 1920	Phalaris
					Scapa Flow
				Nogara b. 1928	Havresac II
					Catnip
			Cecily b. 1940	Cecil br. 1931	Foxlaw
					Star of Blyth
				Matanilla b. 1927	Friar Marcus
					Reef

Allez France 1974

Allez France

The Field

For the first time since her accession to the throne of England in February 1952, Her Majesty Queen Elizabeth II had, in 1974, a runner in the Prix de l'Arc de Triomphe. Her Queen's Hussar filly Highclere had been second to Polygamy on her racecourse début at Newmarket. She reversed that form when finishing a head second to Celestial Dawn in the Princess Margaret Stakes at Ascot in July with Polygamy third. On her final outing as a juvenile, Highclere beat Reigning Grace a neck in the Donnington Maiden Stakes (7f) at Newbury in September.

Highclere did not have the benefit of a preliminary race before the 1000 Guineas but looked very well in the paddock. She took the lead two furlongs out and coming out of the Dip was very strongly challenged by the favourite, Polygamy. Her Majesty's filly, however, fought back with such tenacity and gameness that she held on to win by a short head. Highclere stood almost 16·3 and it was felt that the course at Epsom would not suit her, so it was decided to run her in the Prix de Diane at Chantilly. When Polygamy won the Oaks, hopes that Highclere would become the first filly to win both the 1000 Guineas and the Prix de Diane were considerably raised. It appeared her most dangerous rivals would be Hippodamia and Comtesse de Loir, who had finished first and second respectively in the Critérium des Pouliches the previous autumn. Comtesse de Loir had only managed fifth place behind Paddy's Princess in the Prix de la Grotte at Longchamp on her first outing in 1974 and then was beaten nearly six lengths when third to Rose Bed in the Prix Pénélope. In the Prix Saint-Alary on 26 May, however, she took up the running 400 metres out and decisively beat Hippodamia 2½ lengths. Three weeks earlier Hippodamia had been runner-up to Dumka in the Poule d'Essai des Pouliches.

Her Majesty the Queen came to Chantilly to see Highclere run on 16 June. She was entertained to lunch before the race by Monsieur Marcel Boussac, the Président of the Société d'Encouragement. Highclere was always in the leading group, and on the turn into straight was lying a close third to Hippodamia and the Prix Cléopâtre winner, Tropical Cream. She made her challenge a little over 200 metres out. Comtesse de Loir, who had made rapid progress from the rear, tried in vain to get on terms with Highclere, but the 1000 Guineas heroine increased her advantage to 2 lengths at the winning post with Odisea half a length away third and Hippodamia fifth. The crowd at Chantilly gave the royal filly a tumultuous welcome on her return to the unsaddling enclosure.

Highclere next took on the best in Europe in the King George VI and the Queen Elizabeth Stakes at Ascot. The favourite was Nelson Bunker Hunt's Dahlia, who earlier in July had come back to her best form when beating On My Way in the Grand Prix de Saint-Cloud. Highclere was not in the best position on the turn into the straight; when she got a clear run she finished very strongly to take second place 2½ lengths behind Dahlia with the Prix Lupin winner, Dankaro, a length away third. Buoy, Card King, the surprise Derby winner Snow Knight, and the useful Italian racemare, Orsa Maggiore, were all further behind.

Highclere met Dahlia again in the Benson and Hedges Gold Cup. Dahlia won 2½ lengths from Imperial Prince and Snow Knight, but the Queen's filly ran below her best and finished only sixth. She was, however, running on at the finish and possibly by then required a mile and a half. Since its inauguration the Benson and Hedges Gold Cup had become notorious for its surprise results and its form is normally best disregarded. On her best form Highclere could not be left out of the reckoning for the 1974 Prix de l'Arc de Triomphe.

Comtesse de Loir only ran once between the Prix de Diane and Sunday 6 October. In the Prix Vermeille in September she again had to be content with second place, 4 lengths behind Daniel Wildenstein's much improved three-year-old filly, Paulista, who won unchallenged. Paulista did not have her first race in 1974 until the Prix de Malleret in June when she finished third to Azurella. She then won the Prix de Minerve at Evry and the Prix de Psyché at Deauville. She put up a superb performance in the Prix de la Nonette at Longchamp on 1 September when she beat Pale Ale and Odisea an astonishing 8 lengths and 1½ lengths. There were in fact some people who considered that Paulista might be a very serious danger to her illustrious stable-companion, Allez France, in the Prix de l'Arc de Triomphe.

Allez France and Paulista had a new trainer in 1974. On the retirement of Albert Klimscha the previous autumn, Angel Penna had taken over Daniel Wildenstein's stable. Allez France was unbeaten in the spring and summer of 1974. On 15 April she beat the useful English four-year-old, Ksar, and Mister Sic Top by an impressive 3 lengths and 1½ lengths in the Prix d'Harcourt with Dahlia only fourth. In the Prix Ganay on 5 May, Saint-Martin sent Allez France into the lead 300 metres out and she drew clear in spectacular fashion to beat Tennyson and Gombos 5 lengths and 4 lengths with Card King and Dahlia fourth and fifth. Dahlia had not looked herself in the paddock and normally needed some sun on her back before she came to her best. Nevertheless it was the fifth occasion that Daniel Wildenstein's filly had met her and the fifth she had proved her superior.

In the Prix d'Ispahan at Longchamp on 30 June, Allez France was still 15 lengths behind the leaders entering the straight, but she produced such an electrifying burst of acceleration that she took the lead close home and beat Margouillat and Mister Dip a length and a head. Saint-Martin probably need not have given her such a hard race had he lain a little closer to the leaders. It was intended to run Allez France at Ascot against Dahlia and Highclere, but a few days before the King George VI and the Queen Elizabeth Stakes, Daniel Wildenstein decided against this plan since she was such a nervous traveller. On 8 September Allez France comfortably gave Tennyson and Sang Bleu 3 lbs in the Prix Foy and beat them a length and 4 lengths. She and Paulista started at odds of 5 to 10 on Sunday 6 October and it would have come as no surprise were they to finish first and second in the Prix de l'Arc de Triomphe.

So great was Allez France's reputation that, apart from Highclere, there were only two other challengers from abroad, Lt-Colonel John Chandos-Pole's Proverb and Tim Sasse's Coup de Feu. Proverb won the two-mile Paradise Stakes at Ascot in May and was a three-quarters of a length runner-up to Ragstone in the Ascot Gold Cup a neck in front of Lassalle. He won the Goodwood Cup for the second consecutive year when he beat King Levanstell 3 lengths, and in September defeated the same horse in the Doncaster Cup. He was, however, an out and out stayer and it was almost impossible to envisage him taking a hand in the finish of the Prix de l'Arc de Triomphe.

The five-year-old Coup de Feu had been bought at the 1973 Newmarket autumn sales for 26,000 guineas. He had run at Cagnes-sur-Mer in the early spring but failed to win. He scored his first success in 1974 when carrying top weight to victory in the Newbury Spring Cup in April. He caused a major surprise when, starting at 33 to 1, he beat Ksar 3 lengths in the Benson and Hedges Eclipse Stakes at Sandown in July with Mount Hagen, Toujours Prêt and Giacometti further behind. In his only other race before 6 October Coup de Feu was only a short head and half a length behind Nonoalco and El Toro in the Prix Jacques le Marois at Deauville on 11 August, with Margouillat fifth. Never the most consistent performer, Coup de Feu was best over a mile to a mile and a quarter and his odds of 69 to 1 were probably a fair reflection of his chance at Longchamp.

Card King and Sang Bleu, fourth and sixth to Rheingold the previous season, were again in the field for the Prix de l'Arc de Triomphe in 1974. Card King, however, had failed to win a race in 1974 prior to 6 October. A distant fourth in the Prix Ganay, well behind Dahlia at Saint-Cloud and Ascot, and only sixth to Ashmore in the Grand Prix de Deauville, were not exactly encouraging credentials. He was runner-up to the Deutsches Derby winner, Marduk, in one of the divisions of the Grosser Preis von Baden in September, but clearly was not so good as he had been in 1973. Sang Bleu's chance seemed almost as remote. His best display in 1974 had been to run Orsa Maggiore to half a length in the Gran Premio di Milano in June with Card King a little over 2 lengths away fourth. Récupéré had trounced Sang Bleu in both the Prix de Barbeville and the Prix Jean Prat in April, and Admetus had beaten him into eighth place in the Prix Maurice de Nieuil at Saint-Cloud on 14 July.

Alan Clore's Récupéré had steadily improved in the autumn of 1973 when he won the Prix Henry Delamarre and the Prix du Conseil Municipal. He displayed a magnificent burst of speed to beat Parnell and Filandre 4 lengths and 2 lengths in the Prix de Barbeville on his seasonal reappearance in April, with Lassalle only fourth. He then won the Prix Jean Prat so easily that he started at odds of 1 to 2 for the Prix du Cadran. Saint-Martin, riding almost too confidently, left Récupéré with an enormous quantity of ground to make up on the leader, Lassalle, in the straight. He again displayed remarkable finishing speed and caught Lassalle and beat him a head on the line with Authi 1½ lengths away third. Unfortunately Récupéré was not entered for the Ascot Gold Cup and he was rested until the Grosser Preis von Baden. There were so many entries that the race had to be divided. Récupéré finished third in that won by Meautry. He probably should have won, for his jockey Pyers gave him far too much to do and although he finished much the best of all, the post came too soon. The crowd expressed their displeasure at the Australian jockey's injudicious riding by vociferous booing. Récupéré was certainly the best stayer in Europe in 1974, but against the likes of Allez France, Highclere and Paulista, might find the distance of the Prix de l'Arc de Triomphe on the short side for him.

The 1973 Grand Prix de Paris winner, Tennyson, had a very light campaign as a four-year-old. No match for Allez France in the Prix Ganay, Tennyson was not at all well ridden in the Coronation Cup at Epsom. Buoy was allowed to set up such a long lead that by the straight he had slipped the field. Although Tennyson made up 8 lengths by the winning post, he was still 1½ lengths adrift of Buoy with Dahlia 3 lengths away third. Unfortunately Walter Burmann's colt was sore after Epsom and only ran once more before the Prix de l'Arc de Triomphe when a hard-held Allez France beat him a length in the Prix Foy. It was difficult to envisage Tennyson doing much better in 1974 than he had done in 1973 and his odds of 15 to 1 on 6 October seemed a little ungenerous.

The unpredictable but sometimes brilliant Margouillat had won the Prix Hocquart in 1973. On his seasonal reappearance in the Prix Dollar on 2 June, Margouillat challenged Admetus and Toujours Prêt inside the final 200 metres and beat them three-quarters of a length and half a length. In the Prix d'Ispahan he was probably flattered when only a length runner-up to Allez France. His only other outing had been when fifth to Nonoalco in the Prix Jacques le Marois at Deauville, but a mile was by then patently too short for him. Margouillat, who had always been held in high regard by his connections, had not always enjoyed the best of luck on the racecourse and at odds of 72 to 1 was a possible outsider on 6 October.

Xavier Beau's four-year-old On My Way had won a small twelve furlong race at Saint-Cloud in March and on 11 May ran Admetus and Luengé to 2 lengths and half a length in the Grand Prix d'Evry. He showed much improved form when only beaten a neck by Dahlia in the Grand Prix de Saint-Cloud with Direct Flight a neck away third. Riot in Paris, Un Kopeck and Card King were fourth, fifth and sixth. On My Way may have been a trifle unlucky, however, for he was pocketed on the rails and when he got clear he was finishing the fastest of all. He then disappointed when only ninth to Ashmore in the Grand Prix de Deauville. On My Way then, however, beat Toujours Prêt and Dahlia a neck and a short neck with Riot in Paris 4 lengths away fifth in the Prix du Prince d'Orange in September. On the strength of this victory On My Way was quietly fancied at odds of 29 to 1 on Sunday 6 October.

After Dahlia's defeat in the Prix du Prince d'Orange, Nelson Bunker Hunt decided not to run her in the Prix de l'Arc de Triomphe but to send her to the United States to run in the Man O'War Stakes at Belmont, the Canadian International at Woodbine and the Washington D.C. International at Laurel Park. In her place on 6 October he ran his useful three-year-old colts, Mississipian and Busiris.

As a two-year-old, Mississipian beat Nonoalco a nose in the Grand Critérium and was 2-length runner-up to Apalachee in the Observer Gold Cup at Doncaster with Busiris 10¾ lengths away fourth. In 1974 he was beaten half a length by Moulines in the Poule d'Essai des Poulains after his training schedule had been interrupted. In the Prix Lupin, Mississipian again had to be content with second place when half a length behind Dankaro with Moulines and Riot in Paris 2½ lengths and half a length away third and fourth. In the Prix du Jockey Club, Mississipian was only fourth to Caracolero, Dankaro and Kamaraan, with Un Kopeck sixth and Riot in Paris ninth. In the Irish Sweeps Derby, Mississipian failed to keep a straight course in the final two furlongs and was relegated from third to fourth place behind English Prince, Imperial Prince and Sir Penfro. It was not until the Prix Niel on 8 September that Nelson Bunker Hunt's colt gained his first success of the season when he beat Mount Hagen (gave 2 lbs) and Sagaro (gave 6 lbs) 1½ lengths and a short neck. Although he and Busiris started second favourites at 13 to 2 (couplé), on balance Mississipian had been disappointing as a three-year-old before he ran in the Prix de l'Arc de Triomphe.

Busiris did not make his seasonal début until July when he won a minor race over ten furlongs at Saint-Cloud and another over eleven furlongs at Clarefontaine in August. In the Prix Royal Oak Busiris took the lead early in the straight and ran on strongly to beat the favourite, Kamaraan and Ashmore by three-quarters of a length and 3 lengths. It was possible, nevertheless, that Busiris would be better suited by a distance further than a mile and a half.

The Aga Khan's Kamaraan won the Prix Lagrange (10f) at Saint-Cloud in March, but was then only fifth, a little over 3¼ lengths behind Dankaro, in the Prix Daru on 21 April. He ran Poil de Chameau to a neck in the Prix Hocquart in May and then was third, 1½ lengths and a head behind Caracolero and Dankaro, in the Prix du Jockey Club. He was again third to Sagaro and Bustino in the Grand Prix de Paris before he finished runner-up in the Prix Royal Oak. Although Kamaraan was both genuine and consistent, he seemed likely to be outpaced by some of his faster rivals.

Gerald Oldham's Sagaro won the Prix Canot (12f) at Saint-Cloud in March, but was beaten 3 lengths when fourth to Dankaro in the Prix Daru. Sagaro then easily accounted for his opponents in both the Prix Lovelace (12½f) at Maisons-Laffitte and the Prix de l'Espérance (15f) at Longchamp in May. In the Grand Prix de Paris Sagaro moved smoothly into the lead 400 metres out and won, totally unextended, by 2 lengths and half a length from the subsequent St Leger winner, Bustino, and Kamaraan. It was a very impressive victory but Bustino, who had been fourth to Snow Knight in the Derby, still had improvement in him. Sagaro was not disgraced behind Mississipian in the Prix Niel.

He was accompanied by a pacemaker, Valdo, in the Prix de l'Arc de Triomphe and they started at odds of 37 to 4. The doubt about Sagaro was whether a mile and a half would be too short for him in October 1974.

As a two-year-old, Riot in Paris had been runner-up to Wittgenstein in the Critérium de Maisons-Laffitte and had been third to Mississipian in the Grand Critérium, before he was relegated after a steward' enquiry to fourth place. He was placed third in the Handicap Optional, but he never fulfilled his two-year-old promise. He was well beaten in the Prix de Fontainebleau, the Prix Lupin and Prix du Jockey Club, and was a disappointing fourth when receiving weight from the Premio Parioli winner, Mansfield, in the Prix Eugène Adam at Saint-Cloud on 21 July. His only good performance in 1974 had been to run a close fourth to Dahlia in the Grand Prix de Saint-Cloud. Riot in Paris's odds of 97 to 1 on Sunday 6 October were probably a fair reflection of his chance.

Un Kopeck began his career in the provinces where he won three races over seven, ten and eleven furlongs at Toulouse, Mont de Marsan and Bordeaux, when trained by the Mont de Marsan trainer, René Servain. Un Kopeck joined John Cunnington Jr's stable in May 1974 and won the Prix La Force (10f) by a short head from Sorbi. He was not disgraced behind Caracolero in the Prix du Jockey Club and was a close-up fifth in the Grand Prix de Saint-Cloud. Further success, however, eluded him after the Prix La Force and his prospects at Longchamp on 6 October could not be described as hopeful.

The field for the 1974 Prix de l'Arc de Triomphe was completed by Riboquill, a half-brother to One for All by Northern Dancer and to First Feather, the dam of Run the Gantlet. He had cost $230,000 as a yearling but did not make his race-course début until May when he won a ten-furlong newcomers' event at Saint-Cloud. He defeated Taros 2 lengths over the same course and distance in July, but was beaten a short head by Brigand in the Prix de Solférino at Longchamp in September. Such a record was not one normally associated with a potential winner of Longchamp's most prestigious race.

Neither the first nor the second in the Prix du Jockey Club were in the field for the 1974 Prix de l'Arc de Triomphe. Caracolero, who had only finished eighth in the Irish Sweeps Derby, was reported to have been injured at the Curragh and was retired to stud. Dankaro also suffered a setback after the King George VI and the Queen Elizabeth Stakes and was also retired to stud. The Epsom Derby winner, Snow Knight, was also an absentee having been exported to race in Canada.

The Final Preparations

Ten days before the Prix de l'Arc de Triomphe, Yves Saint-Martin was thrown in the paddock by Native Friend, his mount in a handicap at Maisons-Laffitte. He broke a small bone at the junction of his thigh and his hip, and it seemed at first that he would be out of the saddle for three weeks. Lester Piggott, who had been booked for Dahlia were she to run, was contacted by Daniel Wildenstein and he agreed to ride Allez France. Two days prior to the race, Saint-Martin rode out with Penna's team. His doctor pronounced him fit to ride, and after he had passed an examination by the Société d'Encouragement's doctor it was agreed that he would ride Allez France. Piggott, therefore, switched to Nelson Bunker Hunt's Mississipian while Geoff Lewis took the mount on Busiris.

Angel Penna announced shortly before the Prix de l'Arc de Triomphe that Allez France's campaign in 1974 had been entirely planned with the race as her principal objective. He added: 'She is at the peak of condition and I shall be the most disappointed man in France, apart from Daniel Wildenstein, if she fails to win.' Yves Saint-Martin was equally confident and thought that there was no way she would be beaten. Lester Piggott was hopeful rather than confident about the chance of his mount Mississipian, although he felt the heavy ground would suit him. In fact he considered that Jimmy Lindley's mount, Tennyson, held an outstanding each-way chance.

The day before the Prix de l'Arc de Triomphe, Sir Michael Sobell's Bold Pirate, trained by Highclere's trainer, Dick Hern, won the Prix Saint-Roman which seemed a good omen for the Queen's filly. Nevertheless Hern was concerned since Highclere was going to encounter heavy going for the first time in her life. He was even more worried on the morning of the race when he discovered she was amiss and it was with some reluctance that he decided to allow her to run.

Sunday 6 October was dull, overcast and very cold – more like January than October. Inevitably this horrible weather meant that not so many Parisians ventured to Longchamp as usual, but there were still many visitors from abroad including the usual large contingent from England. The parlous economic state of affairs in England was not evident at the hundred-franc windows of the Pari-Mutuel, where English seemed the predominant language. Président Pompidou was not at Longchamp that afternoon, unlike a cheerful looking Sir Edward Tomkins, the British Ambassador to France. The Président is supposed to have once remarked, 'The English are supposed to be poor, but personally I've never met one who was, not in Paris anyway.' The sons of Sir Christopher Soames, Nicholas and Christopher, were also to be seen that afternoon in company with the Aga Khan and his wife, the Begum. Daniel Wildenstein's son, Alec, looked optimistic before the Prix de l'Arc de Triomphe and said that he and his father had thanked Lester Piggott for standing down from Allez France with 'such good grace'.

The first race of the afternoon, the Prix de l'Abbaye de Longchamp, had been won for the previous six years by English raiders. On this occasion, however, the Aga Khan's Moubariz held the vigorous challenge of the Ace of Aces, ridden by Piggott for Maurice Zilber's stable by a short head. In the Critérium des Pouliches, Lord Rotherwick's Harmonise was never a danger to the winner, Oak Hill.

The Race

Half an hour before the Prix de l'Arc de Triomphe, Yves Saint-Martin was given an injection of the pain-killer novocaine. Unquestionably Allez France was the pick of the paddock and looked superb. Tennyson, a strong well-built colt, and Comtesse de Loir, not a very attractive filly, both looked fit and hard-trained. When Allez France was mounted the normally noisiest section of the paddock crowd was unaccustomably silent, since they knew that she was inclined to be nervous and they did not wish to disturb her. She was still very much on her toes. On the way to the start the crowd cheered the Queen's filly, Highclere, Lester Piggott on Mississipian and Yves Saint-Martin on Allez France.

There were no untoward incidents at the start and as soon as the stalls opened Valdo, Sagaro's pacemaker, went into the lead from Riot in Paris and Coup de Feu. On settling down Valdo was closely tracked by Busiris, Coup de Feu, Riot in Paris and Margouillat, with Sagaro, Paulista, Tennyson and Highclere not far behind. Saint-Martin on Allez France, who had been drawn fifteen, had settled the mare towards the rear of the field with only five, including Comtesse de Loir, behind her. She was about 12 lengths behind the leaders. The first unfortunate incident in the race took place behind the Petit Bois. Paulista, just in front of Highclere, crossed to the rails rather sharply and hampered her. The Queen's filly was almost immediately severely bumped by Sang Bleu so that she lost at least 8 lengths, dropped back to the rear of the field and was never seen in the race with a chance again. At the top of the hill Valdo still led from Busiris, Coup de Feu and Riot in Paris, and the order behind them remained virtually the same with Allez France still amongst the backmarkers on the outside. Coming down the hill the favourite began to make very fast progress on the extreme outside. Lewis on Busiris sent his mount into a narrow lead from the weakening Valdo halfway down the hill. He was tracked by Riot in Paris, Tennyson and Margouillat, with Kamaraan and Paulista close behind them. Allez France took only 200 metres from the time she had begun to improve to pass all her rivals and deprive Busiris of the lead 700 metres from home, a little before the entrance to the straight. Such was her burst of speed that she was nearly 1½ lengths ahead of her nearest rivals Riot in Paris, Margouillat, Kamaraan and Paulista, with Comtesse de Loir improving rapidly on the outside; she swept majestically into the straight clear of the field. Saint-Martin had no choice but to kick for home. Tennyson and Busiris were weakening quickly and none of the other runners appeared to have any chance at all. For 200 metres none of the three disputing second place, Margouillat, Paulista and Kamaraan, were making any impression on Allez France, although Comtesse de Loir was gradually getting on terms with this trio; Riot in Paris meanwhile had shot his bolt and was dropping back rapidly. Two hundred metres out, Kamaraan and Paulista were under pressure and making no progress, but Comtesse de Loir and Margouillat were perceptibly closing on Allez France. With a hundred metres to go, Comtesse de Loir had cut the favourite's

Allez France (rails, white noseband) just holds Comtesse de Loir (19) and Margouillat (8) by a head and three-quarters of a length

advantage to less than a length with Margouillat only half a length behind her. Saint-Martin, strapped up and not fully fit, realised the danger but all he could do was to push out the mare with hands and heels. He was in no condition to ride a strong finish. Comtesse de Loir, however, was in full flight and gaining surely but slowly and the crowd watched with bated breath as the gap separating her and Allez France became smaller and smaller. The two passed the post almost together with Margouillat half a length behind them third. It seemed that Allez France, who had responded in fine style to Saint-Martin's urgings in the final few metres, had held on by a narrow margin. The cheer that greeted the result of the photograph, that Allez France had resisted Comtesse de Loir's challenge by a head, was deafening. Kamaraan was fourth, 4 lengths behind Margouillat, followed by Paulista another length away fifth. She was 3 lengths and half a length clear of Riboquill and Card King sixth and seventh, and then came On My Way, Mississipian and Tennyson.

On dismounting, Yves Saint-Martin confessed that 'the straight was beginning to feel very long'. He explained that he had not drawn his whip because Allez France 'doesn't like it and doesn't deserve it'. Lester Piggott commented, 'If I couldn't win – and Mississipian never showed at all – I am pleased that the mare did because Mr Wildenstein very kindly said he would give me a present for standing down immediately Yves Saint-Martin had recovered if Allez France won.' Jimmy Lindley reported that Tennyson was going well enough entering the straight but 'found nothing at the finish'. For the others, with the exception of Highclere, there were no excuses. Had the Queen's filly not been amiss and not been so severely hampered early on, she must have been thereabouts on her Prix de Diane form. There was no question, however, that Allez France was the best on the day and, had Saint-Martin been fully fit, the result probably would not have been so close.

Angel Penna, who was winning his second Prix de l'Arc de Triomphe in three years, exclaimed that Allez France was quite 'the best I have ever trained'. For Yves Saint-Martin it was also the second time he had won Longchamp's most important race.

Daniel Wildenstein, however, was achieving his first victory in the Prix de l'Arc de Triomphe. Surprisingly, after the race he was not as happy as one might have expected him to be. He thought he had been unfair to his great filly by letting her be ridden by Yves Saint-Martin when he was not completely fit, as she did not deserve to be subjected to such a hard race. He also expressed himself most disappointed that his filly, Paulista, who had been disputing second place early in the straight, had faded to finish fifth. He stated that he had never run her on the soft before and that he was sure 'she is better than this'. He was, of course, well known throughout the art world as one of its major art dealers. Wildenstein had another piece of good fortune in the final race of the afternoon, the Prix de la Concorde. His top-weight Luengé, trained by Jean-Michel de Choubersky and ridden by Philippe Paquet, finished a head second to Fils Ardent. After an objection and a stewards' enquiry, Luengé was awarded the race.

Jean-Claude Desaint also gained some compensation for Comtesse de Loir's narrow defeat when he partnered Cheryl to victory in the Prix de l'Opéra, the race immediately following the Prix de l'Arc de Triomphe. Saint-Martin, however, did not ride again on 6 October, nor, for that matter, until several days later on account of his injury. Later that afternoon he explained that Allez France had been pushed to the outside of the field by one of her rivals early during the descent. When she saw daylight she immediately changed gear and, since he was not able to restrain her, Allez France took the lead much sooner than he had intended.

The Rest of the Season

It was decided to run Allez France in the Champion Stakes on 19 October. Unfortunately she slightly injured herself when being unloaded from the plane at Cambridge Airport the evening before, so she was withdrawn and sent back to France. Because Penna offered no explanation to the stewards the next day as to the reason for her non-appearance, he was fined £100 at a later Jockey Club enquiry. In her absence the Champion Stakes was won by Gaicometti and the sole runner to have taken part in the Prix de l'Arc de Triomphe, Coup de Feu, only managed sixth place. Kamaraan, however, upheld the form of 20 October when he beat the very lightly raced Boussac colt, Labus, by an impressive four lengths in the Prix du Conseil Municipal. Card King and Sang Bleu were only able to finish sixth and seventh to Authi in the Gran Premio del Jockey Club at San Siro, whilst Proverb ran third a short neck and three-quarters of a length behind Forceful and Campo Moro in the Prix Gladiateur on 27 October. There was a French victory in the Washington D.C. International Stakes at Laurel Park, but not the one expected. As a result of winning the Man o'War Stakes and the Canadian International Stakes Dahlia started favourite at 6 to 10, but was beaten three-quarters of a length and the same by Admetus and Desert Vixen, with Margouillat 3 lengths away fifth and Coup de Feu sixth.

Pedigree and Stud Record

Allez France was sired by Sea Bird (see page 11). Certainly she was the very best of all his progeny. She was bred in the United States by the Bieber Jacobs stable and was bought privately as a foal for Daniel Wildenstein by his son Alec for $190,000.

Allez France's dam, Priceless Gem by Hail to Reason, was the joint-top two-year-old filly of her generation in 1965, the other being Moccassin. Priceless Gem won five of her six starts as a juvenile. She beat the champion two-year-old in the United States, Buckpasser, by half a length in the Futurity Stakes at Aqueduct, and she won the Frizette Stakes by 1½ lengths from Lady Pitt, who ran Moccassin to 2½ lengths in the Gardenia Stakes. Regrettably Priceless Gem and Moccassin never met, but they were both rated only 7 lbs behind

Buckpasser and 5 lbs below Graustark in the Experimental Free Handicap. Priceless Gem never ran up to her juvenile form as a three-year-old but still won two races. During her career on the racecourse she won seven times and amassed $209,267 in prize money. In 1970 she changed hands for the then record sum of $395,000. Wildenstein had been underbidder at the sale and when Paul Arnold, the bloodstock agent, told him Jacobs had kept her foal, Allez France, he sent his son Alec to the U.S.A. He bought her entirely on his own judgement as she was so typical of the La Troiènne family.

Allez France was head and shoulders above any of Priceless Gem's other foals. Her first, National Honour by Flag Raiser, won a minor race in the United States on her only start as a three-year-old. Priceless Gem's fourth foal, Kimberley Mine by Tom Rolfe, won his only outing, a maiden event in Ireland in 1975.

Priceless Gem's dam, Searching by War Admiral, won an incredible twenty-five races between the age of three and six worth $327,381, including the Vagrancy Handicap once, and the Gallorette Stakes twice. Searching was also an excellent broodmare, producing seven winners. Apart from Priceless Gem, the best was Affectionately by Swaps, whose twenty-eight victories between two years and five years netted $546,660 in prize-money. Affectionately was the champion two-year-old filly of her generation and won the Sorority, the Spinaways and the Astoria Stakes, in addition to the Vossburgh Handicap and Interborough Handicap twice. She was also the dam of Personality, the winner of the Preakness Stakes and Horse of the Year in the United States in 1970. Another of Searching's foals, Admiring, won seven races worth $184,581, including the Arlington and Washington Lassie Stakes, and became the dam of the stakes winner Glowing Tribute.

Searching's dam, Big Hurry by Black Toney, won four races as a two-year-old and three-year-old worth $29,860. As a juvenile she won the Selima Stakes over a mile at Laurel Park and ran fourth in the Pimlico Futurity Stakes. She was also an excellent broodmare. Amongst her produce were Bridal Flower, the champion two-year-old filly of her generation, Be Fearless, Ambulance, the dam of King of the Castle, Blue Line, and Great Captain.

Big Hurry's dam, La Troiènne by Teddy, was one of the most important broodmares imported into the U.S.A.; she was bought for 1250 guineas by Colonel E.R. Bradley at Newmarket. She was bred by Marcel Boussac and a half sister to his 1934 Prix de Diane heroine, Adargatis by Astèrus, who was also the dam of Ardan, the 1944 Prix de l'Arc de Triomphe winner and Pardal. La Troiènne was the dam of Bimelech by Black Toney who was unbeaten as a juvenile, winning the Pimlico Futurity and Hopeful Stakes, and the hero of the Belmont Stakes and the Preakness Stakes in 1940 as a three-year-old. She was also the dam of Baby League by Bubbling Over, the dam of Busher, of Black Helen who won the Coaching Club American Oaks in 1935, and of Big Event by Blue Larkspur, the dam of Hall of Fame by Shut Out, the 1948 American Derby winner. It is impossible to over-stress the importance of La Troiènne on the American Stud Book.

La Troiènne's dam, Hélène de Troie by Helicon, was bred by Lord Clonmel and bought at public auction by Jefferson Davis Cohn. She won once as a two-year-old and was bought privately by Marcel Boussac before she ended her racing career as a three-year-old. She was the dam of seven individual winners, of which Adargatis and Léonidas by Teddy, who won the 1930 Lincolnshire Handicap, were unquestionably the best on the racecourse. She traced to Paradoxical by Timothy, the grandam of Jest (Oaks and 1000 Guineas 1913), her full brother Absurd by Sundridge (Middle Park Stakes 1911), and also Black Jester by Polymelus (St Leger 1914).

Thus it can be seen that Allez France was bred in the purple and it is hardly surprising that she became probably the most brilliant racemare in France since Marcel Boussac's Corrida. She remained in training as a five-year-old and when she eventually retired, she had won thirteen of her twenty-one races and 6,254,162 francs in France and England.

Regrettably, Allez France, like so many brilliant racemares, has been a dismal failure as a broodmare but that must in no way detract from her brilliance on the racecourse. She lost her first foal to Secretariat when it was stillborn in 1977. Her first live foal, Ave France, a filly by Seattle Sew, was born in 1980 and is in training with Henry Cecil at Newmarket. In 1983 Allez France produced her second live foal, a colt by Exclusive Native.

PRIX DE L'ARC DE TRIOMPHE 2400 metres
at Longchamp on Sunday 6 October 1974. For three-year-olds and upwards. Going: Heavy.
1,200,000 francs to the winner; 480,000 fr. to the 2nd; 240,000 fr. to the 3rd; 120,000 fr. to the 4th. 300,000 fr. to the breeder of the winner; 120,000 fr. to that of the 2nd; 60,000 fr. to that of the 3rd; 30,000 fr. to that of the 4th. Value to the winner: 1,412,000 francs.

ALLEZ FRANCE	b.f.	by Sea Bird – Priceless Gem by Hail to Reason	4.58½k	Daniel Wildenstein	Y. Saint-Martin	1
COMTESSE DE LOIR	b.f.	by Val de Loir – Neriad by Princequillo	3.54k	George Ohrstrom	J.C. Desaint	2
MARGOUILLAT	b.c.	by Diatome – Tita by Tim Tam	4.60k	Paul de Moussac	G. Doleuze	3
KAMARAAN	b.c.	by Tanerko – Diamond Drop by Charlottesville	3.55½k	H.H. Aga Khan	H. Samani	4
PAULISTA	br.f.	by Sea Hawk – Petite Marmite by Babur	3.54k	Daniel Wildenstein	F. Head	5
RIBOQUILL	b.c.	by Ribot – Quill by Princequillo	3.55½k	William R. Hawn	G. Dubroeucq	6
CARD KING	b. or br.c.	by Cardington King – Nantua II by Again II	4.60k	Raphaël Hakim	A. Lequeux	7
ON MY WAY	b.c.	by Laugh Aloud – Gracious Me by Tulyar	4.60k	Xavier Beau	G. Rivases	8
MISSISSIPIAN	b.c.	by Vaguely Noble – Gazala by Dark Star	3.55½k	Nelson Bunker Hunt	L. Piggott	9
TENNYSON	b.c.	by Val de Loir – Tidra by Prince Taj	4.60k	Walter Burmann	J. Lindley	10
SANG BLEU		by Right Royal – Royal Girl by Norseman	4.60k	Mme Cino Del Duca	J. Taillard	
COUP DE FEU		by White Fire – Winning Bid by Great Captain	4.60k	Tim Sasse	P. Eddery	
PROVERB		by Reliance – Causerie by Cagire II	4.60k	Lt. Col. John Chandos-Pole	W. Carson	
RECUPERE		by Reliance – Nelion by Grey Sovereign	4.60k	Alan Clore	W.B. Pyers	
RIOT IN PARIS		by Sir Ribot – Running Account by Tuleg	3.55½k	L.P. Doherty	A. Gibert	
BUSIRIS		by Ridan – Lily Pons by Licencioso	3.55½k	Nelson Bunker Hunt	G. Lewis	
UN KOPECK		by Picqu' Arrière – Macopek by Près du Feu	3.55½k	Jacques Marx	M. Philipperon	
SAGARO		by Espresso – Zambara by Mossborough	3.55½k	Gerald Oldham	P. Paquet	
VALDO		by Salvo – Ileana by Abernant	3.55½k	Gerald Oldham	G. Thiboeuf	
HIGHCLERE		by Queen's Hussar – Highlight by Borealis	3.54k	H.M. Queen Elizabeth II	J. Mercer	

20 runners.
ALLEZ FRANCE trained by A. Penna; bred by Bieber-Jacobs Stable in the U.S.A.
Distances: Head; ¾ length; 4 lengths; 1 length; 3 lengths; ½ length; short neck; 1 length and 2½ lengths. Time: 2 mins 36.90 seconds.
Pari-Mutuel: Win: 1fr. 50. Places: 1fr. 50; 6fr. 60; 10fr. 20.
S.P.: 5 to 10 ALLEZ FRANCE and Paulista (couplé); 13 to 2 Mississipian and Busiris (couplé); 37 to 4 Sagaro and Valdo (couplé); 14 to 1 Kamaraan; 15 to 1 Highclere and Tennyson; 29 to 1 On My Way; 31 to 1 Récupéré; 53 to 1 Comtesse de Loir; 62 to 1 Card King; 65 to 1 Riboquill; 69 to 1 Coup de Feu; 70 to 1 Proverb; 72 to 1 Margouillat; 95 to 1 Sang Bleu; 97 to 1 Riot in Paris and Un Kopeck.

ALLEZ FRANCE bay filly 1970	Sea Bird ch. 1962	Dan Cupid ch. 1956	Native Dancer gr. 1950	Polynesian br. 1942	Unbreakable
					Black Polly
				Geisha gr. 1943	Discovery
					Miyako
			Vixenette ch. 1944	Sickle br. 1924	Phalaris
					Selene
				Lady Reynard b. 1939	Gallant Fox
					Nerva
		Sicalade b. 1956	Sicambre br. 1948	Prince Bio b. 1941	Prince Rose
					Biologie
				Sif br. 1936	Rialto
					Suavita
			Marmelade b. 1949	Maurépas ch. 1936	Athelstan
					Brocéliande
				Couleur b. 1939	Biribi
					Colour Bar
	Priceless Gem br. 1963	Hail to Reason br. 1958	Turn-to b. 1951	Royal Charger ch. 1942	Nearco
					Sun Princess
				Source Sucrée b. 1940	Admiral Drake
					Lavendula
			Nothirdchance b. 1948	Blue Swords b. 1940	Blue Larkspur
					Flaming Swords
				Galla Colors b. 1943	Sir Gallahad
					Rouge et Noir
		Searching b. 1952	War Admiral br. 1934	Man O' War ch. 1917	Fair Play
					Mahu Bah
				Brush Up b. 1929	Sweep
					Annette K.
			Big Hurry br. 1936	Black Toney br. 1911	Peter Pan
					Belgravia
				La Troiènne b. 1926	Teddy
					Hélène de Troie

Star Appeal 1975

Star Appeal

The Field

In 1975 Daniel Wildenstein's Allez France started favourite for the Prix de l'Arc de Triomphe for the second consecutive year. She heavily defeated Card King and Comtesse de Loir by 4 lengths and 1½ lengths in the Prix Ganay on her seasonal reappearance on 4 May with Dahlia 3 lengths away sixth. The Sea Bird mare appeared as good as ever when she comfortably beat Son of Silver and Un Kopeck 2½ lengths and a neck in the Prix Dollar on 1 June with Card King 3 lengths back fifth. Four weeks later, however, Allez France lost her first race since the 1973 Champion Stakes; she could only finish third in the Prix d'Ispahan, half a length and 1½ lengths behind Ramirez, whom she had easily beaten in the Prix Dollar, and the previous year's Grand Critérium winner, Mariacci. It was Allez France's solitary lapse before Sunday 5 October and it was possible that 1,850 metres was too short for her in 1975. After almost three months rest, Allez France gave 3 lbs to Duke of Marmalade and defeated him a very easy 1½ lengths in the Prix Foy at Longchamp. There was certainly no doubt that the 1974 Prix de l'Arc de Triomphe heroine was a worthy 7 to 4 favourite to become the first filly to win in consecutive years since Corrida in 1936 and 1937. Nevertheless her defeat in the Prix d'Ispahan showed that she was not invincible.

Daniel Wildenstein's champion filly was

again opposed by her great rival, Dahlia. In North America the previous autumn Dahlia had beaten Crafty Khale and London Company 2 lengths and a head in the Man O'War Stakes in the very fast time of 2 minutes 26.6 seconds. She then broke the course record when defeating Big Spruce 1½ lengths in the Canadian International Stakes at Woodbine Park. In a slowly-run race for the Washington D.C. International Stakes at Laurel Park, Dahlia was only able to finish third to Admetus and Desert Vixen. In her eight outings before the 1975 Prix de l'Arc de Triomphe, Nelson Bunker Hunt's mare only won once. She was never in contention in the Prix Ganay, the Prix Jean de Chaudenay or even the Gran Premio di Milano. Although she was hampered in the Grand Prix de Saint-Cloud, Dahlia was only fifth to Un Kopeck. In the King George VI and the Queen Elizabeth Diamond Stakes she showed signs of temperament before the start and Lester Piggott had to dismount her before she would consent to enter the stalls. Dahlia finished third, half a length and 5 lengths behind Grundy and Bustino, whose heroic battle meant that 2·36 seconds were lopped off the course record at Ascot. All who saw that tremendous and electrifying race will never forget the courage of both Grundy and Bustino. On My Way was 1½ lengths behind Dahlia in fourth place with Card King fifth and Star Appeal ninth. In the Benson and Hedges Gold Cup on 17 August, Grundy and Dahlia met again but the hero at Ascot was but a shadow of himself; clearly his titanic effort to overhaul Bustino had taken more out of him than was realised. Dahlia led the whole way and beat Card King and Star Appeal 1½ lengths and 5 lengths with Grundy a poor fourth. The fact that Nelson Bunker Hunt's mare broke the course record made some feel she was back to her best, but it must be remembered that there was a long dry spell in late July and August which produced very fast going.

The value of the form of the Benson and Hedges Gold Cup was again drawn into question for the third year out of four when Dahlia could only manage third place, 1½ lengths and a neck behind Duke of Marmalade and L'Ensorceleur in the Grand Prix de Deauville. She was promoted second when Duke of Marmalade was relegated to fifth place after a stewards' enquiry. Dahlia fared no better when half a length and a neck behind Kasteel and Ramirez in the Prix du Prince d'Orange on 21 September. The general consensus of opinion was that in 1975 she had deteriorated and that her three-year-old stable companion, Nobiliary, held a better chance against Allez France in the Prix de l'Arc de Triomphe.

Nobiliary by Vaguely Noble, who was a half-sister to Lyphard, won the Prix de la Grotte on her first outing as a three-year-old and then ran Ivanjica to 1½ lengths in the Poule d'Essai des Pouliches on 4 May, with the best filly in 1974, Broadway Dancer, 6 lengths away third. Nelson Bunker Hunt's filly was so impressive when defeating Lighted Glory, who had run fifth to Nocturnal Spree in the 1000 Guineas, by 2 lengths in the Prix Saint-Alary on 25 May that she ran next in the Epsom Derby. At Epsom Nobiliary beat all her rivals with the exception of the brilliant Grundy, who had 3 lengths to spare over her at the finish. As a result of this excellent display Nobiliary started a short-priced favourite for the Benson and Hedges Eclipse Stakes. She played up in the paddock and at the start, however, and she eventually finished a disappointing fifth to Star Appeal. Nobiliary failed again when a neck and 2 lengths behind Juliette Marny and Tuscarora in the Irish Guinness Oaks on 19 July. She was rested until the Prix Vermeille in September when she ran Ivanjica to 1½ lengths with the Yorkshire Oaks and Park Hill Stakes heroine, May Hill, 8 lengths away third. Nobiliary's trainer, Maurice Zilber, was confident that she would run Ivanjica much closer in the Prix de l'Arc de Triomphe, if not turn the tables, as he felt the race would have brought her on a lot. She and Dahlia started at odds of 35 to 4 couplé on 5 October.

Alec Head's stable was represented in the 1975 Prix de l'Arc de Triomphe by both Ivanjica and Green Dancer; the first was by Sir Ivor, the latter one of the first crop of Nijinsky. In her first race as a three-year-old Ivanjica ran third to Sea Sands and Beautiful Sea in the Prix Vanteaux in early April. She produced a really impressive burst of speed to account for Nobiliary in the Poule d'Essai des Pouliches. Regrettably the Prix de Diane was not run in 1975 as stable lads at Chantilly were occupying the course, so Ivanjica ran instead in the Prix d'Ispahan, in which she suffered an injury and was only fifth, 6 lengths behind Ramirez. In the Prix d'Ispahan she was ridden by R. Jallu, since her normal jockey, Freddie Head, had broken his leg in a car accident earlier in June. In the Prix de la Nonette at Longchamp on 7 September Ivanjica, again partnered by Jallu, produced a scintillating burst of acceleration to beat the Prix d'Astarté and the Prix de Malleret winner, Infra Green, by a length. When Ivanjica won the Prix Vermeille, such good fillies as the Oaks d'Italia heroine, Carnuba, and the Ribblesdale Stakes winner, Gallina, trailed a long way behind her and Nobiliary. There was no question that Ivanjica was the best three-year-old filly in France regardless of Zilber's hopes for Nobiliary and that she represented a serious threat to Allez France on Sunday 5 October.

Her stable-companion, Green Dancer, had run a close second to Mariacci in the Prix des Chênes as a two-year-old before he won the Observer Gold Cup from Sea Break and No Alimony at Doncaster. He reappeared in the Poule d'Essai des Poulains on 27 April and comfortably beat Condorcet a length. In the Prix Lupin on 18 May Green Dancer again met Mariacci, who had won the Prix Greffulhe in April. A battle royal ensued between these two very good colts with Nijinsky's son eventually wearing down his rival by three-quarters of a length with the subsequent Grand Prix de Paris winner, Mattahawk, 6 lengths away third. Green Dancer as a result of this performance started 6 to 4 favourite for the Epsom Derby, but he was only able to finish sixth, 10 lengths behind Grundy, without ever threatening to win. It is quite possible that his extremely hard race with Mariacci was the cause of this below-par performance. Green Dancer was rested until the Prix Niel in September; he was far from disgraced when failing by a neck to give 7 lbs to the Epsom Derby fourth and Irish Sweeps Derby third, Anne's Pretender. A fully recovered Freddie Head chose to ride Ivanjica in preference to Green Dancer in the Prix de l'Arc de Triomphe so Gary Moore rode him instead. He was probably the best French three-year-old colt in the field and the pair started second favourites at 17 to 4 couplé.

An English trained three-year-old colt, who had only finished fourteenth in the Epsom Derby, had come very much into prominence in the five weeks preceding the Prix de l'Arc de Triomphe. On 8 May Bruni by Sea Hawk II won the Wincanton Maiden Plate at Salisbury and then only lost the Predominate Stakes at Goodwood to the Ladbroke Craven Stakes winner, No Alimony, by a short head. Lester Piggott partnered him in the Derby but uncharacteristically got carried wide on Tattenham Corner. After that the grey Sea Hawk II colt had no chance and, even worse, returned home with sore shins and shortly afterwards caught a virus. On 30 August he beat the useful handicapper, Royal Match, by 7 lengths in record time in the Friends of the Variety Club Stakes (10f) at Sandown. Bruni took the lead two furlongs out in the St Leger and turned the race into a procession slaughtering King Pellinore, the 2-length runner-up to Grundy in the Irish Sweeps Derby, by an incredible 10 lengths with Libra's Rib third. Bruni's trainer, Ryan Price, also ran Carolus in the Prix de l'Arc de Triomphe. He had been outclassed in the Derby and St Leger, but had won two minor events at Kempton and Ayr and had run third to Dakota in the Terry's All Gold Ebor Handicap. Carolus had absolutely no chance on Sunday 5 October and started at 99 to 1 but Bruni was extremely well fancied at 9 to 2.

Jean-Michel de Choubersky's stable also had two three-year-old colts in the Prix de l'Arc de Triomphe in 1975, Baron de Zuylen de Nyvelt's Kasteel and Baron de Rédé's Citoyen. Kasteel, a half brother to Amadou, ran third to Speedy Dakota and Wronsky in the Prix Jean Prat at Chantilly on 8 June on his fourth outing in 1975 and was again third to Free Round and Anne's Pretender in the Prix Eugène Adam at Saint-Cloud on 21 July. His steady improvement continued dramatically when he decisively beat L'Ensorceleur and Condorcet 3 lengths and three-quarters of a length in the Prix de la Côte Normande at Deauville on 15 August. He then won the Prix du Prince d'Orange by half a length and a neck from Ramirez and Dahlia. Kasteel was clearly a very useful mile and a quarter colt, but the Prix de l'Arc de Triomphe was his first attempt at a mile and a half.

Baron de Rédé's Citoyen won two minor events at Evry and Maisons-Laffitte over ten furlongs in the spring but was only eighth to Val de l'Orne, Patch and Mariacci in the Prix du Jockey Club. He was much better suited by the longer distance of the Grand Prix du Paris and finished 5 lengths runner-up to Mattahawk with Patch only seventh. He was again second, a neck behind Mosstrooper, in the Prix Kergorlay at Deauville. Citoyen disappointed in the Grand Prix de Deauville, but ran Henri le Balafré to 2 lengths in the Prix Royal Oak a head in front of Olmeto. Unlike Kasteel, Citoyen needed a thorough test of stamina and was likely to be taken off his feet in the closing stages of the Prix de l'Arc de Triomphe for which the pair started at 23 to 1 couplé.

Henri le Balafré had won three races from six outings prior to his victory in the Prix Royal Oak, the most important of which had been in the Grand Prix de Lyon. He had been a well-beaten fifth to Free Round in the Prix Eugène Adam immediately prior to his success over Citoyen, which came, therefore, as a slight surprise. The absence of Mattahawk from the field of the Prix Royal Oak certainly made his task easier, and it is probable that the form did not amount to much. Henry le Balafré was trained by François Mathet, but shortly before the Prix de l'Arc de Triomphe he was bought by Paul de Moussac and transferred to the stable of R. de Mony Pajol.

Gerald Oldham's Olmeto, who beat Val de Fier in the Prix de Fontenoy as a juvenile and was fourth in the Grand Critérium, never really fulfilled his two-year-old promise in 1975. He only finished eighth to Corby in the Prix du Lys at Chantilly in June. He did win the Prix de Saint-Gatien at Deauville, but his close third in the Prix Royal Oak made that form suspect.

There was one three-year-old colt from the U.S.A. in the 1975 Prix de l'Arc de Triomphe, Ogden Mills Phipps's Intrepid Hero. He had won several good races in his native land including the Hollywood Derby and the Secretariat Stakes. He had also only been beaten a neck by the very useful Wajima in the Monmouth Invitational Handicap. Nevertheless Intrepid Hero, who was trained by John Russell, was a few pounds below the best three-year-olds in the United States, and it was likely he would fare no better than the last American challenger, One for All, in 1971.

Besides Allez France and Dahlia, there were twelve other older horses in the field on Sunday 5 October, several of whom had really no chance at all. Fabliau had won two races in Bordeaux and one in Toulouse and was sure to be outclassed. Paddy's Princess won the Prix de Victot (12½f) and the Prix de Pomone (13½f) at Deauville, but these victories were the first signs of good form she had shown since the spring of 1974 and she would have created a major surprise if she were to be concerned in the finish of the Prix de l'Arc de Triomphe. Yves Lelong's mare Ambrellita, who had run fourth in the Prix Vermeille and won the Prix de Royallieu in 1974, was well beaten in the Grand Prix de Saint-Cloud and Grand Prix de Deauville. She did beat Dom Luc at Vichy over thirteen furlongs in early August but was also sure to be out of her depth at Longchamp.

It was also very difficult to make a case for William R. Hawn's Riboquill and Beau Buck. The latter beat Blue Diamond in his only outing in 1974 in a race for newcomers at Longchamp in April. A year later he ran Card King and On My Way to a neck and 2 lengths in the Prix d'Harcourt with Riboquill 2 lengths away fourth. Beau Buck was only fifth to Ashmore in the Prix Jean de Chaudenay at Saint-Cloud on 13 May, and then was last in the Benson and Hedges Eclipse Stakes at Sandown. He did beat Meautry over eleven furlongs at Clairefontaine on 26 August and, although clearly well regarded at home, Beau Buck did not have the credentials one associates with a winner of the Prix de l'Arc de Triomphe. Riboquill had run well to be sixth to Allez France in 1974 and on his reappearance won over twelve and a half furlongs at Saint-Cloud in March. He was only sixth in the Grand Prix de Saint-Cloud and, although promoted to third place in the Grand Prix de Deauville, had not really lived up to his three-year-old promise.

After his victory in the Prix d'Harcourt, Card King, a seven-year-old in 1975, failed to win again before the Prix de l'Arc de Triomphe. He was runner-up in the Benson and Hedges Gold Cup and the Prix Ganay and third in the Grosser Preis von Baden. Nevertheless he was only fifth, both in the Prix Dollar and behind Grundy at Ascot, and eighth in the Grand Prix de Saint-Cloud. He was probably getting a little long in the tooth to take an active part in the finish on 5 October.

Xavier Beau's five-year-old On My Way had run with admirable consistency prior to the 1975 Prix de l'Arc de Triomphe. Although he had not won a race, he had always finished in the frame. He had been runner-up in both the Prix d'Harcourt and La Coupe de Longchamp; he had run a close third to Un Kopeck in both the Grand Prix d'Evry and Grand Prix de Saint-Cloud, when he suffered considerable interference, as he did when fourth in the Grand Prix de Deauville. On My Way was far from disgraced when 7 lengths fourth to Grundy at Ascot and at odds of 79 to 1 could certainly be regarded as a possible outsider.

The outsider who represented the best value, however, was the German-trained five-year-old Star Appeal, whose odds were an incredible 119 to 1. In 1974 he was trained by A.

Pohlkotter and won the Grosser Preis der Badischen Wirtschaft at Baden Baden and the Concentra-Pokal der Deutschen Investment Trust (10f) at Frankfurt. He ran exceptionally well when 4¾ lengths fourth to Giacometti in the Champion Stakes. In 1975 he gained his first victory of the season on his third outing in the Grosser Preis der Badischen Wirtschaft for the second consecutive year. Star Appeal then started at 20 to 1 in the Gran Premio di Milano at San Siro on 15 June. He took the lead early in the straight and beat Duke of Marmalade and Un Kopeck three-quarters of a length and the same with Orange Bay, the Derby Italiano and Premio Emanuele Filiberto hero, Laomedonte, the winner of the Gran Premio d'Italia and Dahlia the next to finish. Un Kopeck in fact was relegated to fifth place. Not merely was it the biggest success Star Appeal had gained in his career but also one of the strongest fields for the race for several years. Nevertheless Star Appeal started at 20 to 1 for the Benson and Hedges Eclipse Stakes at Sandown on 7 July but, taking the lead inside the final furlong, he proved a decisive 2-length winner over Taros with Nobiliary and Beau Buck never in contention. His new trainer Theo Grieper, therefore, had the distinction of saddling the first German-trained horse to win in England since the victory of Turnus in the 1850 Stewards Cup. Star Appeal was no match for Grundy and Bustino at Ascot, finishing ninth, but ran third to Dahlia and Card King at York and fourth in very heavy going to Marduk in the Grosser Preis von Baden. On his best form Star Appeal was about the same horse as Duke of Marmalade and a little better than Un Kopeck.

Duke of Marmalade, the mount of Lester Piggott in the 1975 Prix de l'Arc de Triomphe, had been trained by Robert Armstrong in England the previous season, when he won three races over a mile and a quarter and a mile and a half at Nottingham and Newmarket. He began his four-year-old career in the stable of M. Bertini in Italy and won twice over ten and twelve furlongs before he ran second to Star Appeal in the Gran Premio di Milano. He was then transferred to J.R. Lyon's stable in France where, on his first outing, he ran Record Run and Ramirez to a short head and a short neck in the Prix Gontaut-Biron (10f) at Deauville on 16 August. Duke of Marmalade then beat L'Ensorceleur and Dahlia in the Grand Prix de Deauville, but was disqualified and placed fifth after a stewards' enquiry. He may have been a little flattered by his proximity to Allez France when she defeated him 1½ lengths very comfortably in the Prix Foy at Longchamp. Nevertheless there seemed absolutely no logic as to why Duke of Marmalade should start at odds of 19 to 2 and Star Appeal the complete outsider at 119 to 1 on Sunday 5 October.

The ex-provincial four-year-old, Un Kopeck, was much improved in 1975. He won the Grand Prix d'Evry by a neck and a nose from Dhaubasix and On My Way on 26 April. He then ran Ashmore (received 4 lbs) and Battle Song (received 6 lbs) to 2 lengths and half a length in the Prix Jean de Chaudenay at Saint-Cloud on 19 May and was third to Allez France in the Prix Dollar. After his disqualification from third place in the Gran Premio di Milano, Un Kopeck led throughout to beat Ashmore and On My Way half a length and a short head in the Grand Prix de Saint-Cloud with Comtesse de Loir and Dahlia 3 lengths and the same behind fourth and fifth. On that form alone, at 42 to 1, he seemed to have a reasonable each way chance in the Prix de l'Arc de Triomphe. Un Kopeck was only sixth in the Grosser Preis von Baden but he was unable to handle the exceptionally heavy ground.

Comtesse de Loir, who had run Allez France so close the previous autumn, was lightly raced as a four-year-old. She ran below her best in the spring finishing a well-beaten third in the Prix Ganay, and was never a danger when 7 lengths fourth to Bustino and Ashmore in the Coronation Cup. Although fourth in the Grand Prix de Saint-Cloud, she was still 3½ lengths behind Un Kopeck. On 29 July Comtesse de Loir beat Rosy Ride, subsequently a close third in the Prix Foy, by a short neck in the Prix des Erables over eleven furlongs at Chantilly. This race was only open to horses who had not won a Group Three race for twelve months so it was difficult to judge whether Comtesse de Loir had recaptured her three-year-old brilliance; if she had, she could not be ruled out of contention on 5 October.

The field for the 1975 Prix de l'Arc de Triomphe was completed by the Aga Khan's Kamaraan. He had suffered from leg trouble in 1975 and had only one outing before running for the second consecutive year in Longchamp's most important weight-for-age race. He won the ten-furlong Prix de l'Ecoute s'il peut at Evry on 12 September by 4 lengths and a length from Tom Paine and Admetus, the 1974 Washington

D.C. International Stakes winner. His victory was achieved so impressively that Kamaraan was more than a quiet whisper in Chantilly and Longchamp in the days before the Prix de l'Arc de Triomphe and he started at odds of 16 to 1.

The regrettable absence of the best three-year-old colt in Europe, Grundy, from the field for the 1975 Prix de l'Arc de Triomphe had been long anticipated. Before his famous success at Ascot, the English Levy Board had bought a 75 per cent share in him for £750,000 and it had been announced that he would not be trained for the race at Longchamp. Bustino was an intended runner but regrettably suffered an injury to his near-fore during his preparation and was retired to stud. Val de l'Orne, who had won the Prix Hocquart and the Prix du Jockey Club, and Mattahawk, successful in the Grand Prix de Paris, had both broken down and gone to stud. Another absentee was the 1974 Irish St Leger winner Mistigri, who suffered a slight injury in training and was withdrawn shortly before the race.

The Final Preparations

In 1975 the Prix du Moulin de Longchamp was not run on the same day as the Prix de l'Arc de Triomphe but was transferred to the prior Sunday which previously did not have a Group One race in its programme. In its place was substituted the Prix de l'Opéra, a Group Two race for three- and four-year-old fillies, over 1,850 metres. Prior to the Prix de l'Arc de Triomphe Daniel Wildenstein's son, Alec, was of the opinion that should Allez France be beaten, it would be by a four-year-old. He considered the three-year-olds, with the exception of Grundy, to be a moderate bunch. Bruni's jockey, Tony Murray, said that he was most frightened of the filly Ivanjica, whom he considered had been extremely impressive in the Prix Vermeille. Most French *turfistes* regarded Allez France as the most likely to take the prize but still regarded Bruni as the principal threat to her supremacy. On account of an industrial dispute at Chantilly nearly all the French runners had to be boxed up by six o'clock in the morning; for the 45-kilometre trip to Longchamp the horse-boxes were given a police escort. Thus as Maurice Zilber pointed out the unfortunate horses missed their breakfast at the normal time. Those foreign challengers, like Star Appeal, who were lodged in the racecourse stables at Longchamp, did not have to endure this change in routine.

Président Giscard d'Estaing, elected on the death of Président Pompidou in May, came to Longchamp on Sunday 5 October surrounded by numerous gendarmes, fore and aft; when he made his way from the Président's box to the paddock, he enthusiastically acknowledged the applause of the crowd and appeared to be enjoying himself.

The afternoon commenced very well for Angel Penna and Yves Saint-Martin because Daniel Wildenstein's Lianga won the Prix de l'Abbaye de Longchamp. Yves Saint-Martin quickly completed a double when he won the Critérium des Pouliches on Theia.

The Race

The paddock crowd were rather surprised that Alfred Gibert was not wearing the orange colours with dark blue sleeves of Xavier Beau. Instead he had on the white colours of the Société d'Encouragement. On My Way's connections had ordered a brand new set of colours for Gibert to ride in the Prix de l'Arc de Triomphe. A parcel containing the new jacket had arrived the previous evening but it was not opened until a short time before the race the following afternoon. It contained a dark blue jacket with orange sleeves, the colours of Comtesse Batthyany. As a result Gibert was forced to wear the colours of the Société d'Encouragement.

In the paddock Allez France again looked an absolute picture and was certainly the pick. Green Dancer, a medium-sized attractive colt, also took the eye. Bruni, however, did not look as well as he had done at Doncaster and appeared as if he had gone a little over the top. That actually was the opinion of Ryan Price, his trainer, who would have preferred not to have run him after the St Leger, and in his normal forthright manner said that he had been brainwashed to run the grey colt.

The favourite was the last into the stalls and as soon as they opened Citoyen took the lead and set a good pace to Olmeto. They were closely followed by Nobiliary, Dahlia, who was surprisingly prominent, Carolus, Un Kopeck and Allez France, who was on the inside of Carolus about a length behind him. At this stage the backmarkers were Ivanjica, Comtesse de

Star Appeal (3) beating On My Way (2) and Comtesse de Loir (13) 3 lengths and 2½ lengths

Loir, Star Appeal and Bruni. The order remained almost the same to the top of hill, except that Kasteel made progress to be just on the heels of the leading group. Throughout the first half of the race Lewis on Carolus appeared a little uneasy as he was continually glancing over his shoulder. Shortly after the descent began Dahlia either stumbled or struck into the heels of another runner with the result that she lost 10 lengths and any chance she might have had. Halfway down the hill Citoyen still held a slender lead over Nobiliary, Un Kopeck, Carolus and Allez France. It was at this moment that Lewis virtually pulled Carolus up as if he had broken down. He cannoned into Allez France and they both dropped back almost to last place, interfering and hampering several other runners. This incident was most unfortunate since clearly it put paid to the favourite's chance. At the same time both Bruni and Comtesse de Loir could be seen making significant progress towards the leaders. On the turn into the straight Nobiliary led from Kasteel and Un Kopeck together with Bruni on the outside going very easily. They were tracked by Duke of Marmalade, the improving Comtesse de Loir, On My Way and Kamaraan. Both Allez France and Star Appeal were still amongst the backmarkers however. Four hundred metres out there was little to choose between Nobiliary, Kasteel and Bruni, but all three were visibly beginning to struggle, and on their heels were Un

Kopeck and Comtesse de Loir, who seemed poised to take the lead whenever Desaint gave her her head. On My Way was also going easily just behind her and Star Appeal could be seen wending his way through beaten horses, followed by Allez France. Three hundred metres out Comtesse de Loir moved through on the rails to take the lead from the weakening Nobiliary, Kasteel and Bruni. She was followed by Un Kopeck, On My Way and Star Appeal, who was travelling fastest of all. Allez France was at last running on but looked to have too much to do. A very fortunate gap appeared between Un Kopeck and the weakening Nobiliary 200 metres out through which Starkey steered Star Appeal, and so fast was he going that he had passed Comtesse de Loir in a few strides, drawing away from his rivals to a 3-length victory over the strong finishing On My Way who beat Comtesse de Loir 2½ lengths for second place. Un Kopeck was another 1½ lengths back fourth, only a neck in front of Allez France. She was followed by Nobiliary, Bruni, Green Dancer and Kamaraan with the American challenger Intrepid Hero tenth. There was a stunned silence at Longchamp immediately after the horses passed the post as few had anticipated this result.

It had been an extremely rough race. Daniel Wildenstein complained that Allez France had been knocked about like a punch-bag and she had returned to the unsaddling enclosure minus one shoe and a cut on her near hind leg. The cut was not that serious but clearly the incident with Carolus had not helped her and she did very well to recover to finish fifth. Green Dancer also had a very rough passage coming down the hill, whilst Head blamed the high draw for Ivanjica's total failure. Murray thought that he would win on Bruni turning for home, but added 'he weakened quickly a furlong and a half out'. The St Leger winner had been forced to make his effort on the outside and his owner Charles St George felt that this had probably caused his defeat. Zilber was most critical of the riding of the American jockey Navarro, whom Nelson Bunker Hunt had brought over to ride Dahlia. He blamed the American both for taking on Nobiliary in the early part of the race and for causing Dahlia to stumble at halfway.

All of these recriminations were inclined to obscure the fact that Star Appeal had produced a quite astonishing burst of speed in the straight to win the race strictly on merit. He thus became the first ever German-trained victor of the Prix de l'Arc de Triomphe. Theo Greiper trained near Cologne and his bold policy of taking on the best in Europe in 1975 with Star Appeal had paid off handsomely. He had no English whilst his jockey, Greville Starkey, spoke no German. The orders in the paddock were given by a bilingual stable-lad.

It was also Starkey's first success in the Prix de l'Arc de Triomphe. On Monday 29 September at Goodwood he had been found guilty of careless riding but fortunately the suspension that he had been given was not due to commence until Wednesday 8 October. The 35-year-old jockey had been apprenticed to H. Thomson Jones at Newmarket and in 1957 was the champion apprentice. He had been stable jockey to John Oxley between 1961 and 1969 and to Henry Cecil from 1970 to 1974. In 1975, however, he was riding freelance and unquestionably his success on Star Appeal was the biggest of his career up to that time. Immediately after the race he commented, 'In the King George VI and the Queen Elizabeth Stakes the pacemakers killed my horses, so I decided to drop him out today. I never had to hit him with the stick and only pushed him out with hands and heels.'

Star Appeal's owner Waldemar Zeitelhack, a Nuremburg steel magnate, had the reputation in Germany of being a man who, if he touched something, it turned to gold. This seemed to have happened again after his purchase of Star Appeal.

The Rest of the Season

Both Star Appeal and Allez France ran in the Champion Stakes on 18 October. Daniel Wildenstein's mare gained her revenge over her Longchamp conqueror. She was beaten, however, by the brilliant filly Rose Bowl by 1½ lengths with Ramirez and Star Appeal three-quarters of a length and the same third and fourth. Neither the Prix de l'Arc de Triomphe winner nor Allez France could match the finishing speed of Rose Bowl who was running over ten furlongs for the first time.

The next day Kamaraan beat Beau Buck, who was receiving weight, 1½ lengths in the Prix du Conseil Municipal with Olmeto, Card King and Ambrellita in rear. Duke of Marmalade was only third to Laomedonte and Consul in the Gran Premio del Jockey Club at San Siro, but on

15 November in the Premio Roma he and Henri le Balafré dead-heated for first place 5 lengths ahead of Klairvimy. Dahlia again ran in the Canadian International Stakes at Woodbine, but was given too much to do by her jockey and only finished fourth to the much-improved Snow Knight.

In the Washington D.C. International Stakes at Laurel Park on 8 November, Nobiliary led throughout to beat Comtesse de Loir and On My Way three-quarters of a length and 3 lengths with Star Appeal only fifth and Dahlia, who suffered an injury in running, a disappointing eighth. Considering his strenuous season, and that he was probably not suited by the track, it was no disgrace for Star Appeal, but the fact that three French-trained runners filled the first three places again upheld the form of the Prix de l'Arc de Triomphe. Allez France ran once after the Champion Stakes, in the National Thoroughbred Championship at Santa Anita in November which is run on dirt. That fact was probably why she was a disappointing last of eleven. Considering also that she was such a nervous traveller, it was surprising that she was even sent to run the last race of her career in California.

When Star Appeal retired to the National Stud at Newmarket he had won eleven times and been placed twelve times from thirty-nine outings and had won a total of £263,015. Although foaled in Ireland, he had been bred by Frau Mehl-Mülhens, whose stud, the Gestüt Rottgen, was one of the best in Germany.

Pedigree and Stud Record

Star Appeal's sire, Appiani II by Herbager, had won the Derby Italiano in 1966 and the Premio Presidente della Repubblica in 1967. He was also runner-up to Marco Visconti in the Gran Premio di Milano as a four-year-old and third to Busted and Great Nephew in the Eclipse Stakes and fourth behind Busted in the King George VI and the Queen Elizabeth Stakes. He began his stud career in Ireland but in 1975 stood in France. Star Appeal was much the best of his progeny by 1975, though he had also sired the Galtres Stakes winner Lucky for Me, Gritti Palace (Manchester November Handicap), and Attivo, the very game stayer and hurdler owned by Peter O'Sullevan. Appiani II's dam, Angela Ruccellai, had won the Oaks d'Italia and Premio Regina Elena in 1957.

Star Appeal's dam Sterna was by Neckar. He had won the Deutsches Derby in 1951 and was five times leading sire in Germany prior to his death in 1974. He was also the sire of Tadolina (Oaks d'Italia 1965), Hogarth (Derby Italiano 1968) and Peseta II, the dam of the Oaks winner, Pia. Sterna, however, was not a particularly good racemare; she never won but was placed six times in Germany as a three-year-old. As a broodmare she proved to be very good. Sterna was the dam of eight winners of which the best was Star Appeal. Sternau by Tamerlane won as a three-year-old in Ireland and was the dam of Strong Gale, whose four victories included the Grosser Preis der Bayerischen-Wirtschaft and who was also runner-up to Northern Baby in the Prix Dollar and in the Irish 2000 to Jaazeiro in 1978. Sterna's Larkspur foal, Sternspur, won seven races in England, Ireland, Germany and Belgium, including the Grand Prix de la Ville d'Ostende. Star Wolf by Wolver Hollow won three races in France and Germany including the Allianz Pokal. Star's Salute by Sallust won four races in Ireland and one in France and ran third in the Grosser Hertie-Preis (Group III) and the Desmond Stakes. Stern Regina by Royal and Regal won three times and was third in the Brownstown Stakes.

Sterna's dam, Stammesart by Alchimist, won three races in Germany, including the Deutscher Stutenpreis and the Bayerischer Stutenpreis. She was the dam of seven winners, the most important of which was Stani, who ran third to Ribot and Norman in the 1955 Gran Premio del Jockey Club. Stani's thirteen victories in Germany and Belgium included the Grosser Preis von Baden, von Frankfurt and von Dusseldorf. He also won the Grosser Preis von Hessen twice. Stammesart was also the dam of Santa Cruz by Caran d'Ache whose six victories included the Preis der Diana in 1960, the Schwarzgold Rennen and the Ratibor Rennen. Santa Cruz herself became a very useful broodmare and amongst her progeny were Sawara (Furstenburg Rennen) and Santayana (Turf Club Handicap). Stammesart was also the grandam of Stella Maris (Hamburger Criterium).

Stammesart's dam, Stammesfahne by Flamboyant, won three races including the Oppenheim Rennen. She also was the dam of winners in Germany. This family was one of the best in Germany since the turn of the century.

It is too early to say if Star Appeal has been a

success yet at stud. From his first crop he sired Star Way (Chesham Stakes and 2nd Royal Lodge Stakes), Western Story (Scherping-Rennen). His daughter Madam Gay won the Prix de Diane in 1981 in record time and ran third in the King George VI and the Queen Elizabeth Diamond Stakes, in the Prix Vermeille and in the Arlington Million as well as being runner-up in the Oaks. In 1982 another daughter Star won the Magyar Karcadis (Oaks) in Budapest. Up to October 1982 his progeny had won thirty-four races in England worth £323,353 and abroad during the same period he has sired winners of sixty-nine races worth £194,407.

PRIX DE L'ARC DE TRIOMPHE 2400 metres
at Longchamp on Sunday 5 October 1975. For three-year-olds and upwards. Going: Soft.
1,200,000 fr. to the winner; 480,000 fr. to the 2nd; 240,000 fr. to the 3rd; 120,000 fr. to the 4th. 300,000 fr. to the breeder of the winner; 120,000 fr. to that of the 2nd; 60,000 fr. to that of the 3rd; 30,000 fr. to that of the 4th. Value to the winner: 1,468,150 francs.

STAR APPEAL	b.c.	by Appiani II – Sterna by Neckar	5.60k	Waldemar Zeitelhack	G. Starkey	1
ON MY WAY	b.c.	by Laugh Aloud – Gracious Me by Tulyar	5.60k	Xavier Beau	A. Gibert	2
COMTESSE DE LOIR	b.f.	by Val de Loir – Neriad by Princequillo	4.58½k	George Ohrstrom	J.C. Desaint	3
UN KOPECK	b.c.	by Piqu'Arrière – Macopeck by Près du Feu	4.60k	Jacques Marx	M. Philipperon	4
ALLEZ FRANCE	b.f.	by Sea Bird – Priceless Gem by Hail to Reason	5.58½k	Daniel Wildenstein	Y. Saint-Martin	5
NOBILIARY	ch.f.	by Vaguely Noble – Goofed by Court Martial	3.54k	Nelson Bunker Hunt	W.B. Pyers	6
BRUNI	gr.c.	by Sea Hawk – Bombazine by Shantung	3.55½k	Charles St George	A. Murray	7
GREEN DANCER	b.c.	by Nijinsky – Green Valley by Val de Loir	3.55½k	C. Carratt	G.W. Moore	8
KAMARAAN	b.c.	by Tanerko – Diamond Drop by Charlottesville	4.60k	H.H. Aga Khan	H. Samani	9
INTREPID HERO	b.c.	by Forli – Bold Princess by Bold Ruler	3.55½k	Ogden Mills Phipps	A. Cordero Jr	10
CARD KING		by Cardington King – Nantua II by Again II	7.60k	Raphaël Hakim	R. Jallu	
DUKE OF MARMALADE		by Vaguely Noble – Mock Orange by Dedicate	4.60k	Mme Bruno Bucci	L. Piggott	
BEAU BUCK		by Buckpasser – Blue Northern by Windy City II	4.60k	William R. Hawn	G. Dubroeucq	
RIBOQUILL		by Ribot – Quill by Princequillo	4.60k	William R. Hawn	P. Eddery	
FABLIAU		by Le Fabuleux – Barde by Prince Taj	4.60k	Mme Fernand Ripoli	R. Mongil	
DAHLIA		by Vaguely Noble – Charming Alibi by Honey's Alibi	5.58½k	Nelson Bunker Hunt	N. Navarro	
AMBRELLITA		by Misti – Ambrella by Amber	4.58½k	Yves Lelong	A. Badel	
PADDY'S PRINCESS		by St. Paddy – Russian Princess by Henry the Seventh	4.58½k	Garfield Weston	M. Depalmas	
HENRI LE BALAFRE		by Sassafras – Galoubinka by Tamerlane	3.55½k	Paul de Moussac	A. Lequeux	
OLMETO		by Val de Loir – Vivara by Mossborough	3.55½k	Gerald Oldham	P. Paquet	
CITOYEN		by Above Suspicion – Tomorrow by Citation	3.55½k	Baron de Rédé	J. Taillard	
CAROLUS		by King Log – Stormy Venture by Arctic Storm	3.55½k	Mme W.H.D. Riley-Smith	G. Lewis	
KASTEEL		by King of the Castle – Esme by Le Haar	3.55½k	Baron Th. de Zuylen de Nyevelt	G. Rivases	
IVANJICA		by Sir Ivor – Astuce by Vieux Manior	3.54k	C. Carratt	F. Head	

24 runners.
STAR APPEAL trained by T. Grieper at Köln-Heumar in Germany; bred by Frau Mehl-Mülhens in Ireland.
Distances: 3 lengths; 2½ lengths; 1½ lengths; 1½ lengths; a neck; a short head; ¾ length; 1½ lengths; a nose. Time: 2 mins 33.60 seconds.
Pari-Mutuel: Win: 119fr. 70. Places: 24fr. 10; 16fr. 60; 5fr. 60.
S.P.: 7 to 4 Allez France; 17 to 4 Green Dancer and Ivanjica (couplé); 9 to 2 Bruni; 35 to 4 Dahlia and Nobiliary (couplé); 19 to 2 Duke of Marmalade; 16 to 1 Kamaraan; 21 to 1 Comtesse de Loir; 23 to 1 Citoyen and Kasteel (couplé); 31 to 1 Intrepid Hero; 32 to 1 Henri le Balafré; 42 to 1 Un Kopeck; 49 to 1 Olmeto; 79 to 1 On My Way; 99 to 1 Beau Buck and Riboquill (couplé), Card King, Ambrellita, Paddy's Princess, Carolus and Fabliau; 119 to 1 STAR APPEAL.

STAR APPEAL bay colt 1970	Appiani II b. 1963	Herbager b. 1956	Vandale b. 1943	Plassy b. 1932	Bosworth
					Pladda
				Vanille b. 1929	La Farina
					Vaya
			Flagette ch. 1951	Escamillo ch. 1939	Firdaussi
					Estoril
				Fidgette ch. 1939	Firdaussi
					Boxeuse
		Angela Rucellai b. 1954	Rockefella br. 1941	Hyperion ch. 1930	Gainsborough
					Selene
				Rockfel br. 1935	Felstead
					Rockcliffe
			Aristareta ch. 1947	Niccolo dell' Arca b. 1938	Coronach
					Nogara
				Acquaforte ch. 1936	Blenheim
					Althea
	Sterna br. 1960	Neckar bl. 1948	Ticino br. 1939	Athanasius br. 1931	Ferro
					Athanasie
				Terra br. 1929	Aditi
					Teufelsrose
			Nixe bl. 1941	Arjaman br. 1930	Herold
					Aditja
				Nanon b. 1935	Graf Isolani
					Nella da Gubbio
		Stammesart b. 1944	Alchimist b. 1930	Herold br. 1917	Dark Ronald
					Hornisse
				Aversion br. 1914	Nuage
					Antwort
			Stammesfahne b. 1929	Flamboyant b. 1918	Tracery
					Simonath
				Selika b. 1916	Nuage
					Safety

Ivanjica 1976

Ivanjica

The Field

The French three-year-olds in 1976 were almost certainly the best since the remarkable year of Sea Bird, Reliance and Diatome. All the English classics, with the exception of the 2000 Guineas, won by Wollow, and both the Irish Sweeps Derby and the Irish Guinness Oaks, were captured by French challengers. It is hardly surprising, therefore, that there were only four foreign-trained runners in the 1976 Prix de l'Arc de Triomphe: the four-year-old Bruni and the five-year-olds Duke of Marmalade and Dakota from England and the four-year-old Noble Dancer from Norway.

Charles St George's Bruni reappeared in the Yorkshire Cup (14f) on 13 May. Although he was slow out of the stalls, he beat the useful stayers Mr Bigmore, Sea Anchor and Dakota, with the utmost ease. The 1975 St Leger winner would have made an ideal Cup horse, but the fashion in breeding dictated that Bruni should bypass the Prix du Cadran and the Ascot Gold Cup as a four-year-old and have the principal middle distance races in France and England included in his programme, if he were to become a well-patronised stallion. In the Hardwicke Stakes Bruni once again was very slowly away and lost at least 10 lengths. Orange Bay took the lead shortly before the straight but Lester Piggott drove Bruni past him inside the final hundred yards to go a neck in front. Orange Bay fought back and on the line had a head to spare over Bruni. There was no doubt that the latter was an unlucky loser; had he started on level terms he must have won. His newly acquired bad habit of starting very slowly had put him at a grave disadvantage and was to plague him and worry his connections for the remainder of the season. In the King George VI and the Queen Elizabeth Diamond Stakes on 24 July, Charles St George's St Leger winner again lost ground at the start, though admittedly only 5 lengths. He was 10 lengths behind the leader Pawneese at Swinley Bottom, and was then hampered by the weakening Coin of Gold four furlongs out. On the turn into the straight Bruni appeared to be in a totally hopeless position. When Piggott eventually managed to extricate him and begin his challenge, Pawneese was gone beyond recall. Bruni, however, finished fastest of all and was only a length behind Pawneese at the winning post with Orange Bay and Dakota a short head and three-quarters of a length away third and fourth; Malacate was fifth and Youth only ninth. On 23 September in the Cumberland Lodge Stakes at Ascot Bruni again dug his heels in at the start and lost 10 lengths. Nevertheless he struck the front 150 yards out and so fast was he travelling that he thrashed the useful three-year-old Illustrious Prince 5 lengths.

Bruni could certainly be regarded as an unlucky loser of both the Hardwicke Stakes and the King George VI and the Queen Elizabeth Diamond Stakes; he was to start second favourite for the Prix de l'Arc de Triomphe on Sunday 3 October at odds of 5 to 2, which seemed a little ungenerous in view of his habit of loitering at the start.

Madame Bruno Bucci's Duke of Marmalade was running in the Prix de l'Arc de Triomphe for the second consecutive season. In the spring he had won the Premio Ellington (12f) in Rome and on 27 May he ran Shamsan and Record Run to a length and 2 lengths in the Premio Presidente Della Repubblica. He was beaten 1½ lengths and three-quarters of a length by Rouge Sang and Art Style in the Gran Premio di Milano at San Siro on 20 June and was last to Pawneese at Ascot. He was then transferred to the stable of Nick Vigors in England and ran Ashmore and Diagramatic to 4 lengths and a short neck in the Grand Prix de Deauville on 29 August. It was difficult, however, to envisage Duke of Marmalade taking a hand in the finish on Sunday 3 October and his odds of 66 to 1 reflected this opinion.

Guy Reed's Dakota won the 1975 St Simon Stakes at Newbury and the Ebor Handicap. His performance behind Pawneese at Ascot had surprised most observers for he had only run third, receiving weight, to Orange Bay in the Jockey Club Stakes (12f) in April. After Ascot he resumed his temperamental and mulish behaviour and was a well-beaten fourth to Swell Fellow in the Geoffrey Freer Stakes at Newbury in August and a very disappointing sixth to Gunner B in the Doonside Cup at Ayr in September. His chance was considerably more remote than his odds of 99 to 1 on Sunday 3 October.

Gunnar Schjelderup's Noble Dancer was trained by Gordon Smyth in England in the first part of his juvenile campaign and won over six furlongs at Kempton Park before he was exported to Norway. In 1975 as a three-year-old he won both the Oslo Cup and the Norsk St Leger. Prior to his arrival in Paris Noble Dancer was unbeaten in three outings in Norway including a second and very easy success in the Oslo Cup on 5 September. It was, however, very difficult to assess his chance with any accuracy in the Prix de l'Arc de Triomphe. His trainer, Terjé Dahl, nevertheless had several times brought off successful raids with his hurdlers in England, and the fact that Noble Dancer started at odds as low as 23 to 1 on 3 October showed the confidence that his connections had in the best horse seen in Norway for very many years.

With the exception of Bruni, and there was a question mark about him, the foreign challengers did not appear to represent the usual threat to the French defenders. Their three-year-olds seemed to dominate the Prix de l'Arc de Triomphe in 1976 and the best of these were Nelson Bunker Hunt's Youth and Exceller, Daniel Wildenstein's Pawneese and Crow, and Mme Alec Head's Riverqueen.

Youth, who was trained by Maurice Zilber, had won a débutants' race and been runner-up to Far North in the Prix Saint-Roman as a two-year-old. On his reappearance in the Prix Greffulhe, Youth finished very strongly to beat Velino 1½ lengths on 4 April, and a fortnight later he had 4 lengths to spare over Danestic in the Prix Daru. Youth next won the Prix Lupin on 16 May by three-quarters of a length and a length from Arctic Tern and his stable-companion Empery. The latter was sent to Epsom to contest the Derby as it was felt the course would suit him better; his victory in England's premier classic confirmed this opinion. Youth ran in the Prix du Jockey Club at Chantilly on 6 June; he took the lead half way up the straight and drew away to win unchallenged by 3 lengths and three-quarters of a length from Twig Moss and Malacate with Arctic Tern only ninth. When Malacate beat Empery 2½ lengths in the Irish Sweeps Derby on 26 June, there was a very good case to suggest that Youth was the best three-year-old colt in Europe. He started favourite for the King George VI and the Queen Elizabeth Diamond Stakes, but Freddie Head, who had partnered him in all his previous outings in 1976, had great difficulty in steering him at Ascot. Youth ran so wide on the turn into the straight and forfeited so much ground that he lost any chance he might have had and he finished a dismal ninth, 10 lengths behind Pawneese. Lester Piggott partnered Youth in the Prix Niel on 5 September; Nelson Bunker Hunt's colt beat Arctic Tern, who was receiving 2 kilos, and Malacate by a comfortable three-quarters of a length and 2½ lengths. If the débâcle at Ascot were to be disregarded, Youth clearly held an outstanding chance in the 1976 Prix de l'Arc de Triomphe and he was to start 2 to 1 favourite with his owner's other three-year-old colt Exceller.

Exceller, who was trained by François Mathet, won over nine furlongs at Evry as a juvenile but was disqualified from first place in the Prix Herod and placed third, the position he also occupied in the Prix de Condé. On his first run as a three-year-old Exceller was beaten a length by Malacate in the Prix de Suresnes, but he then had 2½ lengths to spare over Lithidan in the Prix Matchem (10f) at Evry on 19 May. Exceller did not run in the Prix du Jockey Club, but on the same day defeated Brouhaha in the Prix du Lys in a canter by 6 lengths. In the Grand Prix de Paris on 27 June Exceller trounced the subsequent St Leger runner-up, Secret Man, by 4 lengths. In his final outing before the Prix de l'Arc de Triomphe Exceller slammed the runaway Ebor Handicap winner, Sir Montagu, by 4 lengths in the Prix Royal Oak on 15 September. Some felt that a mile and a half might be too sharp for Exceller, but he certainly provided an extremely useful second string to Nelson Bunker Hunt's bow on 3 October.

Daniel Wildenstein's Crow did not run as a two-year-old but on his début in the Prix Northeast (8½f) in March he beat Youth's lead horse Oldfield. He was only fourth in the Prix Lagrange on 15 March and then suffered from a blood disorder and was off the course until the Grand Prix de Compiègne on 6 July when he ran Hunza Dancer, who had been third to Grundy in the 1975 Derby, to half a length. Crow then beat Avaray, the Prix de la Jonchère winner, a short head in the Prix Eugène Adam at Saint-Cloud on July 18th, with Arctic Tern and Vitiges only fifth and sixth. Crow ran Wollow to a length in the Benson and Hedges Gold Cup on 17 August, the pair finishing 5 lengths ahead of their nearest pursuers, including Twig Moss and Trépan. In the St Leger on 11 September Daniel Wildenstein's colt never looked in danger once Saint-Martin sent him to the front; he strode away to win 2 lengths from Secret Man, the Prix de Menneval winner. Crow was certainly improving with each race and would present a serious danger to both Youth and Exceller on Sunday October 3rd.

In midsummer Crow's brilliant stable-companion Pawneese was the best three-year-old in Europe. Pawneese won the Prix la Camargo at Saint-Cloud in the second week of March and on the same course beat Loveliest 2 lengths in the Prix Pénélope on 27 March. In the Prix Cléopâtre at Saint-Cloud in May she gave Elise Ma Belle, runner-up to Theia in the Prix Vanteaux, two kilos and a 2½-length thrashing as a result of which she was an extremely short-priced favourite for the Oaks at Epsom. Taking the lead after only three furlongs, Pawneese was never off the bit and trounced Roses for the Star and African Dancer 5 lengths and 4 lengths. Her stable-companion, Flying Water, had earlier won the 1000 Guineas and it had been intended to run her in the Prix de Diane, but she had suffered a setback. Flying Water's place was taken by Pawneese. Thus but nine days after her success at Epsom, Pawneese faced Madame Alec Head's unbeaten Luthier filly, Riverqueen, at Chantilly.

Riverqueen had won her sole race as a juvenile at Saint-Cloud the previous October. On 11 April she beat the subsequent 1000 Guineas third, Kesar Queen, 2½ lengths in the Prix de la Grotte. She next won the Poule d'Essai des Pouliches by 3 lengths from Suvannée. In the

Prix Saint-Alary on 23 May Riverqueen hit the front 300 metres out and was not unduly pressed to hold the challenge of the previously unbeaten Theia by three-quarters of a length. She therefore started favourite for the Prix de Diane. Pawneese led from the start and held a long lead over Riverqueen at the entrance to the straight. Madame Alec Head's filly tried in vain to get on terms with Pawneese throughout the final 400 metres but was still 1½ lengths adrift at the winning post. The pair were 3 lengths clear of Lagunette with Antrona, Sarah Siddons and Theia occupying the next three places.

Pawneese next took on the best in Europe in the King George VI and the Queen Elizabeth Diamond Stakes. Yves Saint-Martin again led from the outset and set a fast pace. He gave as masterly display of waiting in front as he had done on Match. Pawneese passed the post a comfortable length winner from the unlucky Bruni. Riverqueen had earlier more than redeemed her reputation when she easily beat Ashmore and Tip Moss half a length and 2 lengths in the Grand Prix de Saint-Cloud on 4 July with Beau Buck last. Madame Alec Head's filly thus became the first three-year-old filly ever to win the Grand Prix de Saint-Cloud. Three fillies had won it, Corrida, Banassa and Dahlia, but as four-year-olds.

It was assumed that the Prix Vermeille on 19 September would be a match between these two exceptionally good fillies, but both ran stones behind their best form. Pawneese was nearly 9 lengths behind Lagunette in seventh place with Riverqueen another 8 lengths away ninth. The going that afternoon was very soft, which some offered as an excuse for their abysmal display, but nonetheless it was the biggest surprise of the season, especially as it had been reported that Riverqueen had totally outpointed Ivanjica in a gallop a few days earlier. Some, however, had the impression that both the fillies had gone over the top and, if that were true, it was possible that they would not shine in the Prix de l'Arc de Triomphe. Nevertheless Pawneese and Crow started at 21 to 4 (couplé) on 3 October.

Riverqueen was also accompanied to the post in the 1976 Prix de l'Arc de Triomphe by an illustrious stable-companion, the four-year-old Ivanjica. She had been the best three-year-old filly in France in 1975 but had taken some time to find her form in 1976. She had only managed sixth place behind Liloy in the Prix d'Harcourt on 11 April. Although Ivanjica was a close third to Infra Green and Kasteel in the Prix Ganay on 2 May, she was beaten more than 5 lengths when seventh to Kasteel in the Prix Dollar four weeks later. In the Prix d'Ispahan on 27 June Ivanjica was defeated 2½ lengths when runner-up to Full of Hope. To some it appeared that she might not be as good as she had been the previous season, but in reality it was more likely that the distances of the Prix Dollar and Prix d'Ispahan were on the short side for her. Ivanjica was rested until the Prix du Prince d'Orange over ten furlongs on 19 September, when she ran out a most authoritative 4-length winner from Sea Sands with Trépan and Infra Green three-quarters of a length and a length away fourth and fifth. This impressive victory suggested that Ivanjica was back to the form that had won her the Prix Vermeille the previous year and, were that assumption true, she would represent a danger to all her rivals. She and Riverqueen started at odds of 71 to 10 on Sunday 3 October, and neither, on their best form, could be ruled out of contention.

There were two other three-year-old fillies in the field for the 1976 Prix de l'Arc de Triomphe, Marius Berghgracht's Lagunette and Lawrence Gelb's Floressa. Lagunette had won over a mile and a half at Longchamp on 27 May before finishing third in the Prix de Diane. On 11 July she comfortably beat Sarah Siddons 2 lengths in the Irish Guinness Oaks, with Floressa a well-beaten thirteenth. Sarah Siddons then won the Yorkshire Oaks, but neither she or Lagunette were given much chance again Pawneese and Riverqueen in the Prix Vermeille. In the event, Lagunette and Sarah Siddons, partnered for the first time by Lester Piggott, engaged in a terrific duel throughout the final 200 metres, with Lagunette getting the better of her rival in a very close finish by a nose, with Floressa a length away third. While victory for Lagunette on Sunday 3 October would not constitute a major surprise, she was only the third-best of her sex in France and her odds of 21 to 1 seemed to reflect her chance.

Floressa had won the Prix de Royaumont in June and had run Antrona and Ranimer, who was to win the Sun Chariot Stakes at Newmarket on 30 September, to 2 lengths and three-quarters of a length in the Prix de Malleret on 27 June. Nevertheless she had disappointed at the Curragh, and was only fifth to the deadheaters Sweet Rhapsody and Moquerie in the Prix de Pomone at Deauville. On balance, she looked more than a little out of her depth on Sunday 3

October and she started at odds of 99 to 1.

There were two other French three-year-old colts who might be considered possible candidates for the minor honours on their very best form, Arctic Tern and Twig Moss. Arctic Tern beat the Poule d'Essai des Poulains runner-up, Roan Star, by three-quarters of a length in the Prix de Fontainebleau on 4 April but in the classic itself on 25 April was only fifth to Red Lord and Roan Star. In the Prix Lupin Arctic Tern was runner-up to Youth but he disappointed in the Prix du Jockey Club and was only fifth in the Prix Eugène Adam. He did, however, run second to Youth in the Prix Niel; he ran much his best on very soft ground which he would find at Longchamp in the Prix de l'Arc de Triomphe. That was probably why his odds were as short as 23 to 1.

Twig Moss won his first race of the season over eight and a half furlongs at Cagnes-sur-Mer, and another over a mile and a quarter on the same course. He won the Prix Noailles on 25 April, but could only manage third place to Grandchant in the Prix Hocquart on 9 May. He had no chance with Youth in the Prix du Jockey Club but did well to hold off Malacate by three-quarters of a length for second place. In the Benson and Hedges Gold Cup Twig Moss was only fifth to Wollow and Crow, who was 9 lengths in front of him. Twig Moss started at 54 to 1 for the 1976 Prix de l'Arc de Triomphe and on balance that seemed the correct estimate of his chance.

The remaining three-year-old in the field on Sunday 3 October was Ravi Tikoo's Java Rajah, trained in France by Scobie Breasley. Although he had won a handicap at Deauville in August and the Prix Omnium Handicap at Longchamp on 12 September in very soft going, it was very difficult to enthuse about his chance against opposition of the calibre he would meet on Sunday 3 October. Confidence in Java Rajah was not increased when he was disqualified from first place in the Prix Omnium after traces of caffeine were revealed in his system.

The caffeine that caused most sensation in 1976 was that found in Trépan after his victory over Anne's Pretender in the Prince of Wales's Stakes at Royal Ascot on 15 June in a course record time. Prior to that victory he had won a handicap under topweight at Longchamp and his four victories in 1975 had included La Coupe de Maisons-Laffitte. The Jockey Club enquiry was not held until 12 August and on 3 July Trépan beat Wollow in the Joe Coral Eclipse Stakes once more in record time. He was again disqualified as theobromine, which had also been found in his system after the Ascot race, was discovered in his system after the race. His trainer François Boutin attributed the presence of caffeine and theobromine in Trépan's system after the Prince of Wales's Stakes to a diuretic drug, hepatorenal, which had been mistakenly given to him by a stable employee twenty-four hours before the race. François Boutin was fined £1000. He explained to the Jockey Club Enquiry that Trépan had received no medication between Royal Ascot and the Joe Coral Eclipse Stakes. It was agreed that some traces of theobromine could have remained in Trépan's system and Boutin was only fined £250 for the offence at Sandown Park. Although a lot of space was devoted in the English press to these two disqualifications, both were in reality but technical, if careless, breaches of the Rules of Racing comparable to the disqualification of Rock Roi from the Ascot Gold Cup. Trépan was certainly a very useful performer over nine and ten furlongs but had never run over a mile and a half before the 1976 Prix de l'Arc de Triomphe. He was last of seven to Wollow and Crow in the Benson and Hedges Gold Cup and only fourth to Ivanjica in the Prix du Prince d'Orange which made many feel he had gone over the top. His odds of 19 to 1 rather over-estimated his chance on 3 October.

Another four-year-old who was outstanding over a mile and a quarter but did not truly stay twelve furlongs was Baron de Zuylen de Nyevelt's Kasteel. After running fifth in the Prix d'Harcourt, Kasteel was most unlucky not to win the Prix Ganay on 3 May. He took the lead from the start and looked certain to win with 200 metres to go, when he came away from the rails, with the result that Infra Green came through to beat him a short head with Ivanjica and Rose Bowl a neck and half a length away third and fourth. When his jockey George Rivases was questioned by his trainer Jean-Michel de Choubersky as to why he did not keep a straight course, he replied that he had seen a patch of virgin ground away from the rails that he thought would prove faster. Kasteel then won the Prix Dollar by 2 lengths from El Rastro with Ivanjica only seventh on 30 May. He was totally unsuited by the firm ground when seventh to Wollow in the Joe Coral Eclipse Stakes at Sandown but ran the useful Larkhill, who was receiving 3 kilos, to 1½ lengths in the Prix Gontaut-Biron at Deauville. Kasteel finally gave On My Way 2 kilos and beat him a short head in the Prix Foy on 12 September with Beau Buck 4½ lengths away fourth. Kasteel, nevertheless, started at 64 to 1 because it was likely he would find a mile and a half too far for him.

Xavier Beau's On My Way had only run once in 1976 before to the Prix de l'Arc de Triomphe – in the Prix Foy. His previous race had been when third to Nobiliary and Comtesse de Loir in the Washington D.C. International Stakes in November 1975. To have run Kasteel so close in the Prix Foy had been an extremely promising performance. It was not surprising, therefore, that the runner-up to Star Appeal the previous season started at the short odds of 14 to 1.

Infra Green, the very fortunate winner of the Prix Ganay, had subsequently only managed fifth place to Quiet Fling in the Coronation Cup and she occupied the same place nearly 6 lengths behind Ivanjica in the Prix du Prince d'Orange. As a three-year-old the previous year she had won the Prix Chloë, the Prix de Mallaret and the Prix d'Astarté but it seemed likely she would be out of her depth in the Prix de l'Arc de Triomphe.

William Hawn's five-year-old Beau Buck won over twelve and a half furlongs at Saint-Cloud in April, and on 27 May beat the useful Luengé 1½ lengths in La Coupe at Longchamp. He was unsuited by the good going when last in the Grand Prix de Saint-Cloud but was only fourth in the Prix Foy. The heavy going he would encounter at Longchamp on 3 October would suit Beau Buck; his odds of 56 to 1 was probably a fair estimate of his chance.

The Final Preparations

The weights for the Prix de l'Arc de Triomphe were altered in 1976 in favour of the older horses. Previously four-year-olds and upwards had to carry 60 kilos and three-year-olds 55½ kilos, with mares and fillies being allowed 1½ kilos. In 1976 four-year-olds and upwards had to carry 59 kilos and the three-year-olds 56 kilos, with mares and fillies still being allowed 1½ kilos. Thus the three-year-olds were one and a half kilos worse off with their older rivals.

Bruni, whom Lester Piggott chose to ride in preference to Youth, travelled to Paris by sea and horsebox. His trainer, Ryan Price, was of the opinion that Charles St George's grey was still an immature horse when only seventh to Star Appeal. He was adamant before the 1976 Prix de l'Arc de Triomphe that Bruni had never been better. He added that the only thing that he feared was that the ground might become very heavy which would not suit such a good actioned horse as Bruni.

On Friday 1 October Yves Saint Martin rode both Crow and Pawneese in their final workouts on the sand gallops of 'la route des Lions' at Chantilly. After a long and serious discussion with Angel Penna, he chose to ride Crow who he felt was improving and would be better suited by the heavy going likely to be encountered on Sunday 3 October. The mount on Pawneese, therefore, went to Pat Eddery, whose intended mount Orange Bay was scratched four days before the race on account of the going. The mount on Youth went to the Australian Bill Pyers, who had ridden him in all his final gallops at home in 1976 without partnering him on the day. His trainer Maurice Zilber reported that Youth was 'in splendid condition' and that he too had been working very well. Gerard Dubroeucq, the jockey on Nelson Bunker Hunt's other classic colt, Exceller, was also confident that his mount would go very well, placing a lot of faith on his undoubted stamina, but there were several who felt Exceller would not be suited by the heavy going. Freddie Head chose to ride Ivanjica instead of Riverqueen who was to be partnered by R. Jallu. His father, Alec Head, was certain Ivanjica had come back to her best. He added that 'the heavy going would suit her perfectly' and believed that 'she possessed an excellent chance'. It was also reported that Trépan had done a brilliant gallop on his last serious workout.

The winners of both the Epsom Derby, Empery, and the Irish Sweeps Derby, Malacate, were absentees from the Prix de l'Arc de Triomphe in 1976. Empery and Youth had been syndicated together for $6 million before the end of June. After his defeat by Malacate, Empery did not run again as a result of an injury on the gallops. François Boutin decided to miss another clash with Youth and chose to send Malacate back to Ireland for the Joe McGrath Memorial Stakes at Leopardstown on 25 September when the Irish Sweeps Derby winner beat Mart Lane without difficulty. His charge in the Prix de l'Arc de Triomphe, Lagunette, was put up for sale on the evening before the race but was led out of the sale ring 200,000 francs short of her reserve of 3 million francs.

The Race

The afternoon began with Gentilhombre and Mendip Man deadheating for the Prix de l'Abbaye de Longchamp a head and a neck in front of Raga Navarro and Girl Friend. The British trainer Neil Adam protested to the Stewards at the Judge's verdict but on the five-furlong course there was no mirror to produce a reflected photo-finish image as the sprint course is too wide for a mirror to pick up an image. The Stewards therefore upheld the decision of the Judge since Mendip Man's head was virtually hidden by that of Gentilhombre.

The crowd did not include Président Giscard d'Estaing who was to leave the following morning for an official visit to Iran. There was, however, the usual cosmopolitan gathering, including a huge number of English who even numbered among them their former heavyweight boxing champion Henry Cooper. The Mexican film-star Maria-Felix Berger looked notably relieved that her Rolls-Royce, which had disappeared from the car park for an hour the previous day while unguarded by her chauffeur, had been found by the police fifty metres from where it had been parked.

Alec Head's stable had a boost to their confidence in Ivanjica and Riverqueen when Jacques Wertheimer's Fashion Lad, partnered by Freddie Head, won the second race, the Prix des Champs-Elysées, by a nose from the Zilber-trained Imogène. The Critérium des Pouliches went to the 30 to 1 outsider Kamicia by a very easy 3 lengths from Doha.

In the paddock both Youth and Ivanjica appeared particularly well and were certainly the pick. Bruni was quiet and relaxed but Pawneese seemed to have run up a bit light and gave the impression that she might have gone over the top. Noble Dancer, the Norwegian challenger, was extremely fractious and as soon as Geoff Lewis mounted him, he deposited him on the ground. Fortunately the colt did not get loose and Lewis was soon reunited with his mount.

At the start Dakota was troublesome and most reluctant to enter the stalls. When they opened it was with a great sigh of relief that all the supporters of Bruni saw that the elegant grey got away on level terms with his rivals. Kasteel immediately took up the running from Pawneese and this pair quickly established a clear lead over Youth, Crow and Bruni whom Piggott unaccustomedly kept up with the leaders. They were followed by a group consisting of Trépan, Floressa and Infra Green, tracked by Beau Buck and Exceller. At this stage Ivanjica, who, with Dakota, was last away, tracked across to the rails and this pair was last a little behind Java Rajah, Riverqueen and Noble Dancer. Approaching the top of the hill, Kasteel and Pawneese had drawn almost 12 lengths clear of Youth, Crow and Bruni. Pawneese took over the lead from Kasteel at the beginning of the descent and the chasing group of Youth, Crow and Bruni gradually began to close the gap between them and the leaders. Freddie Head still had Ivanjica on the rails amongst the backmarkers. Noble Dancer and Duke of Marmalade began to improve their positions coming down the hill but Exceller, Lagunette, Riverqueen and, surprisingly, Arctic Tern were all struggling in the heavy going before the straight was reached. Pawneese was still just in front entering the straight but quickly weakened and gave way briefly again to Kasteel. He quickly surrendered the lead to Youth with Crow and Bruni poised to challenge him at his quarters. Ivanjica meanwhile had had a dream of a run through on the rails to take eighth place with 400 metres to go whilst Noble Dancer and Infra Green were also on the heels of the leaders and Beau Buck was beginning a run on the outside. Youth led from Crow and Bruni 300 metres out when Saint-Martin unleashed the St Leger winner who quickly went a length clear of Youth and crossed over to the rails. It was clear by then that Bruni was beaten and Crow looked certain to chalk up another success for Daniel Wildenstein. With only 250 metres to go Freddie Head smoothly put Ivanjica through a gap between the weakening Infra Green and Kasteel and accelerating superbly quickly passed first Bruni and then Youth. Meanwhile the Norwegian challenger, Noble Dancer, looked full of running but was boxed in on the rails behind the weakening Pawneese. Ivanjica then unleased a scintillating burst of speed to catch Crow and flash past him. So fast was she travelling that she had 2 lengths to spare over Daniel Wildenstein's colt who was 3 lengths ahead of the one-paced Youth. The latter was very fortunate to hold on to third place by a short

neck from Noble Dancer, for Lewis had to switch his mount from the rails to the outside 200 metres out to avoid running into Youth; the Norwegian challenger was finishing the fastest of all with the exception of the winner. Bruni ran on at one pace to finish fifth, 1½ lengths behind Noble Dancer, and half a length in front of Beau Buck in sixth place. They were followed by Infra Green, Java Rajah, On My Way, Kasteel and Pawneese.

The drama of the 1976 Prix de l'Arc de Triomphe was not over, however. A smiling and jubilant Freddie Head was waving with his right hand to the crowd in the stand on his return to the unsaddling enclosure when Ivanjica unceremoniously deposited him on the track when she shied at some photographers. She galloped away alone down the course and collided with a plastic rail before she was caught. Fortunately she only suffered superficial cuts but there was considerable delay before she and her jockey were re-united and returned to the unsaddling enclosure. For a few minutes there was some anxiety that Head would not weigh in within the statutory twenty minutes after the finish of the race, which would have entailed automatic disqualification. Happily he made the time limit.

For Freddie Head it was his third victory in the Prix de l'Arc de Triomphe. He rated Ivanjica better than either Bon Mot or San San and 'one of the best I've ever ridden, certainly in the Pistol

Ivanjica (12) has 2 lengths and 3 lengths to spare over Crow (16) and Youth (13) with Noble Dancer (6) fourth, Bruni (grey) fifth, Beau Buck (3) sixth and Infra Green (9) seventh

Packer class'. He added, 'I knew we had won by the middle of the straight.' It was also a third success for his father, Alec Head, for he had trained Nuccio and Saint Crespin to win in 1952 and 1959. Alec Head immediately confirmed that Ivanjica would run in the Washington D.C. International Stakes.

There were in reality no hard-luck stories for the vanquished with the possible exception of Noble Dancer, whose jockey Geoff Lewis said: 'We ran up the backside of Youth, otherwise we could have finished third or perhaps even second.' Lester Piggott reported 'Bruni just wasn't good enough. I had to make too much use of him early on to obtain a prominent position.' Bill Pyers, Youth's rider, said 'I came to win my race but Youth floundered in the mud. With better ground it might have been another story.' Pat Eddery also stated that Pawneese did not handle the ground and that 'as soon as I let her down she had gone'. However, her trainer, Angel Penna, was delighted with Crow's performance.

Considering how long his family had been associated with French racing, it was surprising that Jacques Wertheimer was its first member to win the Prix de l'Arc de Triomphe. Pierre Wertheimer, who died in 1965, had owned such excellent horses as Epinard, Mesa the 1000 Guineas winner in 1935, Lavandin the Derby hero in 1956, Vimy who won the King George VI and the Queen Elizabeth Stakes in 1955, and Chingacgook, victorious in the 1955 Grand Prix de Saint-Cloud. After his death, his widow Madame Pierre Wertheimer won the Poule d'Essai des Poulains with Riverman in 1972, the Prix du Jockey Club with Roi Lear in 1973, the Prix Royal Oak with Bourbon in 1971, the Grand Critérium with Satingo in 1971 and the Poule d'Essai des Poulains and Prix Lupin in 1975 with Green Dancer. Since the early 1950s Alec Head had trained all these winners. He had begun training in 1947 and from that time Alec Head had sent out a continuous stream of classic winners for the Aga Khan and Prince Aly Khan until the early 1960s and for the Wertheimer family on both sides of the Channel. During that thirty-year period only François Mathet and Etienne Pollet could rival his record. This success was not unexpected as the Head family have been training in France for nearly a century. As a teenager Alec Head was sent to school near Slough in England. In May he thought the French raider, Hunter's Moon, was sure to win and tipped him to his headmaster, who won £50. As a result he took the young Alec Head to Epsom to see Bois Roussel win the 1938 Derby.

The Rest of the Season

Ivanjica was accompanied to Laurel Park by Youth, Noble Dancer and On My Way. Youth had already won in North America; he had beaten Improviser and Effervescing 4 lengths and 7 lengths in the Canadian International Stakes at Woodbine Park on 23 October. Effervescing had won the Man o'War Stakes, so it had been a very impressive performance. The course at Laurel Park did not suit Ivanjica on account of its very short straight of only one and a half furlongs. Youth beat On My Way and the Prix de l'Arc de Triomphe heroine 10 lengths and a nose with Noble Dancer fourth and Rose Bowl fifth. When Youth took the lead, he cut very sharply across On My Way who was forced to check. The stewards, however, permitted the result to stand after careful perusal of the film. Ivanjica was certainly not disgraced. It is surprising that in fact no horse who has won the Prix de l'Arc de Triomphe has ever succeeded at Laurel Park, but that is almost certainly due to the tightness of the track which is not totally suitable for such an important middle distance race.

On My Way had earlier won the Prix du Conseil de Paris by 1½ lengths from Beau Buck with Arctic Tern fourth and Noble Dancer seventh on 17 October. The same afternoon at San Siro, Infra Green beat Sierra Morena and Duke of Marmalade 1½ lengths and a head in the Gran Premio del Jockey Club. Duke of Marmalade later won the Premio Roma. As normal, therefore, the form of the Prix de l'Arc de Triomphe was upheld by the results in the late autumn. Crow, however, had not looked well in the Champion Stakes on 13 October and along with Wollow was unplaced to Vitiges. The Champion Stakes almost invariably comes too soon after the Prix de l'Arc de Triomphe for any horse who has been concerned in the finish at Longchamp.

Pedigree and Stud Record

Ivanjica was bred in the United States by the Claiborne Farm, and was bought as yearling for $180,000. Her sire was Sir Ivor whose brilliance on the racecourse has already been related. He has been almost as successful as a sire with the one proviso that in general his fillies are much better than his colts. His best progeny include Godetia (Irish Guinness Oaks), Lady Capulet (Irish 1000 Guineas), Sweet Alliance (Kentucky Oaks), Optimistic Gal (Frizette Stakes and Selima Stakes), Equanimity (Fantasy Stakes), Cloonara (Phoenix Stakes), Malinowski (Craven Stakes), Cavo Doro (2nd Epsom Derby), and such good horses as Land Girl, Fascinating Girl, Sir Ivor Again, Golden Reserve, Sir Penfro, Padroug, Ercolano, Realty, Flashy, Val's Girl and Istiea, to mention but a few. In North America up to September 1982 his offspring had won 334 races worth $5,865,702; in England and Ireland in the same period they had won 149 races worth £708,137; and in France 70 races worth 6,924,575 francs.

Ivanjica's dam, Astuce by Vieux Manoir, was exported to the United States from France in 1968. She was trained by John Cunnington Jr who also had the best three-year-old filly in France in 1967, Gazala, in his yard. As a juvenile Astuce won once, was second in the Prix Saint-Roman and was given 8 stone in the Handicap Optional, 18 lbs behind the best filly in 1966, Silver Cloud. Astuce did not win as a three-year-old but ran in the highest class. She was third to Tidra and Fix a Date in the Prix Saint-Alary, fourth to Lionel in the Grand Prix de Deauville, and also fourth to Modeste in the Prix de Pomone on the same course. She was no match for her stable-companion Gazala in the Prix de Diane, but was far from disgraced when seventh of nineteen to Casaque Grise in the Prix Vermeille. Ivanjica was Astuce's fourth foal; her first a colt by Bold Ruler, Lawgiver, won as a three-year-old in the U.S.A. Her next two, both colts by Sir Gaylord, Sir Smarts and the unraced Sir Arthur, were nothing exceptional; Sir Smarts was unable even to reach the frame as a three-year-old in Ireland but won a hurdle race at the age of four. Nevertheless Astuce ensured her success as a broodmare when she foaled Ivanjica in 1972. Her colt, Kentucky Frenchman by Le Fabuleux, won two races as a juvenile in the U.S.A. in 1976.

Astuce's dam, Ashleen by Alizier, won two races worth 12,260 francs but her fame really rests with her progeny. She foaled four winners the best of which unquestionably was the filly Princeline by Princely Gift who won the Prix Morny in 1968 and was runner-up to Koblenza in the Poule d'Essai des Pouliches and also was second in the Prix Jean Prat.

Ashleen's dam was the very useful racemare Asheratt by Sunny Trace. As a two-year-old Asheratt won the Prix de la Salamandre, was runner-up in the Prix de la Forêt, was third to Ad Astra in the Prix Morny, and was only beaten a short head, a head and a neck by Gossip, Cillas and Blue Star in the Grand Critérium. In the 1937 Handicap Optional she was given 59 kilos, only 3 kilos behind the top-weight Cillas. As a three-year-old Asheratt's victories included the Prix Vanteaux and the Prix la Rochette. She was also third to Victrix and Dadji in the Prix du Président de la Republique. In all Asheratt won five races worth 225,050 francs. Inevitably the war interrupted Asheratt's career as a broodmare. Including Ashleen she produced five winners of nine races on the flat and two hurdle races, the best of which was the Brantôme filly Ash Tray.

Asheratt's dam, Fée Esterel by Cadum, won one race as a two-year-old and two as a four-year-old worth 67,620 francs. Her best produce as a broodmare was unquestionably Asheratt, but she also got two winners on the flat and Fehlheimer by Téléférique who won two hurdle races and five steeplechases between 1949 and 1951 worth 1,419,525 francs.

Fée Esterel's dam, Reine Mab, was a half-sister to Sly Fox by Flying Fox, the winner of the Prix la Rochette (1000 metres) and to Saint Just by St Frusquin who won the Prix de Condé. She was a prolific and most influential broodmare producing eleven winners of fifty-two races. Her daughter Orianne by Sans Souci, was the dam of Reine Isaure by Blandford, successful in the Prix Vanteaux and runner-up to Mary Tudor in the Prix Vermeille in 1934 and the dam of Cranach by Coronach, the very useful German colt and later a good stallion. Another of Reine Mab's daughters, Brocéliande by La Farina, won seven races and was the dam of Maurépas by Athelstan, the winner of the Grand Prix de Paris in 1940 and also of the Prix du Cadran, Grand Prix de Saint-Cloud and the Prix des Sablons in 1942. Reine Mab was also the ancestress of Dictaway (Poule d'Essai des Pouliches 1955 and dam of Diatome), Timandra (Prix de Diane and Poule d'Essai des

Pouliches 1960), and Dragon Blanc (Grand Critérium 1952).

Reine Mab's dam, Justitia by Le Sancy, won the Prix Morny and was a half-sister to the Prix de la Forêt winner Chalet. The family eventually traced to May Bell. Although Ivanjica was foaled in the United States, she came from a very well-established French family and it was very appropriate that she should gain her brilliant victories in France where her dam had been born and raced.

Ivanjica was sold at Keeneland in November 1981 for $2,150,000, then a world record price for a broodmare. It was broken the following year at the Keeneland November sales when Royal Honorée was sold for $3,800,000.

Her only foal up to 1983 has been Figure Libre by Northern Dancer. She won the Prix de la Brie (1800 metres) at Evry in April 1982 and the Prix de Bois Josselyn (2100 metres) the following December. She also ran third in the Prix de Meudon (2000 metres) and fourth in the Prix de Saint-James (2000 metres). Figure Libre won a total of 145,200 francs as a three-year-old, nowhere near the 1,533,800 francs that her much more illustrious dam amassed during her career on the racecourse.

PRIX DE L'ARC DE TRIOMPHE 2400 metres
at Longchamp on Sunday 3 October 1976. For three-year-olds and upwards. Going: Heavy.
1,200,000 francs to the winner; 480,000 fr. to the 2nd; 240,000 fr. to the 3rd; 120,000 fr. to the 4th. 300,000 fr. to the breeder of the winner; 120,000 fr. to that of the 2nd; 60,000 fr. to that of the 3rd; 30,000 fr. to that of the 4th. Value to the winner: 1,200,000 fr.

IVANJICA	b.f.	by Sir Ivor – Astuce by Vieux Manoir	4.57½k	Jacques Wertheimer	F. Head	1
CROW	ch.c.	by Exbury – Carmosina by Right of Way	3.56k	Daniel Wildenstein	Y. Saint-Martin	2
YOUTH	b.c.	by Ack Ack – Gazala by Dark Star	3.56k	Nelson Bunker Hunt	W.B. Pyers	3
NOBLE DANCER	b.c.	by Prince de Galles – Helen Traubel by Sing Sing	4.59k	Gunnar Schjelderup	G. Lewis	4
BRUNI	gr.c.	by Sea Hawk – Bombazine by Shantung	4.59k	Charles St George	L. Piggott	5
BEAU BUCK	ch.c.	by Buckpasser – Blue Norther by Windy City II	5.59k	William R. Hawn	J.C. Desaint	6
INFRA GREEN	ch.f.	by Laser Light – Greenback by Fric	4.57½k	Mme John Pochna	M. Depalmas	7
JAVA RAJAH	ch.c.	by Jacinto – Sabana by Bryan G.	3.56k	Ravi N. Tikko	F. Durr	8
ON MY WAY	b.c.	by Laugh Aloud – Gracious Me by Tulyar	6.59k	Xavier Beau	A. Gibert	9
KASTEEL	b.c.	by King of the Castle – Esme by Le Haar	4.59k	Baron de Zuylen de Nyevelt	G. Rivases	10
DUKE OF MARMALADE		by Vaguely Noble – Mock Orange by Dedicate	5.59k	Mme Bruno Bucci	G. Doleuze	
DAKOTA		by Stupendous – Ardneasken by Right Royal	5.59k	Guy Reed	A. Barclay	
TREPAN		by Breakspear II – Quiriquina by Molvedo	4.59k	Rodolph Schafer	J. Mercer	
TWIG MOSS		by Luthier – Top Twig by High Perch	3.56k	Richard Brooke	A. Murray	
EXCELLER		by Vaguely Noble – Too Bald by Bald Eagle	3.56k	Nelson Bunker Hunt	G. Dubroeucq	
ARCTIC TERN		by Sea Bird – Bubbling Beauty by Hasty Road	3.56k	Mme John S. Knight	M. Philipperon	
PAWNEESE		by Carvin – Plencia by Le Haar	3.54½k	Daniel Wildenstein	P. Eddery	
LAGUNETTE		by Val de Loir – Landerinette by Sicambre	3.54½k	Marius Berghgracht	A. Lequeux	
FLORESSA		by Sassafras – Florentia by Relic	3.54½k	Lawrence M. Gelb	J.P. Lefèvre	
RIVERQUEEN		by Luthier – Riverside by Sheshoon	3.54½k	Mme Alec Head	R. Jallu	

20 runners.
IVANJICA trained by A. Head at Chantilly; bred by Claiborne Farm in the U.S.A.
Distances: 2 lengths; 3 lengths; short neck; 1½ lengths; ½ length; ¾ length; 5 lengths; ¾ length; head. Time: 2 mins 39.4 seconds.
Pari-Mutuel: Win: 8fr. 10. Places: 3fr. 30; 2fr. 90; 2fr. 30.
S.P.: 2 to 1 Youth and Exceller (couplé); 5 to 2 Bruni; 21 to 4 Crow and Pawnesse (couplé); 71 to 10 IVANJICA and Riverqueen (couplé); 14 to 1 On My Way; 19 to 1 Trépan; 21 to 1 Lagunette; 23 to 1 Noble Dancer and Arctic Tern; 54 to 1 Twig Moss; 56 to 1 Beau Buck; 64 to 1 Kasteel; 66 to 1 Duke of Marmalade; 75 to 1 Java Rajah; 99 to 1 Dakota; Floressa and Infra Green.

IVANJICA bay filly 1972	Sir Ivor b. 1965	Sir Gaylord b. 1959	Turn-to b. 1951	Royal Charger ch. 1942	Nearco
					Sun Princess
				Source Sucrée b. 1940	Admiral Drake
					Lavendula
			Something Royal b. 1952	Princequillo b. 1940	Prince Rose
					Cosquilla
				Imperatrice b. 1938	Caruso
					Cinquepace
		Attica ch. 1950	Mr Trouble ch. 1947	Mahmoud gr. 1933	Blenheim
					Mah Mahal
				Motto br. 1932	Sir Gallahad
					Maxima
			Athenia b. 1943	Pharamond br. 1925	Phalaris
					Selene
				Salaminia ch. 1937	Man o' War
					Alcibiades
	Astuce b. 1964	Vieux Manoir b. 1947	Brantôme b. 1931	Blandford br. 1919	Swynford
					Blanche
				Vitamine b. 1924	Clarissimus
					Viridiflora
			Vieille Maison b. 1936	Finglas b. 1923	Brûleur
					Fair Simone
				Vieille Canaille b. 1930	Zionist
					Ficelle
		Ashleen b. 1947	Alizier b. 1936	Téléferique ch. 1934	Bactériophage
					Beauté de Neige
				Alizarine b. 1939	Coronach
					Armoise
			Asheratt b. 1935	Sunny Trace b. 1925	Abbot's Trace
					Sunny Moya
				Fée Esterel b. 1929	Cadum
					Reine Mab

Alleged 1977

The Field

There were nine foreign challengers for the 1977 Prix de l'Arc de Triomphe, for which there was a field of twenty-six runners. Alleged, Sarah Siddons and Panamint were trained in Ireland; Her Majesty the Queen's filly Dunfermline, Orange Bay and Balmerino were from English stables, though Orange Bay had raced as a juvenile and as a three-year-old in Italy, whilst the majority of Balmerino's racing had been in his native New Zealand and Australia. The mare Vivi came from West Germany, Cunning Trick from the United States, and Mia was the first-ever challenger from Argentina. The favourite at odds of 38 to 10 was Robert Sangster's Hoist the Flag colt Alleged, whilst Dunfermline and Orange Bay started co-second favourites with Crystal Palace at 8 to 1.

Alleged had won his only race as a juvenile at a seven-furlong event at the Curragh in November. On his seasonal début he easily won the Baldoyle Stakes (10f) at Leopardstown on 20 April. He was the total outsider for the Royal Whip Stakes at the Curragh on 13 May for which his stable-companion Valinsky, partnered by Lester Piggott, was odds-on favourite. To general amazement Alleged beat Valinsky by a length. The Hoist the Flag colt showed his victory was no fluke when he won the Gallinule Stakes by a length from Orchestra at the Curragh at the end of May. Alleged was rested until the Great Voltigeur Stakes and put up a superb performance to demolish the Irish Sweeps Derby third, Classic Example, by an effortless 7 lengths having led from the start. It was certainly the most impressive display by a three-year-old up to that point in the season. Robert Sangster's colt appeared to have the St Leger at his mercy; the only conceivable danger was Her Majesty the Queen's Dunfermline.

In the St Leger Dunfermline had a pacemaker Gregarious who set a strong pace to the turn with Alleged close behind. The pacemaker, however, prevented Piggott on Alleged from waiting in front and dictating the pace. The Hoist the Flag colt took the lead early in the straight but Dunfermline loomed up menacingly two furlongs out, and in a terrific struggle to the line she eventually outstayed Alleged and won by 1½ lengths, the pair finishing 10 lengths clear of their nearest pursuer, Classic Example. Most people felt that the Irish-trained colt would reverse the form in the Prix de l'Arc de Triomphe over the shorter distance of a mile and a half.

Dunfermline was by the hitherto disappointing sire Royal Palace out of the St Paddy mare Strathcona. Her dam was Stroma by Luminary whom the Queen had personally decided to purchase; she had cost 1150 guineas at the 1956 Doncaster Sales. Although she did not win as a two-year-old in three outings Dunfermline ran with immense promise and finished runner-up to Triple First in the May Hill Stakes at Doncaster in September and she was also second in the Argos Star Stakes at Ascot. At Doncaster she had looked immature in the paddock, so her performance was even more meritorious. In her first race as a three-year-old Dunfermline had 4 lengths to spare over the subsequent Irish Guinness Oaks winner Olwyn in the Pretty Polly Stakes (10f) at Newmarket in April. Her task in the Oaks was made easier when the favourite Durtal, who had deadheated behind the brilliant Madelia in the Poule d'Essai des Pouliches, and had been the best filly in England in 1976, had to be withdrawn. Her saddle slipped in front of the stands as Lester Piggott began to canter to the start. Durtal panicked and Piggott's foot was caught in a stirrup; he was dragged over a hundred yards before getting free. Durtal, however, injured herself so badly that she only ran once again – disappointingly. Dunfermline came from behind in the Oaks to join Freeze the Secret at the furlong pole and ran most gamely to win by three-quarters of a length with Vaguely Deb 3 lengths away third. This victory very appropriately came at the end of the week which marked the Silver Jubilee celebrations of the Queen's accession to the throne. In the Yorkshire Oaks in August Dunfermline was easily beaten by Busaca and Royal Hive. It was, however, no fault of hers; the race was run at a farcical pace for the first six furlongs, which did not suit the Oaks winner at all. It was partly for this reason that she had Gregarious as a pacemaker in the St Leger. Although stamina was Dunfermline's strongest suit, she had won the Oaks and over ten furlongs in the spring. It was by no means certain that Alleged would prove her superior at Longchamp over a mile and a half. Her trainer, Dick Hern, decided not to run a pacemaker for Dunfermline in the Prix de l'Arc de Triomphe as he knew that the race was always run at a fast pace.

Dr Carlo Vittadini's Orange Bay had won

the Premio Presidente della Repubblica and the Derby Italiano in 1975, before he joined Peter Walwyn's stable in England. As a four-year-old he won the Jockey Club Stakes and the Hardwicke Stakes and was third to Pawneese and Bruni in the King George VI and the Queen Elizabeth Diamond Stakes. In 1977 Orange Bay won the Aston Park Stakes at Newbury but finished an abysmal last to Meneval in the Hardwicke Stakes. He was blinkered for the first time in the King George VI and the Queen Elizabeth Diamond Stakes. He was headed by The Minstrel, the winner of both the Derby and the Irish Sweeps Derby, inside the final furlong, but fought back with such courage and tenacity that he only failed by a short head with Exceller, the winner of the Coronation Cup, third, Crystal Palace fourth and Crow unplaced. When 4 lengths and 2 lengths third to Relkino and Artaius in the Benson and Hedges Gold Cup, Orange Bay found the ten furlongs too sharp for him. It was another occasion when the Benson and Hedges Gold Cup turned the form book upside down; in the Sussex Stakes at Goodwood Artaius had beaten Relkino by 7½ lengths. Sarah Siddons and Malacate were fourth and fifth three-quarters of a length and 3 lengths behind Orange Bay. In his final outing before travelling to Paris Orange Bay slammed Meneval by 4 lengths in the Cumberland Lodge Stakes at Ascot. Although he was co-second favourite at 8 to 1 he was reported to be a bad traveller and on the really big occasion normally found one or two rivals too fast for him.

The other English-trained challenger, R.K. Stuart's Balmerino, in fact looked a greater danger to Alleged and Dunfermline. In the 1975/1976 season in New Zealand and Australia Balmerino had been the outstanding three-year-old, winning fourteen of his eighteen races including the Brisbane Cup (16f) and the Rawson Stakes (9f) and the New Zealand Derby and 2000 Guineas. In the 1976/7 season he won four races from nine starts in Australia and was then sent in early 1977 to race in California, where he won an eleven-furlong allowances race but was unplaced in three other valuable handicaps. This was almost certainly due to the fact that he was still becoming acclimatised. He arrived at John Dunlop's Arundel stables in the late summer and immediately created a very big impression on the gallops. Indeed Lavinia Duchess of Norfolk went so far as to say she had never seen a horse go better on the Downs near Arundel, and that he made 'the others look like goats'. He had only one race in Europe in the Valdoe Stakes at Goodwood in September before travelling to Longchamp. He produced a scintillating burst of acceleration to slam Lucent 5 lengths in a most authoritative fashion. Lucent herself was no slouch; she had earlier won the Lingfield Oaks Trial Stakes but she was totally outpointed by Balmerino. He started at 12 to 1 and looked very good value at those odds on Sunday 2 October.

Another challenger from the southern hemisphere was the three-year-old Mia from Argentina who was allowed 2 kilos as she was so immature compared to her contemporaries in Europe. Mia had been the best two-year-old filly in Argentina and won the Clasico Eliseo Ramirez and the Clasico Jorge de Atucha, the equivalent of the Critérium des Pouliches. As a three-year-old, Mia won the Pola de Potrancas and the Argentine 1000 Guineas, and her victory in the Gran Premio Seleccion was spectacularly effortless. It was a very brave venture of the Ecurie Los Sauces to send Mia to Paris, but it seemed likely that the age difference and lack of time to acclimatise would make her chance rather remote.

It was difficult to enthuse about the chances of any of the other four foreign challengers. The German racemare Vivi had been a useful three-year-old filly in 1975 but was rated 4½ kilos behind the best of her generation in Germany, Princess Eboli. As a four-year-old and five-year-old she had not made any significant improvement and her chance at Longchamp was little better than Mia's. Sarah Siddons had been the best three-year-old filly in England and Ireland in 1976 winning the Irish 1000 Guineas and the Yorkshire Oaks and finishing runner-up to Lagunette in both the Prix Vermeille and the Irish Guinness Oaks. She had been unplaced to Arctic Tern in the Prix Ganay in April and she was only fourth to Relkino in the Benson and Hedges Gold Cup. There was a strong suspicion that Sarah Siddons was not such a good four-year-old as she had been a three-year-old, although she had been a close third in the Gladness Stakes in April. The Irish three-year-old colt Panamint did beat Transworld a length in the Blandford Stakes at the Curragh in August, but was only eighth to him in the Irish St Leger. Earlier in the spring he had been a well-beaten third to Alleged and Valinsky in the Royal Whip Stakes, so his odds of 99 to 1 were a fair reflection of his chance. The American challenger Cunning Trick won the Gallant Fox Handicap and the Arlington Handicap, run on turf over twelve furlongs, on 5 September. Nevertheless he was some way behind the best in the United States and seemed likely to find several of his opponents too good for him at Longchamp.

The French defenders against this very strong foreign challenge for the 1977 Prix de l'Arc de Triomphe did not seem as formidable as normal. Baron Guy de Rothschild's Crystal Palace won the Prix de Courcelles by 2 lengths from Montcontour on 11 April. He then ran Pharly to three-quarters of a length in the Prix Lupin with Carwhite and Amyntor fourth and sixth at Longchamp on 15 May. Pharly had run second to Blushing Groom in the Poule d'Essai des Poulains, so this defeat was no disgrace. He then had half a length to spare over Artaius in the Prix du Jockey Club with Carwhite, Guadanini and Pharly, who did not stay, all unplaced. Since his dam was the 1971 Prix de Diane winner Hermières, Crystal Palace clearly was better suited by twelve furlongs than the ten and a half furlongs of the Prix Lupin. Remarkably it was the first victory for the famous blue and yellow Rothschild colours in the Prix du Jockey Club since that of Sardanaple in 1914. Crystal Palace then ran fourth, 4½ lengths behind The Minstrel and Orange Bay at Ascot. He was in fact given far too much to do in the straight and might well have finished closer had his jockey ridden a more judicious race. In the Prix Niel on 11 September, Crystal Palace gave Paico and Vagaries 3 kilos and impressively beat them 2½ lengths and three-quarters of a length with Amyntor only fifth. If Baron Guy de Rothschild's colt's performance at Ascot were ignored, it seemed certain that Crystal Palace would be involved in the finish on Sunday 2 October.

Jacques Wertheimer's Carwhite won the Prix Daru at Longchamp on 17 April but was 1¾ lengths behind Crystal Palace when fourth to Pharly in the Prix Lupin. Though only seventh to Baron Guy de Rothschild's grey colt in the Prix du Jockey Club, Carwhite was merely 2½ lengths behind him. After finishing runner-up to Casaque in the Prix Eugène Adam, Carwhite beat Gairloch and Waya a head and 2 lengths in the Prix du Prince d'Orange on 18 September, with Montcontour unplaced. It was a reasonable trial for 2 October but the best Carwhite could hope for was one of the minor places.

When Monseigneur beat the much-vaunted General in the Prix de Suresnes at Longchamp on 1 May, his trainer François Boutin decided to run him at Epsom where he finished fourth to The Minstrel, Hot Grove and Blushing Groom. Monseigneur was only fifth to The Minstrel in the Irish Sweeps Derby on 25 June. He then was third to Gairloch in the Prix de la Côte Normande at Deauville and was again third to Pharly and River Dane in the Prix du Rond Point at Longchamp on 4 September. The week before the Prix de l'Arc de Triomphe Monseigneur was 1½ lengths runner-up to Pharly in the Prix du Moulin de Longchamp. It was his best effort since the Derby and at odds of 78 to 1 he was perhaps the best outsider on Sunday 2 October.

Montcontour, the runner-up to Crystal Palace in the Prix de Courcelles in April, beat Guadanini 1½ lengths in the Prix Hocquart on 8 May, but was off the course due to illness till running unplaced in the Prix du Prince d'Orange and it was difficult to envisage him taking an active part in the finish on Sunday 2 October. The same applied to Guadanini who had won the Prix Yong-Lo at Saint-Cloud in April and Prix Brisecoeur on the same course in July. He had defeated Sissoo and Shafaraz in the Grand Prix de Vichy, but was well beaten by Crystal Palace at Chantilly and only fifth to Funny Hobby and Valinsky in the Grand Prix de Paris with Dom Alaric sixth.

Dom Alaric was owned by Jean-Paul Van Gysel and trained by François Mathet until after his three-quarters of a length victory over Midshipman, third in the Grand Prix de Paris, in the Grand Prix de Deauville. He was then bought by Mme Daniel Malingue and transferred to the stable of Olivier Douieb. Although he had won the Prix du Raincy (10f) at Evry in April and the Prix du Moulin-à-Vent (12f) at Chantilly in July, and had been runner-up in both the Prix de l'Espérance and Prix de Menneval, Dom Alaric looked as if he would be outclassed on Sunday 2 October. The same such fate was probable for Yelpana, the winner of the Prix Berteux at Chantilly in July and runner-up to Solaro in the Prix Kergorlay.

Marcel Boussac's Amyntor had been 4 lengths second to Blushing Groom a head in front of J. O. Tobin in the 1976 Grand Critérium. As a three-year-old he was a disappointment. He was beaten 7¾ lengths by Rex Magna in the Prix Greffulhe on 4 April and was a neck behind Catus in the Prix Noailles. He only finished sixth in the Prix Lupin and was off the course till finishing fifth to Crystal Palace in the Prix Niel. His starting price of 65 to 1 fairly reflected his chance in the Prix de l'Arc de Triomphe.

There were two French three-year-old fillies in the 1977 Prix de l'Arc de Triomphe. Madame Rabatel's Kamicia had been an impressive winner of the Critérium des Pouliches as a juvenile. After finishing third to Hartebeest in the Prix La Carmago at Saint-Cloud on 7 March and second to the same filly in the Prix de la Grotte in April, Kamicia was a disappointing sixth to the brilliant Madelia in the Poule d'Essai des Pouliches on 1 May. She never showed with a chance behind Madelia and Trillion in the Prix de Diane de Revlon which was the first time a classic was sponsored in France. Kamicia was third in the Prix de Malleret in July but when she beat Edinburgh a nose in the Prix de la Nonette on 4 September there seemed a possibility that she was coming back to her best. She then won the Prix Vermeille by 1½ lengths and a neck from Royal Hive and Fabuleux Jane on 18 September with Busaca only fourth. It seemed as if she was always better in the autumn than in the spring and she started at odds of 14 to 1 in the Prix de l'Arc de Triomphe.

Fabuleux Jane started at the same odds as Kamicia on Sunday 2 October. She had not looked as fit in the paddock but she finished fastest of all in the Prix Vermeille. It was her first outing since finishing third a head and 2½ lengths behind Madelia and Trillion in the Prix de Diane de Revlon. Madelia, who had also won the Prix Saint-Alary in brilliant style, broke down in the straight and most courageously struggled through the final hundred metres to the winning post. The injury was so serious that she was never able to run again. Had she done so she would have been a severe threat to Alleged and Dunfermline. Fabuleux Jane had run rather too freely when sixth to the Queen's filly in the Oaks at Epsom. Prior to that Fabuleux Jane had won the Prix des Belles Filles at Longchamp on 8 May. It was probable that her chance in the Prix de l'Arc de Triomphe was not quite as good as that of Kamicia.

Daniel Wildenstein's four-year-old Crow had had a disappointing season prior to the Prix de l'Arc de Triomphe. He was unplaced to Arctic Tern in the Prix Ganay in May. He did not run again until the Grand Prix de Saint-Cloud when he finished fourth a neck, half a length and 2 lengths behind Exceller, Riboboy and Iron Duke. He did meet some interference in the straight and might have finished closer to Exceller. Crow was strongly fancied to take his revenge on Exceller in the King George VI and the Queen Elizabeth Diamond Stakes at Ascot but he was never going well and finished a poor ninth to The Minstrel. In spite of these poor performances Crow started at the very short odds of 19 to 2 for the Prix de l'Arc de Triomphe. On his 1977 form it was difficult to give him much chance but if his trainer Angel Penna had got him back to his best he would be a threat to all his rivals.

The 1976 Irish Sweeps Derby winner Malacate was sold to Tim Rogers of the Airlie Stud in the autumn by his owner, Maria-Felix Berger. He turned out to be only semi-fertile and it was decided to put him back into training with François Boutin. He trounced Fulgus 5 lengths in the Prix des Erables (8f) at Chantilly on 29 July, but disappointed when fifth to Relkino in the Benson and Hedges Gold Cup. In the Prix Foy, however, on 11 September Malacate beat On My Way and Ranimer three-quarters of a length and a short head with Arctic Tern three-quarters of a length away fourth and Shafaraz sixth and last. Although not the best colt of his generation Malacate had been more than useful. While his starting price of 12 to 1 seemed a little ungenerous only the imprudent would rule him totally out of contention on Sunday 2 October.

The Prix Foy was the first race of the season for the seven-year-old On My Way, and it was a fine performance to run second to Malacate. He was a proven autumn performer but at the age of seven was probably not quite as good as he had been when runner-up to Star Appeal. His starting price of 12 to 1, therefore, also seemed a little cramped.

The erratic Sea Bird's four-year-old Arctic Tern had begun his campaign by finishing third to Kasteel in the Prix d'Harcourt on 11 April. He then put up possibly the best performance of his career when beating Exceller and Infra Green half a length and a neck in the Prix Ganay on 1 May and he was far from disgraced when failing by three-quarters of a length to give Trépan 3 kilos in the Prix Dollar. In the Joe Coral Eclipse Stakes at Sandown he was upset by the journey, was slowly out of the stalls and he finished only third to Artaius having never looked like getting on terms. If he consented to show his very best form Arctic Tern might spring a surprise but on balance it seemed unlikely.

The five-year-old mare, Infra Green, was unplaced to Kasteel in the Prix d'Harcourt before she finished third in the Prix Ganay. She was three-quarters of a length runner-up to Sirlad, the champion Italian three-year-old, in the Gran Premio di Milano on 17 June but was beaten about 2 lengths into sixth place in the Prix Gontaut-Biron by Iron Duke at Deauville. Her odds of 57 to 1 probably represented her chance at Longchamp on 2 October.

The four-year-old Iron Duke had won La Coupe de Maisons-Laffitte and the Prix de la Côte Normande as a three-year-old when he was some pounds behind the best of his generation. On his seasonal reappearance he ran fourth to Mittainvilliers in the Prix de Ris-Orangis (8f) at Evry on 31 March but could only finish fifth to Lucky Wednesday in the Westbury Stakes (10f) at Sandown Park. Iron Duke was runner-up 3 lengths behind Diagrammatic in the Prix Jean de Chaudenay in May and was an excellent third, a neck and half a length behind Exceller and Riboboy, in the Grand Prix de Saint-Cloud. He then beat Diagrammatic a head in the Prix Gontaut-Biron at Deauville on 15 August. When competing at the highest level Iron Duke normally found several of his rivals too good for him and this seemed likely to occur again in the Prix de l'Arc de Triomphe.

The field was completed by the four-year-old Shafaraz. In the spring he was owned by the Aga Khan and trained by François Mathet and won two Gentleman-Riders' events at Saint-Cloud in March. He was then sold to Mme Albert Mina and trained by Patrick-Louis Biancone. He won another Gentleman-Riders' race, the Prix des Centaurs at Longchamp in May, after which he was also successful there in the Prix de l'Esplanade – a handicap over a mile and a half. Shafaraz then won the Prix du Carrousel at Longchamp on 26 June but was only third in the Grand Prix de Vichy and last in the Prix Foy. In spite of the number of his victories, by no stretch of the imagination could he be said to have a chance in the Prix de l'Arc de Triomphe.

Apart from the absence of Madelia through injury, two of the best horses in Europe were not in the field for the 1977 Prix de l'Arc de Triomphe, Exceller and The Minstrel. Both had been sent to the United States, Exceller to race and The Minstrel to go to stud. There had been an outbreak of contagious metritis in Europe in the summer and both had been hastily shipped to America before an imposition of import restrictions was placed on European horses. Alleged was more than able deputy for The Minstrel for Robert Sangster and in fact was probably a better colt, but Nelson Bunker Hunt had no runner in 1977. Neither did the Aga Khan whose brilliant Blushing Groom had palpably failed to stay when third to The Minstrel in the Derby. The non-staying Pharly also sidestepped the Prix de l'Arc de Triomphe and had run instead in the Prix du Moulin de Longchamp the previous Sunday. Relkino was another absentee. He had been second to Empery in the Derby in 1976 and after his success in the Benson and Hedges Gold Cup, his trainer Dick Hern decided to aim for the Champion Stakes at Newmarket as he required a sound surface, not frequently encountered at Longchamp in October. Funny Hobby, the rather undistinguished winner of the Grand Prix de Paris, was another absentee through injury.

Lester Piggott was asked before the St Leger if he would win on Alleged at Doncaster and he replied, 'I don't know but I'll win the Prix de l'Arc de Triomphe'. Before the Doncaster Classic, Alleged had been owned in partnership by Robert Sangster and the Americans Mrs Shirley Taylor and Robert Fluor. Since it was one of Robert Sangster's great ambitions to win Longchamp's most prestigious race, he acquired Robert Fluor's share before the St Leger. At Chantilly, Kamicia's trainer Jean Laumain was extremely optimistic about her chance. He reported that she did not even lose a pound in weight after her success in the Prix Vermeille and that she had been going even better on the gallops since her victory. Crystal Palace, who was not the easiest of rides, was not to be partnered by his habitual jockey Gerard Dubroeucq, he had lost his position as François Mathet's stable jockey. Instead he was to be ridden by Alain Badel which might be slightly to the colt's disadvantage.

The Final Preparations

Before racing commenced at Longchamp on 2 October, Lester Piggott told Alleged's trainer Vincent O'Brien that he was so confident of winning that he was contemplating making all the running. Even O'Brien was a little surprised, because it was not at all normal for a runner to attempt to lead from the start to the finish in the Prix de l'Arc de Triomphe. The afternoon's racing began with the English sprinter Gentilhombre beating Madang and his own stable-companion Haveroid 4 lengths and 2 lengths in the Prix de l'Abbaye de Longchamp. It was not necessary, therefore, for the judge to have recourse to the photo-finish as he had done the previous year, which had caused such controversy when a dead-heat between Gentilhombre and Mendip Man was declared. Sir Michael Sobell's Cistus, who was trained by Dick Hern, was only beaten a length by Tarona in the Critérium des Pouliches. Even though Tarona was trained by François Boutin, she was owned by the Englishman Gerald Oldham.

The Race

In the paddock Alleged was cool and relaxed and was without question the pick of the field. Dunfermline and Orange Bay also looked very well and Balmerino, in spite of his globetrotting throughout the year, was difficult to fault. On the other hand Kamicia was rather a plain filly and Mia, the Argentinian challenger, was dwarfed by some of her rivals. With so many runners in the race, the paddock was very crowded and even at Longchamp that afternoon there was some momentary confusion while trainers looked from one end to the other to find their charges so that they could be mounted by their respective jockeys.

Yelpana, Dunfermline and Alleged were first out of the stalls. Carson eased the Queen's filly back into about sixth place and Yelpana set a very leisurely pace to Alleged with the rest of the field closely bunched behind them and Cunning Trick and Mia bringing up the rear. None of the jockeys appeared to want to make the running and for the first time for many years there was not the usual blistering gallop from the outset of the Prix de l'Arc de Triomphe. The penetrometre used to gauge the going recorded 2·9; on the day San San equalled Mill Reef's course record it was 2·8 so one might have anticipated a near record time. As the field disappeared behind the Petit Bois, Lester Piggott took Alleged into the lead from Yelpana followed by Crystal Palace, Carwhite, Crow, Dunfermline, Orange Bay and Amyntor, with Balmerino, Infra Green and Malacate in the mid-division and Cunning Trick and Mia last. The fact that Piggott had been able to take Alleged into the lead so early, considering his draw at nineteen, and that Crystal Palace, drawn twenty-three, was third, was a further demonstration of the slow pace in the first 400 metres. The English jockey then proceeded to give a masterly and superb demonstration of the tactic of waiting in front. At the top of the hill Alleged was still lobbing along in front of Crystal Palace, Crow and Yelpana, with Orange Bay close behind, Dunfermline on the rails and Carwhite and Amyntor a little behind her. Coming down the hill Piggott still did not increase the leisurely pace he was setting on Robert Sangster's colt who appeared totally in command. Crow did move up to take second place from Crystal Palace but did not make any serious challenge to the leader. Balmerino did, however, make progress from the mid-division to be on the heels of Yelpana, Orange Bay and Dunfermline. As Alleged swung round the turn into the straight, Piggott suddenly increased the pace. Within 200 metres Robert Sangster's colt had increased his lead over Crow and Crystal Palace from 1 length to 4 lengths. He had set up an unassailable advantage. It was one of Alleged's great assets to be able to produce a burst of acceleration which left his rivals powerless to respond in time and he displayed it again that afternoon at Longchamp. With 300 metres to go, Alleged still held a decisive advantage over Crystal Palace and Crow. Dunfermline on the rails was at this point hampered by the rapidly weakening Yelpana, and Balmerino was beginning a challenge from behind Crow. In fact the colt who had travelled from New Zealand looked the only possible danger to the leader at this point since the Queen's filly was still trapped on the rails. Alleged, however, was still going like a winner. Although Balmerino flashed past Crystal Palace inside the distance and finished fastest of all, Alleged had a very comfortable 1½ lengths to spare over him with the one-paced Crystal Palace plugging on 2 lengths behind in third place. Carson had eventually managed to extricate Dunfermline who finished very well when she got clear and only failed to take third place by a short neck. Crow was another neck away fifth with the fast-finishing Monseigneur 1½ lengths behind sixth. He was followed by Infra Green, Malacate, Yelpana and Guadanini who crossed the line almost together.

There was no absolutely question that Alleged, aided by the masterly tactics of Lester Piggott, had totally dominated the race. He had

Alleged (20) beating Balmerino (2) and Crystal Palace (19) 1½ lengths and 2 lengths with Dunfermline (23) a close fourth

outgalloped his rivals in a manner reminiscent of Ribot. The time of 2 minutes 30·6 seconds was, however, nearly 2 seconds slower than that of Mill Reef and San San. The race had not been run to suit those runners such as Crow and Dunfermline who required a strong test of stamina. The Queen's St Leger winner, Dunfermline, returned to the unsaddling enclosure having spread a plate when hampered by Yelpana, but Willie Carson did not use this as an excuse; he felt the lack of pace had been the cause of her downfall. Afterwards Dick Hern felt it might have been a mistake not to have run a pacemaker as he had done in the St Leger. Had he done so, and the pacemaker taken on Alleged, it would have been much more difficult for Piggott to have totally dictated the tactics and pace of the race. Nevertheless the 1977 Prix de l'Arc de Triomphe was the story of a superb demonstration of waiting in front by Piggott on an exceptional colt. Dunfermline, however, remained the sole racehorse to have beaten him as a three-year-old.

Alleged and Piggott received a tremendous ovation when they returned to the unsaddling enclosure. Robert Sangster and his wife and Lester Piggott were presented to Président Giscard d'Estaing. It was then that Yves Saint-Martin commented on the triumphant jockey, 'Lester is the supreme master of his craft'. It was a second success, both for Piggott and for Alleged's brilliant trainer Vincent O'Brien, in the Prix de l'Arc de Triomphe. O'Brien, however, had had to wait nineteen years since Ballymoss had triumphed in 1958. He had to endure watching Sir Ivor and Nijinsky both finishing second during that time. Although he trained in Ireland, O'Brien had saddled five winners of the Derby in the previous fifteen years: Larkspur, Sir Ivor, Nijinsky, Roberto and The Minstrel. In addition, he had won the Oaks twice with Long Look and Valoris, the 2000 Guineas on two occasions with Nijinsky and Sir Ivor, and the St Leger twice with Nijinsky and Boucher. In 1977 he had virtually dominated English racing, setting a new record by winning £439,124 in prize money. Amongst his horses that year in addition to Alleged and The Minstrel, who also won the Irish Sweeps Derby, were Artaius, who won the Joe Coral Eclipse Stakes and the Sussex Stakes and was second in the Prix du Jockey Club; Be My Guest, the winner of the Waterford Crystal Mile; Meneval, hero of the Hardwicke Stakes; and Robert Sangster's champion two-year-old, Try My Best, who won the Dewhurst Stakes. O'Brien also trained Valinsky, runner-up in the Grand Prix de Paris. It is difficult to remember a year when one trainer ever won so many important races in England, Ireland and France in one season as did Vincent O'Brien in 1977.

The Rest of the Season

Alleged did not run again in 1977 but remained in training as a four-year-old. Monseigneur, who beat Paico 3 lengths in the Prix du Conseil Municipal on 16 October, paid a handsome tribute to Robert Sangster's colt. Arctic Tern, however, could only finish fourth to Flying Water and Relkino in the Champion Stakes at Newmarket. In 1977 the Société d'Encouragement had decided to move the Prix Royal Oak from the middle of September to the last meeting in October. Dunfermline seemed to have the race at her mercy and started at 6 to 10. She could only manage third place, however, 1½ lengths and half a length behind the very lightly raced Prix Hocquart winner Rex Magna and Trillion, with Amyntor 4 lengths away fourth. Dunfermline had gone in her coat in the paddock and was probably over the top. Balmerino had earlier contested the Gran Premio del Jockey Club on 16 October. He passed the post a short head in front of Stateff but the San Siro stewards found that he had interfered with the runner-up in the final hundred metres. Balmerino was disqualified and placed second. This did not prevent his continuing his globetrotting, for he ran in the Washington D.C. International Stakes on 5 November. The favourite was Nelson Bunker Hunt's Exceller, who on his first outing in the United States had been runner-up to Majestic Light in the Man o'War Stakes. He had then beaten Majestic Light and Johnny D a length and 3½ lengths in the Canadian International Stakes at Woodbine Park on 25 October. At Laurel Park the placings were reversed with Johnny D beating Majestic Light 2½ lengths, with Exceller third, Balmerino fourth, Crow fifth and Monseigneur sixth. Considering the amount of travelling Balmerino had done since the summer, his performance at Laurel Park, a course not ideal for his style of running, could not be described as a disgrace. In the Premio Roma on 14 November Yelpana was 1½ lengths runner-up to Montorselli, who had won the Prix de l'Espérance in May, with On My Way only fifth. The latter had earlier run Campero to three-quarters of a length in the Grand Prix de Marseille on 29 October with Montorselli 2 lengths away third.

Pedigree

Alleged, a May foal, was bred in the United States by Mrs June McKnight. He was sold as a yearling to the Californian Monty Roberts for $34,000. Roberts sent him to the California two-year-old sales and he was bought for $175,000 by Robert Sangster in partnership with Mrs Shirley Taylor and Robert Fluor. Sangster was offered $¼ million dollars for the colt immediately after the sale but did not hesitate for a moment in refusing the offer.

Alleged's sire was the Tom Rolfe colt Hoist the Flag. He was the champion two-year-old in the United States in 1970, winning five of his six races. He would have been unbeaten but for crossing in the first three furlongs of the Champagne Stakes and being demoted to third place. He won the Cowdin Stakes and beat Droll Role and Jim French in the Bay Shore Stakes. During his preparation for the Kentucky Derby, Hoist the Flag fractured his off-hind pastern. Fortunately it was not necessary to put him down; the veterinary surgeon was able to graft a piece of hip-bone on to the break and save him.

Alleged came from his second crop, from which Hoist the Flag also got Sensational, the champion two-year-old filly in the United States in 1976. Alleged is still the best horse got by Hoist the Flag to date, but since 1975 his progeny also include True Colours (Round Table Handicap, Fort Marcy Handicap), Fifth Marine (American Derby), Flag Officer (Illinois Derby), A Thousand Stars (3rd Poule d'Essai des Pouliches) and Ginistrelli (Ladbroke's Derby Trial Stakes).

Alleged's dam, Princess Pout by Prince John, was a very useful racemare. She won thirteen times from thirty-one starts from the ages of three to five and earned $260,265 in prize money. She was bred, like Alleged, by Mrs June McKnight and ran in her colours. Princess Pout won the Sheepshead Bay Handicap twice, the New York Handicap, and the Benjamin F. Lindheimer Handicap (9½f) at Arlington Park. She was also runner-up in the Chrysantheum Handicap. As a five-year-old in 1971 Princess Pout was rated level with Fort Marcy in the official handicap for grass horses in America, only 1 lb behind the season's top-rated grass horse, Drumtop. Alleged was Princess Pout's first foal; she was barren in 1975 and produced a filly by Sir Ivor, Allegedly in 1976. Her next foal, the Bold Bidder colt Benefactor, cost $700,000 as a yearling and was trained in Ireland by Vincent O'Brien. In Ireland he only managed to win the Cork Maiden Plate (12f) at Naas in July 1980. Benefactor raced in the United States as a four-year-old and won two races worth $33,900.

Princess Pout's dam, Determined Lady by Determine, won two minor races from eleven starts as a three-year-old worth $8,625. She was a much better broodmare than she had been on the track; from six foals she produced three other winners besides Princess Pout: Duchess of Malfi won one race as a four-year-old worth $23,980 and was the dam of Triple Try; Goose Bay and South American Way won nine minor races between them in the United States.

Determined Lady's dam, Tumbling by War Admiral, won six times from forty-one starts and earned $35,340 in prize-money. She was also third in the Hollywood Oaks and La Centinela Stakes. She was a prolific broodmare, producing thirteen foals, among them Proud Kenn, who earned $199,719 in prize money from his twenty-three victories, which included the Princeton Handicap, the Whirlaway Handicap and the Germantown Handicap, and Falling Evidence, the victor of the Bicentennial Stakes.

Tumbling's dam, Up the Hill by Jacopo, was a useful filly whose victories included the Ladies Handicap at Belmont Park. In addition to Tumbling she was also the dam of Pail of Water by Jack High who won the Arlington Lassie Stakes in 1948.

Up the Hill's dam, Gentle Tryst by Sir Gallahad, was out of Cinq à Sept by Roi Hérode. In 1927 Cinq à Sept won five consecutive races including the Irish Oaks and Park Hill Stakes. Secretariat, Nonoalco and Elocutionist all trace back to Cinq à Sept. On his breeding, therefore, it was no great surprise that Alleged should have turned out to be a champion.

PRIX DE L'ARC DE TRIOMPHE 2400 metres
at Longchamp on Sunday 2 October 1977. For three-year-olds and upwards. Going: Good.
1,200,000 francs to the winner; 480,000 fr. to the 2nd; 240,000 fr. to the 3rd; 120,000 fr. to the 4th. 300,000 fr. to the breeder of the winner; 120,000 fr. to that of the 2nd; 60,000 fr. to that of the 3rd; 30,000 to that of the 4th. Value to the winner: 1,200,000 fr.

ALLEGED	b.c.	by Hoist the Flag – Princess Pout by Prince John	3.56k	Robert Sangster	L. Piggott	1
BALMERINO	b.c.	by Trictrac – Dulcie by Duccio	5.59k	R.K. Stuart	R. Hutchinson	2
CRYSTAL PALACE	gr.c.	by Caro – Hermières by Sicambre	3.56k	Baron Guy de Rothschild	A. Badel	3
DUNFERMLINE	b.f.	by Royal Palace – Strathcona by St. Paddy	3.54½k	H.M. Queen Elizabeth II	W. Carson	4
CROW	ch.c.	by Exbury – Carmosina by Right of Way	4.59k	Daniel Wildenstein	Y. Saint-Martin	5
MONSEIGNEUR	ch.c.	by Graustark – Brown Berry by Mount Marcy	3.56k	Arthur Seeligson	A. Murray	6
INFRA GREEN	ch.f.	by Laser Light – Greenback by Fric	5.57½k	Mme John Pochna	G. Doleuze	7
MALACATE	b.c.	by Lucky Debonair – Eyeshadow by My Babu	4.59k	A.D.D. Rogers	P. Paquet	8
YELPANA	ch.c.	by Yelapa – Zampana by Reliance	3.56k	Werner Heimberg	G. Starkey	9
GUADANINI	ch.c.	by Luthier – Ilrem by Prudent	3.56k	Joseph Kaïda	W.B. Pyers	10
ON MY WAY		by Laugh Aloud – Gracious Me by Tulyar	7.59k	Xavier Beau	A. Gibert	
ORANGE BAY		by Canisbay – Orange Triumph by Molvedo	5.59k	Carlo Vittadini	P. Eddery	
ARCTIC TERN		by Sea Bird – Bubbling Beauty by Hasty Road	4.59k	Mme John S. Knight	M. Philipperon	
SHAFARAZ		by Levmoss – Asharaz by Sicambre	4.59k	Mme Albert Mina	J. Taillard	
CUNNING TRICK		by Buckpasser – Intriguing by Swaps	4.59k	Ogden Phipps	J. Cruguet	
IRON DUKE		by Sicambre – Insulaire by Aureole	4.59k	Gérard Tournier	H. Samani	
VIVI		by Luciano – Vitesse by Panaslipper	5.57½k	Mme Isolde Wilhelms	S. Léonardos	
SARAH SIDDONS		by Le Levanstell – Mariel by Relko	4.57½k	Mme J.R. Mullion	J. Mercer	
AMYNTOR		by Sir Gaylord – Crepellana by Crepello	3.56k	Marcel Boussac	G. Rivases	
PANAMINT		by Run the Gantlet – Panaview by Panaslipper	3.56k	Bertram R. Firestone	W. Swinburn	
MONTCONTOUR		by Luthier – Moskvitchka by Mossborough	3.56k	Mme H. Hausmann	G. Dubroeucq	
DOM ALARIC		by Sassafras – Ordenstreue by Orsini	3.56k	Mme Daniel Malingue	E. Lellouche	
CARWHITE		by Caro – White Paper by Honeyway	3.56k	Jacques Wertheimer	F. Head	
KAMICIA		by Kashmir II – Micia by Mincio	3.54½k	Mme Henri Rabatel	A. Lequeux	
FABULEUX JANE		by Le Fabuleux – Native Partner by Raise a Native	3.54½k	Ralph C. Wilson	J.C. Desaint	
MIA		by Climber – Arana by Tobago	3.52½k	Ecurie Los Sauces	C. Peszoa	

26 runners.
ALLEGED trained by M.V. O'Brien at Cashel, Ireland; bred by Mrs J.H. McKnight in the U.S.A.
Distances: 1½ lengths; 2 lengths; short neck; neck; 1½ lengths; ½ length; neck; short head; short head. Time: 2 mins 30.6 seconds.
Pari-Mutuel: Win: 4fr. 90. Places: 3fr. 60; 5fr. 90; 3fr. 70.
S.P.: 38 to 10 ALLEGED; 8 to 1 Dunfermline, Crystal Palace and Orange Bay; 19 to 2 Crow; 12 to 1 Balmerino and Malacate; 14 to 1 Kamicia and Fabuleux Jane; 17 to 1 On My Way; 20 to 1 Carwhite; 23 to 1 Arctic Tern; 57 to 1 Infra Green; 58 to 1 Cunning Trick; 59 to 1 Sarah Siddons; 60 to 1 Iron Duke; 65 to 1 Amyntor; 66 to 1 Mia; 78 to 1 Monseigneur; 82 to 1 Vivi; 99 to 1 Yelpana, Guadanini, Shafaraz, Dom Alaric, Montcontour and Panamint.

ALLEGED bay colt 1974	Hoist the Flag b. 1968	Tom Rolfe b. 1962	Ribot b. 1952	Tenerani b. 1944	Bellini
					Tofanella
				Romanella ch. 1943	El Greco
					Barbara Burrini
			Pocahontas b. 1955	Roman b. 1937	Sir Gallahad
					Buck Up
				How br. 1948	Princequillo
					The Squaw II
		Wavy Navy b. 1954	War Admiral b. 1934	Man o' War ch. 1917	Fair Play
					Mahubah
				Brushup b. 1929	Sweep
					Annette K
			Triomphe b. 1947	Tourbillon b. 1928	Ksar
					Durban
				Mélibée b. 1938	Firdaussi
					Métairie
	Princess Pout b. 1966	Prince John ch. 1953	Princequillo b. 1940	Prince Rose b. 1928	Rose Prince
					Indolence
				Cosquilla b. 1933	Papyrus
					Quick Thought
			Not Afraid b. 1948	Count Fleet br. 1940	Reigh Count
					Quickly
				Banish Fear br. 1932	Blue Larkspur
					Herodiade
		Determined Lady b. 1959	Determine gr. 1951	Alibhai ch. 1938	Hyperion
					Teresina
				Koubis gr. 1946	Mahmoud
					Brown Biscuit
			Tumbling b. 1953	War Admiral b. 1934	Man o' War
					Brush Up
				Up the Hill b. 1938	Jacopo
					Gentle Tryst

Alleged 1978

Alleged

The Field

Alleged was a 14 to 10 favourite to win the Prix de l'Arc de Triomphe for the second consecutive season in 1978, and so to emulate Ribot, Tantième, Corrida and Ksar. At the beginning of the year it had seemed that he would carry off nearly all the important middle-distance races in England and France. He made his seasonal début in the Royal Whip Stakes at the Curragh and beat Irish Riddle by 2½ lengths without ever coming off the bit. Robert Sangster's colt was withdrawn from the Coronation Cup on account of the firm going at Epsom but he was then stricken with the virus. Although it was not a severe attack and he was recovering well by the Joe Coral Eclipse Stakes at the beginning of July, he was not fit enough to run in the King George VI and the Queen Elizabeth Diamond Stakes and Vincent O'Brien very wisely adopted a policy of patience with him. Alleged did not run again until the Prix du Prince d'Orange at Longchamp on 17 September. Piggott took him into the lead at the entrance to the straight and the issue was settled there and then. He slammed Louksor 2½ lengths with Pevero and Pyjama Hunt third and fourth, in the process lowering the Longchamp course record by half a second. Piggott had even been able to take a long look over his shoulder at his toiling opponents. As Robert Sangster immediately remarked: 'It has been a brilliant feat on Vincent's part to get Alleged ready after that bad attack of the virus.' All who saw Alleged win that afternoon in such scintillating fashion felt that he was back to his best and that he would be very difficult to beat in the Prix de l'Arc de Triomphe on Sunday 1 October.

The principal threats to Robert Sangster's Hoist the Flag colt appeared to be the Aga Khan's Acamas and Jacques Wertheimer's pair, Dancing Maid and Gay Mecene. Acamas was owned by Marcel Boussac as a juvenile and in the spring and summer of 1978. On his racecourse début he won the Prix Vaublanc (1100 metres) by 4 lengths at Chantilly on 21 July but was then well beaten by Vific in the Prix Robert Papin. Acamas ran third, only defeated three-quarters of a length and a nose by Super Concorde and Pyjama Hunt in the Grand Critérium; this was a fine performance with much promise for the future. Acamas reappeared in the Prix Greffulhe on 9 April but failed by half a length to overhaul Naasiri, who had the benefit of a race at Evry earlier in the spring which he had won. In the Prix Lupin on 14 May Acamas challenged Pyjama Hunt 200 metres out and drew clear for an effortless 1½ length victory with Turville and Gay Mecene half a length and 2½ lengths away third and fourth. On the afternoon of the Prix du Jockey Club there was a terrific thunderstorm and a violent downpour during the parade before the race. On the turn into the straight Acamas received an almighty bump from one of his rivals; he completely lost his action and dropped back almost last. Yves Saint-Martin even considered pulling him up. Entering the straight, he seemed to be in an impossible position but then he regained his action and flew past all his rivals to head Frère Basile in the last strides and beat him a nose, with Turville and Noir et Or a length and a neck back third and fourth and Dancing Master only sixth. It was a truly outstanding performance. It was the twelfth time that Marcel Boussac, in his ninetieth year in 1978, had won the Prix du Jockey Club. Sadly his textile empire had been struggling for several years and most of his personal fortune had been swallowed up in attempts to keep the mills rolling. Shortly before the victory of Acamas, Marcel Boussac's business affairs were placed in the hands of the Receiver. The Receiver ordered that all the Boussac bloodstock should be sold and a deal was concluded with the Aga Khan who, after some hesitation, bought all the stallions, broodmares and horses in training for a sum reported to be between $8 million and $9 million. It was agreed, however, that when Acamas ran in the King George VI and the Queen Elizabeth Diamond Stakes, the Prix du Jockey Club winner should run for the last time in the famous orange colours and grey cap of the man who had so dominated European racing in the 1930s, 1940s and 1950s. Marcel Boussac, accompanied by Jean Romanet, came to watch Acamas run at Ascot, his last visit to an English racecourse. In the paddock he was standing under the large oak tree nearest the stand with Jean Romanet, and other connections of Acamas. Her Majesty the Queen, whose Dunfermline was already walking round the paddock, saw him come in after her and immediately left her trainer Dick Hern and her racing manager Lord Porchester and went to greet him. It was a most gracious gesture to a man who had contributed so much to international racing.

In the King George VI and the Queen Elizabeth Diamond Stakes itself, Yves Saint-Martin gave Acamas a great deal to do entering the short straight. Coming from almost last the Prix du Jockey Club winner was finishing fastest of all, but he failed by 1½ lengths to catch Ile de Bourbon with Hawaian Sound and Montcontour a neck and 1½ lengths behind third and fourth, Exdirectory sixth and Dunfermline and Trillion further behind. Many felt that had Acamas been closer to the leaders on the turn into the straight he would have given the winner much more to do and might well have won, though he did hang first left and then right in the final furlong. It was the last time Acamas ran in Marcel Boussac's colours. Sadly he was later disqualified as traces of a prohibited drug were found in his system. He started second favourite at odds of 15 to 4 and looked a very serious danger to Alleged.

Very good cases could also be made out for Jacques Wertheimer's pair Gay Mecene and Dancing Maid. Although only fourth in the Prix Lupin to Acamas, Gay Mecene was unbeaten in his four other outings. He won his only race as a two-year-old, the Prix Dagor (1600 metres), at Maisons-Laffitte in November. He reappeared in the Prix de Guiche and convincingly defeated Peloponnes half a length on 16 April with the Grand Critérium winner Super Concorde a disappointing fifth. Gay Mecene did not run after the Prix Lupin until the Prix Eugène Adam at Saint-Cloud on 16 July when he beat Rusticaro and Pyjama Hunt three-quarters of a length and a short neck with the 2000 Guineas winner Roland Gardens 2½ lengths away fourth. In the Prix Niel at Longchamp on 10 September Gay Mecene got the better of Noir et Or, after a terrific struggle in the final 200 metres, by a short head, with Frère Basile and Dancing Master 5 lengths and a length away third and fourth. Strictly on the book Gay Mecene should not reverse the Prix Lupin form with Acamas but he was improving and might well take a hand in the finish on 1 October.

His stable-companion, the three-year-old filly Dancing Maid, looked an even livelier threat to Alleged and Acamas. As a two-year-old, Dancing Maid ran fourth in the Prix de la Loubère (1500 metres) at Saint-Cloud in September and then won the Prix de Précy (1700 metres) at Longchamp on 28 October. On her début as a three-year-old, Dancing Maid comfortably beat Reine Mère 1½ lengths in the Prix Vanteaux at Longchamp on 16 April. In the Poule d'Essai des Pouliches on 30 April Jacques Wertheimer's filly went clear at the entrance to the straight, and totally outclassing his rivals trounced Fruhlingstag and A Thousand Stars by 5 lengths and a short head. In the Oaks at Epsom, Dancing Maid looked the assured winner when clear of the remainder at the distance but Fair Salinia produced a devastating burst of speed to catch her on the line and beat her by the shortest of short heads. Alec Head, her trainer, felt she should have won and she probably would have done had she not shortened her stride fifty yards out, perhaps due to the roar of the crowd. On 12 July Dancing Maid comfortably held Tingtagel by half a length in the Prix Chloë at Evry. In the Prix Vermeille on 17 September Jacques Wertheimer's filly found her way blocked by Lester Piggott on Amazer 300 metres out, but Freddie Head calmly switched her to the outside, took the lead at the distance and comfortably held the Irish filly Relfo by a length with Amazer a length back third and Fair Salinia 1½ lengths away fifth. She thus avenged her defeat at Epsom and was probably the better of the Alec Head trained pair, who started at odds of 4 to 1 (couplé), in the Prix de l'Arc de Triomphe.

Jean-Pierre Binet's Frère Basile, the runner-up to Acamas at Chantilly, ran second to Gracias in the Prix Noailles in soft going on 16 April, with Dancing Master third. In the Prix Hocquart in very heavy going on 7 May Frère Basile beat Naasiri 8 lengths with Gracias only fourth. After the Prix du Jockey Club he was rested till the Prix Niel when he ran third to Gay Mecene and Noir et Or. He certainly benefited from that race and was expected by his connections to run very well in the Prix de l'Arc de Triomphe. His odds of 14 to 1, however, seemed rather ungenerous as Frère Basile was undoubtedly at his best in heavy going.

Paul de Moussac's Noir et Or was runner-up in his début as a juvenile in the Prix de Vauville at Clairefontaine, but then beat Galiani 2 lengths in the Prix des Aigles (1600 metres) at Chantilly on 14 September. When Noir et Or won the Prix Saint-Roman by a length from Orange Marmelade at Longchamp on 1 October, he looked to have classic potential. On his reappearance as a three-year-old, however, he made very heavy weather of winning the Prix de Suresnes by a short head on 30 April and then finished a disappointing seventh in the Prix Lupin. After his close fourth in Prix du Jockey Club, Noir et Or finished very well to take third place behind Guadanini and Trillion in the Grand Prix de Saint-Cloud. In the Prix Niel he looked the winner when clear of his rivals shortly after entering the straight but was unable or unwilling to fight back when challenged by Gay Mecene. Noir et Or gave the impression that he did not relish a struggle and the very best his connections could hope for was one of the minor places.

Mme Eugène Cashman's Turville won the Prix de Louvre (1400 metres) at Longchamp in October 1977 from four outings as a juvenile. He had failed, however, to win in six runs before the Prix de l'Arc de Triomphe in 1978. Third both in the Prix Lupin and the Prix du Jockey Club, Turville ran Midshipman to three-quarters of a length in the Prix Maurice de Nieuil at Saint-Cloud on 14 July. He was, however, a well-beaten seventh to Gay Mecene in the Prix Niel. Turville's odds of 55 to 1 on 1 October were a fair reflection of his chance.

It was even more difficult to envisage Dancing Master taking a hand in the finish of the 1978 Prix de l'Arc de Triomphe. His solitary success had been in the Prix de Ferrières at Longchamp on 2 April. He ran third to Gracias and Frère Basile in the Prix Noailles but was only sixth in the Prix du Jockey Club and a well-beaten seventh in the Prix Eugène Adam. Although a close fourth to Tempus Fugit in the Grand Prix de Vichy on 7 August, he was again only fourth to Gay Mecene in the Prix Niel and it appeared unlikely that Dancing Master would reach the frame on 1 October.

Galiani, runner-up in the Prix Ksar (2400 metres) at Saint-Cloud on 23 May, won the minor Prix du Champs-de-Mars (2200 metres) at Longchamp on 1 June. He then sprang a major surprise in the Grand Prix de Paris when, at odds of 95 to 1, he beat Roi de Mai and Whitstead half a length and a short head. In the St Leger he finished fifth 3½ lengths behind Julio Mariner. It seemed sure that Galiani would find many of his rivals on 1 October too speedy for him.

Marcos Lemnos's Julio Mariner was the sole English-trained challenger in the 1978 Prix de l'Arc de Triomphe. As a two-year-old, this brother to the Oaks winner Juliette Marny easily won the Leyburn Stakes at York, and many considered him an unlucky loser when a neck second to Dactylographer in the William Hill Futurity at Doncaster. In the Mecca Dante Stakes, Julio Mariner was runner-up to Shirley Heights, and since it was his first outing as a three-year-old it was hoped he would reverse the form at Epsom. He finished a never dangerous sixth 8 lengths behind Shirley Heights in the Derby and was a very disappointing seventh to Ile de Bourbon in the King Edward VIII Stakes at Royal Ascot. In the Monkgate Stakes (8f) at York on 14 July Julio Mariner easily accounted for mediocre opposition, but then finished a very moderate eighth to Hawaian Sound in the Benson and Hedges Gold Cup at York on 23 August. Not surprisingly, he started at 28 to 1 for the St Leger but ran on strongly to hold the challenges of Le Moss and M-Lolshan by 1½ lengths and a short head with Galiani fifth and Ile de Bourbon sixth. The St Leger start, however, was delayed for more than twenty minutes due to an accident in the stalls to Easter King, a French challenger trained by François Boutin. Several of the runners were visibly upset at having to watch the stricken horse writhing on the ground and in very harrowing circumstances being dragged away from the back of the stalls. Whether some of the runners were as sickened and angered at the inept and slow removal of the dying Easter King as the spectators at Doncaster is impossible to know, but the delay at the start cannot have helped those with a nervous temperament and the form had to be suspect. Anyway, as far as Julio Mariner was concerned it was probable that he needed further than a mile and a half and his odds of 17 to 1 on 1 October seemed cramped.

There was also a three-year-old challenger from Ireland – Mrs J.R. Mullion's Exdirectory, trained by Paddy Prendergast. He easily won the Ballymoss Stakes at the Curragh by 1½ lengths from Strong Gale. In the BMW Nijinsky Stakes at Leopardstown on 5 May he was very badly hampered in the turn into the straight, and was unlucky to be beaten 3 lengths by Slaney Idol. He was totally unsuited by the firm going at Epsom when a disappointing thirteenth to Shirley Heights in the Derby, and in addition he was severely hampered after three furlongs. The good going at the Curragh in the Irish Sweeps Derby on 1 July suited Exdirectory much better and he put up a fine performance to run Shirley Heights to a head with Hawaian Sound a neck away third. In the King George VI and the Queen Elizabeth Diamond Stakes the going was again firm and Exdirectory was never striding out with any fluency when sixth to Ile de Bourbon. On his running against Shirley Heights, Mrs Mullion's colt could be given an outsider's chance in the Prix de l'Arc de Triomphe and he started at 24 to 1.

Edward Stephenson's four-year-old mare Trillion was, like Galiani and Montcontour, trained by Maurice Zilber. In 1977 when in the stable of François Mathet she had been one of the best three-year-old fillies in France, winning four times including the Prix de Minerve at Evry and Prix de Royallieu at Longchamp. She had also been runner-up to Madelia in the Prix de Diane and to Rex Magna in the Prix Royal Oak. In the Prix d'Harcourt on 2 April Trillion went down by a short neck to Monseigneur, but she gained her revenge in brilliant fashion when leading all the way in the Prix Ganay and beating Monseigneur and Sirlad 3 lengths and a short head with Balmerino 2 lengths away fourth. After a narrow defeat in the Prix Jean de Chaudenay she defeated Carwhite and Monseigneur in the Prix Dollar on 28 May. In the Grand Prix de Saint-Cloud Trillion had no answer to Guadanini's challenge and went down by a length, with Noir et Or and Dom Alaric 1½ lengths and a length away third and fourth. Trillion was only twelfth to Ile de Bourbon at Ascot; it was reported she had not travelled well and possibly the firm going did not suit her. It was her sole bad performance of the year. In the Prix Foy on 10 September she again led from the start and was always in command, impressively beating Monseigneur, Dom Alaric and Guadanini, 2 lengths, a nose and 4 lengths. It was impossible to ignore this consistent race mare, especially at odds of 16 to 1, but although she had won over a mile and a half, there was a possibility that she was better over ten and eleven furlongs.

Trillion's stable companion Montcontour had won the Prix Hocquart in 1977 but had suffered from warbles which restricted his campaign as a three-year-old thereafter. He caused a surprise in the Hardwicke Stakes at Royal Ascot when, as a 25 to 1 outsider, he beat Dunfermline and Balmerino 1½ lengths and a head. He was a disappointing seventh in the Grand Prix de Saint-Cloud but ran very well to finish fourth in the King George VI and the Queen Elizabeth Diamond Stakes at Ascot, a little over 3 lengths behind Ile de Bourbon. He occupied the same position a little over 2 lengths behind Valour in the Grosser Preis von Baden on 2 September. Maurice Zilber was quite confident about his chance on 1 October, but it was more realistically represented by his odds of 53 to 1.

After his success in the Prix d'Harcourt, Arthur Seeligson's Monseigneur had been a little disappointing, finishing behind Trillion in the Prix Ganay, Prix Dollar and Prix Foy. He was beaten 11½ lengths when fourth to Crow and Balmerino in the Coronation Cup at Epsom on 7 June. It appeared that Monseigneur had deteriorated as a four-year-old and that it was unlikely he would have a hand in the finish on 1 October.

Joseph Kaida's Guadanini beat Fabuleux Jane half a length in La Coupe at Longchamp on 4 May. In the Prix Jean de Chaudenay at Saint-Cloud on 15 May he took the lead a hundred metres out from the fading Trillion and held the fast finishing Dom Alaric by a head with Trillion a short head away third. Admittedly he was meeting the mare on 2 kilos better than weight for age, but Guadanini confirmed the form in the Grand Prix de Saint-Cloud. He was, however, only seventh to Ile de Bourbon at Ascot and over the shorter distance of the Prix Niel was a well-beaten fourth. The extra hundred metres of the Grand Prix de Saint-Cloud certainly suited him better than Trillion and it was difficult to enthuse about his chance in the 1978 Prix de l'Arc de Triomphe.

Dom Alaric's chance on 1 October seemed even more remote, as he appeared to be held by both Trillion and Guadanini. He did beat Tip Moss a neck in the Grand Prix Prince Rose at Ostende on 21 July but was likely to be outclassed in the Prix de l'Arc de Triomphe.

Shafaraz's best performance in 1978 was to win the Prix Kergorlay by 3 lengths and 4 lengths from Santalino and Campero. He was, however, well-beaten in the Grand Prix de Deauville and in the Prix du Cadran and was unlikely to fare any better than he had done the previous year.

The field for the 1978 Prix de l'Arc de Triomphe was completed by the German five-year-old, Lido. He had won once on 27 June over a mile and a half at Hamburg and been runner-up in the Spreti-Rennen at Baden-Baden in August. Nevertheless he was only sixth to Valour in the Grosser-Preis von Baden and it was clear that he was tackling opposition at Longchamp that would prove much too strong for him.

The Final Preparations

Alleged was brought over to Paris on Thursday 28 September by Vincent O'Brien's travelling head-lad Gerry Gallagher and he was accompanied by no less than a hundredweight of oats since he had such a huge appetite. Alleged's travelling companion, Solinus, who was unbeaten in five top sprints in England and Ireland, including the William Hill July Cup and the William Hill Sprint Championship at York, was confidently expected to begin the afternoon of 1 October on a successful note for his trainer Vincent O'Brien and Lester Piggott.

Freddie Head had chosen to ride Dancing Maid in preference to Gay Mecene who was partnered by Jean-Claude Desaint. The great American jockey, Willie Shoemaker, who had ridden Hawaiian Sound for Robert Sangster into second place behind Shirley Heights in the Derby, was engaged for Trillion whom Lester Piggott had ridden in all her victories in 1978.

The absence of the first two in the Epsom Derby, Shirley Heights and Hawaiian Sound, from the field of the 1978 was regrettable but understandable. After his victory in the Irish Sweeps Derby Shirley Heights was trained for the Great Voltigeur Stakes and the St Leger. He injured a tendon so badly in early August, however, that he was retired to stud immediately, having been syndicated for £1,600,000, with the Royal Studs a prominent shareholder. After the Irish Sweeps Derby, Robert Sangster's Hawaiian Sound was runner-up to Gunner B in the Joe Coral Eclipse Stakes and won the Benson and Hedges Gold Cup.

Sensibly, he did not clash with Alleged in Paris and had the Champion Stakes as his autumn objective. Even after his victory at Ascot it had never been intended to run Ile de Bourbon in the Prix de l'Arc de Triomphe. He was a very late foal, born on 7 June, and after Ascot his only classic objective had been the St Leger. Although Crow, Balmerino and Dunfermline remained in training in 1978, all three were also absentees on Sunday 1 October. Crow had training troubles after the Coronation Cup and did not race again; this was a great pity as it seemed he was back to his best in the spring. Dunfermline disappointed when only fourth to Ile de Bourbon in the Geoffrey Freer Stakes at Newbury in August and was withdrawn from training. Balmerino was seldom in the mood as a six-year-old to show his proven ability and had been returned to New Zealand in the late summer to stand as a stallion.

The morning before the Prix de l'Arc de Triomphe the world was shocked to hear of the sudden death of Pope John Paul I, whose Papacy had lasted only thirty-three days. But his death had no visible effect on the crowd at Longchamp on Sunday 1 October. As usual it was both very cosmopolitan and huge, and it arrived despite a strike on the French railways which had reduced traffic to less than 30 per cent on that day. France was hit by a wave of strikes that autumn and a postal strike had all but crippled the mail.

It was not a pleasant afternoon at Longchamp on the day of the 1978 Prix de l'Arc de Triomphe. It was windy, cloudy, and rather cold with intermittent periods of light rain and an occasional glimpse of a weak autumn sun. There was a great variety of hats but most were hidden by umbrellas which were more often opened than rolled. The majority of the women were well-wrapped up in cashmere pullovers and knee-length boots, though an exception was Princess Caroline of Monaco whose dress paid little heed to the weather.

The Race

The going was soft and the expected success of the Vincent O'Brien trained sprinter, Solinus, was thwarted in the Prix de l'Abbaye de Longchamp by Mme Alec Head's brilliant two-year-old filly Sigy, trained by her daughter Christiane Head. Sigy shot out of the stalls into a 2-length lead and it was clear 400 metres out that Solinus was under pressure and not going to catch her. She eventually won by a comfortable 3 lengths from Solinus with Double Form half a length away third. This victory, the first by a two-year-old in the race for twelve years, was not unexpected as Sigy had broken the course record for 1000 metres at Deauville, formerly held by Nonoalco, in August, and had won the Prix d'Arenberg effortlessly by 4 lengths. Alec Head described her as 'a phenomenon', whilst Freddie Head reckoned she was the fastest filly or colt he had ever ridden. Colonel Sir Douglas Clague, whose fortune had been built in Hong Kong, won the Critérium des Pouliches with Pitasia, who accounted for Jacques Wertheimer's Minstrel Girl by 1½ lengths. Pitasia was trained for Colonel Sir Douglas Clague in Chantilly by Aage Paus. Pitasia had earlier won the Prix Robert Papin and been second to Irish River in the Prix Morny.

The rain-softened ground had caused several misgivings amongst the connections of Alleged, and Lester Piggott remarked reflectively that he had never run on ground like that before. In the paddock Alleged looked quiet, relaxed and very fit. Trillion was turned out in superb condition by Maurice Zilber, and both Dancing Maid and Frère Basile were difficult to fault. Acamas, however, did not look as well as he had done at Chantilly and there was a suspicion that his coat was about to break. In the parade in front of the stands Julio Mariner's quarters were still covered by his rug which was only taken off when he turned to canter to the start.

Acamas was quite troublesome before the start and eventually was last into the stalls. Trillion, drawn one, broke first and led for the first 400 metres from Dom Alaric, Alleged, Frère Basile, Dancing Maid and Turville. They were followed by Noir et Or, Acamas and Guadanini, with Gay Mecene, Montcontour and Julio Mariner among the backmarkers, and Exdirectory, Lido and Shafaraz bringing up the rear. Shoemaker steadied Trillion behind the Petit Bois and Dom Alaric showed the way to Frère Basile, Alleged, Dancing Maid and Turville with little change in the order behind them except that it was clear that Exdirectory was already struggling to keep in touch. At the top of the hill Dom Alaric still led from Frère Basile and Alleged, with Trillion, Turville, Dancing Maid, Acamas and Noir et Or close behind. The order remained the same coming

Alleged (6) has 2 lengths and the same to spare over Trillion (8) and Dancing Maid (18) with Frère Basile (rails) fourth

down the hill when Montcontour and Guadanini began making progress from the rear. On turning into the straight Dom Alaric was still in front of Frère Basile and Alleged who were followed by Trillion, Dancing Maid, Turville, Noir et Or and Acamas on the wide outside. It was already very evident that Alleged was going much the best and that Piggott was poised to take the lead whenever he wanted. On the other hand Acamas was already coming under pressure and seemed beaten, and Gay Mecene was far too many lengths behind to threaten the leaders. With 300 metres to go, Piggott unleashed Alleged and he quickly and decisively accelerated into the lead, going two lengths clear of Trillion and Dancing Maid with Frère Basile and Noir et Or on their heels and Montcontour and Guadanini beginning to close on them. Meanwhile, Acamas was rapidly dropping back and, all too late, Gay Mecene was making a forward move from the rear. Trillion got the better of Dancing Maid 200 metres out, but she and Willie Shoemaker could make no impression on Alleged, who was cruising to victory. Lester Piggott looked round to his right and then to his left with a hundred metres to go for any possible dangers, but none was evident. Alleged passed the post a very comfortable 2 lengths in front of Trillion with another 2 lengths and three-quarters of a length to Dancing Maid and Frère Basile. Guadanini and Montcontour finished fast to take fifth and sixth places half a length and a head further back. They were followed by Noir et Or, Gay Mecene, Dancing Master and Lido.

Alleged thus became the first colt since Ribot to win the Prix de l'Arc de Triomphe in consecutive years, a feat only previously accomplished by Ksar, Corrida and Tantième. A tremendous cheer erupted when Alleged and a smiling Lester Piggott returned to the unsaddling enclosure. Vincent O'Brien, who had been immediately congratulated by Jean Romanet when he entered the unsaddling enclosure, raised his hat to acknowledge the tumultuous applause of the crowd. On dismounting, Lester Piggott said, 'Alleged is as good a horse as I've ever ridden. You can do anything with him and it was very easy today.' He added that Vincent O'Brien had achieved a great feat of training to nurse Alleged so patiently back to health, a compliment echoed by Robert Sangster. A huge mob of photographers converged on Alleged and Piggott on their return to the unsaddling enclosure and in their rush one ended on his back on the ground. When the dual Prix de l'Arc de Triomphe was led back to the stables the enormous crowd clapped him out.

Maurice Zilber, the trainer of Trillion, was quite critical of Shoemaker's riding; he felt that he was too small and unable to get the best out of her. This criticism was possibly a little unjust, although several other trainers felt that had the jockeys on the first two been reversed the finish might have been a little closer. The time for the race was, nevertheless, 6 seconds slower than in 1977 and even taking into account the soft going that was not exactly fast. Zilber's criticism that the jockeys should have forced the pace to make it a real test of stamina possibly had some justification. Edward Hide afterwards confirmed that the slow pace had not suited the disappointing St Leger winner Julio Mariner. Freddie Head had no excuses for Dancing Maid and commented, 'She was just not good enough'. The real disappointment of the race was Acamas whom, Yves Saint-Martin reported, 'had no action'. He felt that the trip to Ascot 'must have taken too much out of him'. The Aga Khan took the defeat of Acamas philosophically but added that François Mathet, Yves Saint-Martin and he had thought the colt 'was back to his best and worth running' and that 'he couldn't understand what had happened'.

Alleged's victory was a third success in the Prix de l'Arc de Triomphe for Lester Piggott and Vincent O'Brien and the second for owner Robert Sangster. The *objet d'art* presented to him was a gold soup tureen. Later in the afternoon he and his wife Susan and various friends celebrated the triumph by drinking champagne out of the tureen beside the bar at the far end of the paddock, and Mrs Sangster was seen to offer some to the Begum Aga Khan to compensate her for her disappointment at the running of Acamas.

Robert Sangster, who is the son of Vernon Sangster, the very successful football pools promoter in Liverpool, had come into racing in the early 1960s, encouraged by his friend Nicholas Robinson, the grandson of Sir Foster Robinson and now proprietor of *Pacemaker* magazine. His first horses were trained by Eric Cousins in Cheshire and included the useful handicapper Water Skier. As his enthusiasm grew, Robert Sangster increased and diversified his string. His horses were trained by Barry Hills, Vincent O'Brien, Ryan Price and Jeremy Hindley. In 1977 he was the leading owner in England and from the mid-1970s became one of the biggest buyers at the Keeneland and Newmarket Yearling Sales. In 1978 his racing empire was spread worldwide; he had horses in training and studs in Australia and California besides his stud in Ireland, and in France his horses were trained by Miguel Clément, until the latter's tragic death in a car accident. In 1978 he was already a member of the Jockey Club and had become one of the dominant figures in international racing and breeding.

The French had to be content with second place again in the Prix de l'Opéra when Sir Michael Sobell's Cistus, who had been runner-up to Reine de Saba in the Prix de Diane, held the fast-finishing Reine Imperiale by a head. Her trainer, Dick Hern, had been reluctant to run Cistus in the soft going and had left the decision to her owner. Clearly the Irish and English played up their winnings from Alleged on the filly for she was supported from 35 to 4 to 27 to 10. The last race was won by Natchitoches trained by Bernard Sécly, who thus received some compensation for Frère Basile's defeat in the Prix de l'Arc de Triomphe.

The Rest of the Season

There was some talk of Alleged running in the Champion Stakes but eventually it was decided to retire him to stud. He had won six of his seven races in England and Ireland and had been second in the St Leger for a total prize money of £49,272. In France he had been unbeaten in three races and had collected 2,520,000 francs. He went to the Walmac Farm Stud in Lexington Kentucky to take up his duties as a stallion.

Only one week after the Prix de l'Arc de Triomphe Trillion ran in the Preis von Europa at Cologne on 8 October but could only finish third half a length and 2 lengths behind the Russian colt Aden and Tip Moss. Clearly the race came much too soon for her as this was several pounds below her best form. She then travelled to the United States and was far from disgraced when going down by a neck and a nose to Waya, who had won the 1977 Prix de l'Opéra, and to Tiller, in the Turf Classic at Belmont Park. The first two had earlier fought out the finish of the Man o'War Stakes. The previous day on 22 October, Mac Diarmida beat Dom Alaric 1½ lengths in the Canadian International Stakes at Woodbine Park. In the Washington D.C. International Mac Diarmida beat Tiller and Waya a head and 3 lengths with Trillion 1½ lengths away fourth and Frère Basile last of eight. It was, however, Trillion's fourth race in five weeks and that must have taken its toll.

Dancing Maid ran in the Champion Stakes, so often the graveyard of those concerned in the finish of the Prix de l'Arc de Triomphe. In the paddock at Newmarket she was beginning to break in her coat; she was beaten before the Dip and finished well behind Swiss Maid who defeated Hawaiian Sound by 1½ lengths. Noir et Or, however, upheld the Prix de l'Arc de Triomphe form when beating Air Peruvian a head in the Prix de Conseil de Paris with Dancing Master fifth. Dancing Master later was runner up to Campero in both the Grand Prix de Marseille and the Prix Max Sicard at Toulouse in the late autumn.

Stud Record

It is much too early to pass any judgment on Alleged as a stallion. His first crop only ran as two-year-olds in 1982 and one would expect his progeny to do better as three-year-olds. Nevertheless Heron Bay won the Larkspur Stakes at Leopardstown and the Lee Maiden Stakes at the Curragh. His filly Alligatrix won the Trident Maiden Stakes at Newmarket and ran third to Acclimatize in the Hoover Fillies Mile at Ascot, whilst Ghaiya won the Goldener Oktober Stakes at Goodwood. If Alleged becomes as good a stallion as he was a racehorse, he will be very good indeed. He was never seriously extended over a mile and a half; he galloped his adversaries into the ground in a manner reminiscent of Ribot, and while comparisons are invidious possibly only Sea Bird, Mill Reef or Nijinsky at his very best would have proved his superior. That has to be a matter of speculation, however, since he never met a colt of that calibre over a mile and a half.

PRIX DE L'ARC DE TRIOMPHE 2400 metres
at Longchamp on Sunday 1 October 1978. For three-year-olds and upwards. Going: Good to soft.
1,200,000 francs to the winner; 480,000 fr. to the 2nd; 240,000 fr. to the 3rd; 120,000 fr. to the 4th. 300,000 fr. to the breeder of the winner; 120,000 fr. to that of the 2nd; 60,000 fr. to that of the 3rd; 30,000 fr. to that of the 4th. Value to the winner: 1,200,000 fr.

ALLEGED	b.c.	by Hoist the Flag – Princess Pout by Prince John	4.59k	Robert Sangster	L. Piggott	1
TRILLION	b.f.	by Hail to Reason – Margarethen by Tulyar	4.57½k	Edward L. Stephenson	W. Shoemaker	2
DANCING MAID	b.f.	by Lyphard – Morana by Val de Loir	3.54½k	Jacques Wertheimer	F. Head	3
FRERE BASILE	br.c.	by Djakao – Pola by Hard Sauce	3.56k	Jean-Pierre Binet	P. Paquet	4
GUADANINI	ch.c.	by Luthier – Ilrem by Prudent	4.59k	Joseph Kaïda	H. Samani	5
MONTCONTOUR	br.c.	by Luthier – Moskvitchka by Mossborough	4.59k	Mme H. Hausmann	J.L. Kessas	6
NOIR ET OR	b.c.	by Rheingold – Pomme Rose by Carvin	3.56k	Paul de Moussac	M. Philipperon	7
GAY MECENE	b. or br.c.	by Vaguely Noble – Gay Missile by Sir Gaylord	3.56k	Jacques Wertheimer	J.C. Desaint	8
DANCING MASTER	gr.c.	by Dancer's Image – Polana by Botticelli	3.56k	Prosper Elmoznino	G. Dubroeucq	9
LIDO	b.c.	by Expresso – List by Wattcan	5.59k	P. Erik Pramm	A. Klimscha Jr	10
SHAFARAZ		by Levmoss – Asharaz by Sicambre	5.59k	Yurk Skalka	M. Planard	
DOM ALARIC		by Sassafras – Ordenstreue by Orsini	4.59k	Mme Daniel Malingue	A. Badel	
MONSEIGNEUR		by Graustark – Brown Berry by Mount Marcy	4.59k	Arthur Seeligson	A. Murray	
ACAMAS		by Mill Reef – Licata by Abdos	3.59k	H.H. Aga Khan	Y. Saint-Martin	
GALIANI		by Luthier – On the Wing by Tanerko	3.59k	André Ben Lassin	A. Lequeux	
TURVILLE		by Lyphard – Tia Bella by Le Fabuleux	3.56k	Mme Eugène C. Cashman	A. Gibert	
JULIO MARINER		by Blakeney – Set Free by Worden	3.56k	Marcos Lemnos	E. Hide	
EXDIRECTORY		by Ballymore – Regal Bell by Princely Gift	3.56k	Mme J.R. Mullion	W. Carson	

18 runners.
ALLEGED trained by M.V.O'Brien at Cashel in Ireland; bred by Mrs J.H. McKnight in the U.S.A.
Distances: 2 lengths; 2 lengths; ¾ length; ½ length; head; 1½ lengths; 2½ lengths; ¾ length; 1½ lengths. Time: 2 mins 36.50 seconds.
Pari-Mutuel: Win: 2fr. 40. Places: 1fr. 50; 3fr. 50; 1fr. 90.
S.P.: 14 to 10 ALLEGED; 15 to 4 Acamas; 4 to 1 Dancing Maid and Gay Mecene (couplé); 14 to 1 Frère Basile; 16 to 1 Trillion; 17 to 1 Julio Mariner; 24 to 1 Exdirectory; 29 to 1 Noir et Or; 36 to 1 Guadanini; 53 to 1 Montcontour; 55 to 1 Turville; 56 to 1 Monseigneur; 60 to 1 Lido; 64 to 1 Galiani; 69 to 1 Dancing Master; 72 to 1 Shafaraz; 80 to 1 Dom Alaric.

Three Troikas 1979

Three Troikas

The Field

For the third consecutive year there was a foreign-trained colt as favourite for the Prix de l'Arc de Triomphe in 1979. Sir Michael Sobell's Troy, who started at 8 to 10, was one of nine foreign challengers, seven from England, one from Germany and one from the United States. Not merely was Troy the virtually unanimous selection of the English and Irish press, but nineteen of the twenty-five racing correspondents of the most important French newspapers thought he would win, despite his never having run in France.

As a two-year-old, Troy was runner-up on his début in the Shrewton Maiden Stakes at Salisbury and then won the Plantation Maiden Stakes (7f) at Newmarket in July. He beat a green Ela-Mana-Mou by 2½ lengths in the Lanson Champagne Stakes at Goodwood. In the Royal Lodge Stakes at Ascot, Troy had a terrific struggle with Lyphard's Wish and got the better of him inside the final furlong but was unable to hold Ela-Mana-Mou's late challenge and went down by three-quarters of a length. Troy's first appearance at three years was in the Classic Trial Stakes at Sandown on 28 April when he held Two of Diamonds by a neck. He looked fitter in the Predominate Stakes at Goodwood on 23 May and proceeded to trounce Serge Liafar and Galaxy Libra. In the Derby, Sir Michael Sobell's colt, whom Carson had elected to ride in preference to Milford and Niniski, was only eighth on the rails with two furlongs to go, but as soon as Carson pulled him out to challenge on the outside it was clear that Troy was going much better than any of his rivals. He drew away from his adversaries in the final furlong and had 7 lengths and 3 lengths to spare over Dickens Hill and Northern Baby, with Ela-Mana-Mou 2 lengths away fourth. It was a truly spectacular victory and no colt had won the Derby by such a wide margin since Manna in 1925. Troy again impressively accounted for Dickens Hill by an easy 4 lengths in the Irish Sweeps Derby on 30 June with Fabulous Dancer, the winner of the Prix du Lys, a disappointing ninth. Unfortunately the field for the King George VI and the Queen Elizabeth Diamond Stakes only attracted seven runners and was not as strong as usual. Troy beat Gay Mecene, the easy winner of the Grand Prix de Saint-Cloud, by 1½ lengths with Ela-Mana-Mou 3 lengths away third and Telescopio fifth. After his victory at Ascot it was announced that Troy's autumn objective would be the Prix de l'Arc de Triomphe with a preliminary race in the September Stakes at Kempton or the Valdoe Stakes at Sandown. This plan was changed, however, and the Derby winner ran instead in the Benson and Hedges Gold Cup at York. Troy beat Crimson Beau and Lyphard's Wish by three-quarters of a length and 4 lengths but had to struggle throughout the final furlong to hold on to his advantage. He had been a long way behind the leaders three furlongs out and Carson had to work very hard to take him into the lead. It was clear that Troy was much better suited by a mile and a half than by ten and a half furlongs and this hard race coming at the end of a strenuous spring and summer campaign cannot have helped his chance seven weeks later at Longchamp. Troy had been victorious in all his six races, but he had had no rest since 28 April. The question arose, therefore, as to whether seven weeks would be sufficient time after the Benson and Hedges Gold Cup to bring him back to his peak in the Prix de l'Arc de Triomphe. His brilliant and scintillating victories fully justified his position as favourite on 7 October, but the narrowness of his success at York meant that he could not be considered invincible in what was likely to be the toughest test of his career. He was accompanied to the post in the Prix de l'Arc de Triomphe by two pacemakers: Rivadon, trained by Dick Hern; and Player, trained by John Cunnington Jr.

Sir Philip Oppenheimer's Ile de Bourbon had done very well physically from three to four when he reappeared in the Clive Graham Stakes (10f) at Goodwood on 23 May. He had no difficulty in outclassing Crimson Beau by 4 lengths; this was a fine display considering it was his first race since his mystifying loss of form in the St Leger in 1978. The Coronation Cup had to be delayed by twenty-four hours due to a torrential downpour and the going was very heavy. Ile de Bourbon led from the start and when John Reid asked him to set sail for home rounding Tattenham Corner, he proceeded to draw further and further away from his rivals. At the winning post he had Frère Basile and Gay Mecene trailing 7 lengths and 3 lengths in his wake with M-Lolshan only fourth. It was a victory possibly even more impressive than that of Troy's in the Derby. Ile de Bourbon was then stricken by a virus infection which laid him low for several weeks and forced him to miss a clash with Troy in the King George VI and the Queen Elizabeth Diamond Stakes. He was indeed so ill in July that there were doubts about his ever running again in 1979. He was thought to have recovered sufficiently to run in the September Stakes at Kempton on 8 September but he was defeated a short head by Cracaval who had been a distant eighth in the Derby but had won the Chester Vase. The race was run in record time, however, and it was hoped that the month before the Prix de l'Arc de Triomphe would allow Ile de Bourbon to make a full recovery from his illness. If he could regain the form that he had shown in the Coronation Cup and in the 1978 King George VI and the Queen Elizabeth Diamond Stakes, Sir Philip Oppenheimer's colt would present an enormous threat to Troy and all his other rivals on 7 October, but in view of his severe illness his odds of 9 to 1 seemed rather ungenerous.

Crimson Beau, who was trained by Paul Cole and had won the Prix de la Côte Normande in 1978, ran Spring in Deepsea to a head in the Diomed Stakes at Epsom after his second to Ile de Bourbon at Goodwood. In the Prince of Wales's Stakes, Crimson Beau led throughout and gamely held Lyphard Wish's challenge by a length with Pevero 8½ lengths away fifth. In the Joe Coral Eclipse Stakes at Sandown on 7 July Crimson Beau ran Dickens Hill to 2 lengths with Northern Baby a length away third. Prior to running Troy so close at York, Crimson Beau was third 1½ lengths behind Noir et Or in the

Grand Prix Prince Rose (2200 metres) at Ostende. He faded in the final 200 metres, however, and it was clear that ten furlongs was his best distance. For that reason he was permitted to start at 82 to 1 on 7 October.

There were three other challengers from England, Raymond Guest's Noble Saint, Antony Shead's Two of Diamonds and George Ward's Valour, but it was difficult to envisage any of them taking an active part in the Prix de l'Arc de Triomphe. Noble Saint's best performance in 1979 was to beat Buttress and Milford 1½ lengths and 5 lengths in the Great Voltigeur Stakes at York in soft going. He had earlier won a maiden race at Newmarket in July. He was certainly a considerably better horse in soft or heavy going but on balance was likely to be outclassed in Paris as he had been the previous autumn in the Grand Critérium. After his close second to Troy in April, Two of Diamonds beat Main Reef in the Dee Stakes at Chester, but was a well-beaten fourteenth in the Derby. In his only other race before the Prix de l'Arc de Triomphe, he ran the previously disappointing four-year-old, Fordham, owned by Robert Sangster, to three-quarters of a length in the Joe McGrath Memorial Stakes at the Curragh on 24 September. The odds on favourite Dickens Hill finished only fifth of six but he ran stones behind his best form. Two of Diamonds's odds of 83 to 1 on balance seemed a fair reflection of his chance at Longchamp. George Ward was represented by his colt Valour in the Prix de l'Arc de Triomphe. Valour had caused a big surprise when he beat Noir et Or a neck in the Prix Jean de Chaudenay at Saint-Cloud on 4 June, but the runner-up was not always prepared to struggle when challenged. George Ward's colt was a well-beaten eighth in the Grand Prix de Saint-Cloud and failed to win in five outings in England in 1979. He was to start one of the total outsiders on 7 October at 99 to 1.

The challenger from the United States was the seven-year-old ex-Norwegian champion, Noble Dancer, who had run fourth to Ivanjica in 1976 before going to America to race. In the United States Noble Dancer had enjoyed great success; in 1978 he had won six times including the United Nations Handicap and the San Luis Rey Stakes. Prior to coming to Longchamp, Noble Dancer had been successful in four of his six outings in 1979. He won the San Luis Rey Stakes by 4 lengths from Tiller and the Pan-American Handicap at Gulfstream Park. He beat Dom Alaric 2 lengths in the United Nations Handicap at Atlantic City on 1 September and in the spring had run Bowl Game to three-quarters of a length in the Hialeah Turf Cup Handicap. Although no seven-year-old had won since Motrico in 1932, it was surprising that Noble Dancer was permitted to start at 99 to 1, seeing that he had won twenty-two races from forty-two outings in his career and more prize money than any of his adversaries at Longchamp.

It was much more difficult to envisage the German-trained but Polish-bred five-year-old Pawiment playing any significant part in the finish of the Prix de l'Arc de Triomphe. He had won once from five outings in 1979 prior to 7 October; he had taken the Grosser Preis von Dortmund on 3 June, but he had only managed fifth place to Nebos and Obraztsovy in the Grosser Preis von Berlin in July. As a three-year-old he had won the Polish 2000 Guineas and run second in the Polish Derby and had been successful in Czechoslovakia as well as Germany before coming to Paris.

The Aga Khan's High Top colt Top Ville was considered to be the greatest French danger to Troy in the Prix de l'Arc de Triomphe. As a two-year-old he had won three times including the Prix Saint-Roman and the Prix de Condé, and had been rated only 7 lbs behind Irish River in the Handicap Optional. On his first outing as a three-year-old Top Ville beat the very useful Bellypha by half a length in the Prix de Guiche on 16 April. He then convincingly beat Sharpman and Irish River, who failed to stay, by a length and three-quarters of a length in the Prix Lupin, in record time on 15 May. In the Prix du Jockey Club the Aga Khan's colt took the lead 400 metres out and quickly went 3 lengths clear of Le Marmot, but the latter fought back gamely. At the winning post, Top Ville still had an advantage of a length over Le Marmot, the pair being 5 lengths clear of Sharpman. He had also broken the course record and looked to be a very good colt. The going at Chantilly was good, but in his preliminary race before the Prix de l'Arc de Triomphe, the Prix Niel, Top Ville was faced by soft going. He could only manage fourth place behind Le Marmot, Fabulous Dancer and his own pacemaker Kamaridaan, though he was beaten less than 2 lengths. It was, however, three months since he had seen a racecourse and, although he had appeared on edge before the race, his trainer François Mathet had four weeks to bring him back to his peak by 7 October. Top

Ville's defeat in the Prix Niel did not exactly advertise his chance of beating Troy and his odds of 23 to 4 (couplé) were not exactly tempting. His pacemaker, Kamaridaan, was by no means useless. He had won a small race in heavy going at Maisons-Laffitte in March where his victims included the subsequent St Leger winner Son of Love, and then had run third to High Sierra and Son of Love in the Prix Noailles. He was only beaten a neck and a head by Le Marmot and Fabulous Dancer in the Prix Niel, although he was admittedly receiving 2 kilos from the winner.

Rodolph Schafer's Le Marmot won twice as a juvenile, including the Prix la Rochette, before finishing a disappointing eighth to Irish River in the 1978 Grand Critérium. On his reappearance on 8 April at Longchamp, Le Marmot trounced Wolverton and subsequent Grand Prix de Paris winner, Soleil Noir, 4 lengths and a length. He was very well suited by the heavy going that afternoon and he encountered similar conditions in the Prix Hocquart on 6 May when he beat High Sierra 3 lengths with Son of Love 9 lengths away fourth. If he had had similar conditions in the Prix du Jockey Club, it is conceivable that he might have beaten Top Ville, whose burst of acceleration 400 metres out left him unable to respond immediately. He did however, run on to finish a creditable and close second. In the Prix Niel he was receiving a kilo from Top Ville so there should have been little between them on 7 October, but he was giving a kilo to Fabulous Dancer and could certainly expect to confirm that form. There was little to choose between Le Marmot and Top Ville, and Rodolph Schafer's colt started at the slightly longer odds of 29 to 4.

Mme Alec Head's Fabulous Dancer won a maiden race at Saint-Cloud over a mile in April on his racecourse début and then took the Prix Matchem at Evry. He beat Northern Baby a head in the Prix La Force on 20 May and on his fourth outing had two lengths to spare over Scorpio in the Prix du Lys at Chantilly on 10 June. Fabulous Dancer ran well below his best in the Irish Sweeps Derby, but finished very fast when a neck behind Le Marmot in the Prix Niel. He was trained by Christiane Head, the daughter of Alec Head. She also saddled the best French three-year-old filly, Three Troikas, in the Prix de l'Arc de Triomphe. Three Troikas, who was owned by Mme Alec Head, won her only outing as a two-year-old at Saint-Cloud in November. She beat Dunette by an effortless 3 lengths in the Prix Vanteaux at Longchamp on 16 April, and two weeks later Three Troikas was most impressive when accounting for Nonoalca, the Prix de la Grotte winner, by 2½ lengths in the Poule d'Essai des Pouliches. In the Prix Saint-Alary Mme Head's filly easily defeated Pitasia and Salpinx by 2 lengths and half a length, with Dunette a well-beaten fifth after running far too freely. Three Troikas, in fact, beat the course record in the Prix Saint-Alary which had been set by Solitude in 1961. The afternoon of the Prix de Diane de Revlon was particularly unpleasant; it was raining continually. Three Troikas did not seem to be moving as well as she had done at Longchamp and had quite a struggle to get the better of Nonoalca and Producer 200 metres out, but she had no answer to Dunette's late challenge and went down by a nose with Producer and Pitasia 2½ lengths and half a length away third and fourth. Christiane Head explained this defeat by saying that Three Troikas had a bruised foot, which may well have contributed to it. Dunette, however, had a strong dislike of Longchamp in addition to being a difficult ride, but on 10 June she had settled well and there was no doubt she performed better at Chantilly. Three Troikas was rested until the Prix Vermeille on 16 September where she was opposed by the Oaks winner Scintillate. Three Troikas mastered Pitasia inside the distance and Freddie Head did not push her out so as not to give her a hard race. As a result she was nearly caught by Salpinx, who was fifth in the Prix de Diane de Revlon but better suited by the longer distance of the Prix Vermeille. On the line, Three Troikas held Salpinx by a short head with Pitasia and Dunette half a length and 1½ lengths back third and fourth and Scintillate only ninth. Mme Alec Head's filly certainly could have won by a greater margin and she and Fabulous Dancer started at 88 to 10 (couplé) for the Prix de l'Arc de Triomphe. She might easily have been unbeaten before 7 October when she seemed sure to be one of Troy's principal rivals.

Colonel Sir Douglas Clague's Pitasia had won three of her four outings as a juvenile including the Critérium des Pouliches and the Prix Robert Papin, and had been runner-up to Irish River in the Prix Morny. After winning the Prix Imprudence at Maisons-Laffitte, Pitasia was a disappointing seventh to One in a Million in the 1000 Guineas. She came back to her best when runner-up in the Prix Saint-Alary and was far from disgraced when failing by a neck behind

Young Generation in the Prix Jean Prat at Chantilly on 3 June. A week later she was fourth in the Prix de Diane de Revlon. Pitasia then beat Producer and Salpinx in the Prix de Malleret at Longchamp on 24 June. In the Prix de la Nonette at Longchamp in September she again proved Salpinx's superior with Producer third. She had never finished in front of Three Troikas, however, in their three meetings and was probably flattered by her proximity to Mme Head's filly when third in the Prix Vermeille. The very best that Pitasia could hope for in the Prix de l'Arc de Triomphe was one of the minor places. Pitasia's stable-companion and pacemaker, Princess Redowa, won two races over a mile at Sandown when trained in England by John Winter in the early summer, after which her owner brought her to France.

In addition to Le Marmot, François Boutin's stable also ran another three-year-old, Northern Baby, in the 1979 Prix de l'Arc de Triomphe. He began his three-year-old campaign by winning at Saint-Cloud and twice finishing runner-up, first to Wolverton in the Prix de Suresnes on 29 April and then a head behind Fabulous Dancer in the Prix la Force. Although 10 lengths behind Troy at Epsom, he beat Ela-Mana-Mou three-quarters of a length for third place and he was a good third to Dickens Hill in the Joe Coral Eclipse Stakes. Northern Baby impressively won the Prix de la Côte Normande at Deauville in August and then ran Rusticaro and Trillion to half a length and a head in the Prix du Prince d'Orange at Longchamp on 16 September. Despite his consistency he was allowed to start at the extremely generous odds of 90 to 1, possibly because it was considered that a mile and a quarter was his best distance.

The five-year-old mare Trillion, owned in partnership by Edward Stephenson and Nelson Bunker Hunt, won the Prix d'Harcourt from Fatusael and Frère Basile on 1 April, but the latter gained his revenge by 1½ lengths in the Prix Ganay four weeks later with Pevero a short head away third. Trillion beat Pevero 2½ lengths in the Prix Dollar on 27 May, but this turned out to be her last victory before the Prix de l'Arc de Triomphe. After finishing third, 5 lengths behind Irish River, in the Prix d'Ispahan, Trillion was disqualified and placed fifth and she was well beaten in the Grand Prix de Saint-Cloud. A rejuvenated Rusticaro then beat her 2 lengths in the Prix Gontaut-Biron at Deauville. Trillion was runner-up 1½ lengths behind Pevero in the Prix Foy on 9 September with Gay Mecene half a length away third, and she could only manage second place behind Rusticaro in the Prix du Prince d'Orange a week later. It appeared that she was not as good as she had been in the previous two seasons and her odds of 16 to 1 on 7 October were to a large extent based on her past reputation and performances.

When Gerald Oldham's Pevero won the Prix de Condé and Prix de Nexon in 1977 as a juvenile, he looked a possible classic colt. He was extremely badly hampered when challenging Naasiri and Acamas in the Prix Greffulhe the following April. This seemed to affect his confidence in the remainder of his races as a three-year-old; he did finish fifth in the Prix du Jockey Club but on balance was a disappointment in 1978. As a four-year-old he was third in the Prix Ganay and runner-up in the Prix Dollar but only fifth to Crimson Beau in the Prince of Wales's Stakes at Ascot. When Pevero won the Prix Foy, however, he produced a burst of acceleration in the final 200 metres that he had not shown since his two-year-old career and it seemed that he was a possible outsider on 7 October.

Mahmoud Fustok's Telescopio had won five consecutive races in Argentina in 1978 and was imported into France in the middle of the summer of 1979. He was a well-beaten eighth on his début in Europe in the Prix Maurice de Nieuil at Saint-Cloud on 14 July. He was not disgraced when 6 lengths fifth to Troy at Ascot, but was easily beaten by Shafaraz at Longchamp on 2 September. He did beat Rayanly 8 lengths in the Prix du Saint-Laurent at Maisons-Laffitte on 1 October. Although his victories in Argentina had included the Gran Premio Internacional Carlos Pellegrini, the Gran Premio Nacional and the Gran Premio Jockey Club, the form he had shown in France and England did not entitle him to start at such short odds of 17 to 1 for the Prix de l'Arc de Triomphe and it was difficult to envisage him taking any part in the finish.

The field was completed by the four-year-old Jeune Loup who had won the Prix de Reux at Deauville. He had been promoted to second place to First Prayer in the Grand Prix de Deauville because Piggott on African Hope had snatched Alain Lequeux's whip from him in the final furlong as he had dropped his own earlier in the race. It seemed very likely that Jeune Loup would find several of his adversaries much too fast for him on 7 October and he was one of the total outsiders at 99 to 1.

Alec Head decided not to run Gay Mecene in the Prix de l'Arc de Triomphe in 1979 after his defeat in the Prix Foy, preferring to run him in the Preis von Europe instead. Dunette was also an absentee because she was coughing, while Dickens Hill had also been found to be coughing after the Joe McGrath Memorial Stakes. Frère Basile was reported to be injured after the Prix Foy. Ela-Mana-Mou had pulled a muscle at Ascot and his autumn objective was the Champion Stakes. Barry Hill's stable was badly affected by the virus in 1979, and Hawaiian Sound, after disappointing in the Prix Ganay and in the Prince of Wales's Stakes at Ascot, never ran again. Dancing Maid also only had one run in 1979 when a poor sixth in the Prix Ganay. In spite of these absentees Troy still had a formidable task at Longchamp.

The Final Preparations

François Boutin felt that Le Marmot had the best chance of the three from his stable. He thought that Rodolph Schafer's colt had not been at his best when winning the Prix Niel and was very pleased with the rain that had fallen in Paris on the Thursday preceding the Prix de l'Arc de Triomphe. He also quietly fancied Pevero. Christiane Head was also hopeful that both Three Troikas and Fabulous Dancer would acquit themselves well. The Prix Vermeille heroine had been particularly impressive in her final gallop with Gay Mecene and Fabulous Dancer earlier in the week. Dick Hern thought that Troy was as well as he had been on the day of the Derby, but added that the pacemakers Player and Rivadon had to set a fast pace. Almost everyone expected Troy to notch up his seventh consecutive success in the Prix de l'Arc de Triomphe when they arrived at Longchamp on Sunday 7 October.

The week prior to the Prix de l'Arc de Triomphe had seen a call for a commission of enquiry into France's relations with the regime of the recently deposed Emperor Bokassa of the Central African Republic called by the opposition leader François Mitterrand, but it was an unruffled and relaxed Président Giscard d'Estaing who was present at Longchamp that afternoon. The presence of the best Derby winner for several years meant that the number of English visitors in the cosmopolitan crowd was even greater than normal.

The first race on Sunday 7 October was the Prix des Champs-Elysées which was won by Jacques Wertheimer's top weight Brenneville, partnered by Freddie Head. The brilliant winner of the Prix de l'Abbaye de Longchamp in 1978, Sigy, had not been as successful as a three-year-old and had finished a dismal last to Thatching in the William Hill July Cup at Newmarket. However, her trainer Christiane Head was of the opinion that Sigy was back to her best. In the Prix de l'Abbaye de Longchamp Sigy again proved a failure and a thrilling struggle was fought out by Double Form, Kilijaro and Greenland Park, who finished in that order separated only by a head and a nose. Double Form's trainer Fulke Johnson Houghton and his jockey John Reid had won the Cambridgeshire at Newmarket the previous day with George Ward's Smartset. If they managed to capture the Prix de l'Arc de Triomphe with Ile de Bourbon, it would be a remarkable treble. The favourite for the next race, the Critérium des Pouliches, was Sir John Astor's Green Dancer filly The Dancer trained by Dick Hern. She ran unaccountably badly and the race went to another Green Dancer filly, Aryenne, who beat the Danish-trained outsider Pom Poes a head with Teacher's Pet, another challenger from England, third.

During the early races several hundred motor cyclists and owners of mobilettes staged a noisy demonstration on the road beside the back straight. They were protesting against a new tax that had been levied on their machines. The policemen were able to move them on however, before the start of the Prix de l'Arc de Triomphe.

The Race

Like Fulke Johnson Houghton, Dick Hern's stable had achieved significant success the previous afternoon at Newmarket when he saddled the two-year-old filly Bireme who had run out a very impressive winner of the Westley Maiden Stakes. In the parade ring Troy looked superb and was so relaxed that he actually appeared to yawn a couple of times. Top Ville, however, seemed upset by some of the rowdier sections of the crowd. The owner of the Banstead Manor Stud, Hugo Morriss, who had visited Ile de Bourbon when he was ill, was

Three Troikas (22) passes the post 3 lengths clear of Le Marmot (15)

pleased to see how well he had recovered, since the colt was to stand at his stud. There was, however, little question that Ile de Bourbon had looked much better at Epsom. Three Troikas was as impressive as she had been before the Poule d'Essai des Pouliches and it was difficult to fault Le Marmot. Pevero, his stable-companion, was not on his best behaviour and had to be led all the way to the post. Pitasia and Noble Dancer also looked in peak condition. Crimson Beau was fractious at the start and more than a little reluctant to enter the stalls.

Pevero dwelt at the start and lost about 5 lengths. Alan Bond on Troy's pacemaker, Rivadon, immediately set a strong gallop, closely attended by Player, Princess Redowa, Kamaridaan and Crimson Beau. They were followed by Northern Baby, Top Ville, Le Marmot, Fabulous Dancer and Trillion. Carson settled Troy towards the rear of the field, while Head on Three Troikas tacked over from her number seventeen draw towards the inside rails and Pevero was clearly last. After the Petit Bois, Rivadon still was at the head of affairs but Crimson Beau and Fabulous Dancer had moved up to dispute second place with the other three pacemakers, but the order behind them remained virtually unchanged. Before the field reached the top of the hill, Geoff Baxter on Crimson Beau took the lead and to everyone's surprise went clear of the pacemakers and Fabulous Dancer. Had the race been over a mile it might have been a sensible and enterprising tactic but it put paid to any chance Crimson Beau might have had. Beginning the descent Crimson Beau was still a long way in front; the pacemakers and Fabulous Dancer began to weaken and he was followed by Northern Baby.

He was tracked by a group consisting of Pitasia, Le Marmot, Top Ville and Trillion. It was at this stage that Three Troikas was switched off the rails past Fabulous Dancer and began to make significant progress on the outside. Meanwhile Troy, who had had only three behind after the Petit Bois, had moved up to the middle of the field where he joined Ile de Bourbon. Two hundred metres from the turn into the straight, Crimson Beau was coming back to Northern Baby and Le Marmot; Pitasia and Trillion came next with Three Troikas going well just behind. Troy and Ile de Bourbon still had about ten horses in front of them and Sir Philip Oppenheimer's colt was already struggling, as was Top Ville a little in front of him. Entering the straight, Crimson Beau was at the end of his tether and Northern Baby took up the running-tracked by Le Marmot, Pitasia, who was under pressure, and Three Troikas on the outside going very easily. Trillion was on their heels in front of Troy, who was beginning a run on the outside, and Pevero, who had made up a tremendous amount of ground coming down the hill on the inside. Le Marmot passed Northern Baby about 400 metres out but Three Troikas was poised to challenge on his outside. Even wider but still only seventh, Troy was making ground but he needed to produce the acceleration he had shown at Epsom if he were to win. With 200 metres to go, Three Troikas swept past Le Marmot and, lengthening her stride impressively, she had put 3 lengths between herself and Rodolph Schafer's colt when she crossed the line. Troy ran on to take third place another length away, but only a neck in front of the fast-finishing Pevero, who was 4 lengths ahead of Trillion. Northern Baby had faded quickly in the last 300 metres and was half a length back sixth. He was followed by Jeune Loup, Noble Saint, Pawiment and Crimson Beau. There was absolutely no question that Three Troikas was the best horse on the day and the manner of her victory was both authoritative and clear-cut.

It was a remarkable family success. Three Troikas was owned by Mme Alec Head, trained by her daughter Christiane and ridden by her son Freddie, who was gaining his fourth victory in the Prix de l'Arc de Triomphe, thus equalling Jacques Doyasbère's record for the race. Although the going was good to soft, the three-year-old filly had recorded a time only six-tenths of a second outside Mill Reef's and San San's record. It was the first time that a woman trainer had won the Prix de l'Arc de Triomphe, although Miss Brita Strokirk had saddled the Swedish challenger Lycaste II in 1959. A delighted Christiane Head, only in her second season as a trainer, welcomed Three Troikas and her brother on the course and led them into the unsaddling enclosure. Afterwards she modestly commented: 'I owe everything to my mother for giving me all the horses and to my father who taught me everything.' She also revealed that it was not until after a gallop the previous Tuesday that Freddie Head had decided to ride Three Troikas in preference to Fabulous Dancer. She added that she thought Three Troikas 'had real class and, but for a bruised foot, she would have won the Prix de Diane de Revlon and been unbeaten'. Her brother's post-race comment was: 'From the moment we turned into the straight I knew I could beat the horses in front of me but I was afraid that Troy might produce the same turn of speed that he had done at Epsom. Fortunately it was not forthcoming this time.'

Dick Hern put a brave face on Troy's first defeat as a three-year-old saying, 'He ran a great race.' Willie Carson also had no excuses, but added that he had been 'on the go for a long time and if we had met the others when fresh we would have killed them'. Inevitably the English began an inquest into Troy's failure. Significantly, that very able trainer Ryan Price had remarked after Troy's narrow victory in the Benson and Hedges Gold Cup that he had showed signs of going over the top. The most likely reason for the failure of Troy and other English colts was that in the 1970s it had become fashionable to run the best classic horses in the Derby, the Irish Sweeps Derby, the King George VI and the Queen Elizabeth Stakes and in the Benson and Hedges Gold Cup. These races all come in a period of between ten and eleven weeks, and the last, inaugurated in 1972, is run over the shorter distance of ten and a half furlongs in the middle of August. Naturally, a trainer tries to bring his Derby colts to a peak for Epsom and it is virtually impossible to keep a colt in top condition for more than two months. If a colt has won the Derby and the King George VI and the Queen Elizabeth Diamond Stakes, or the Eclipse Stakes if his stamina is suspect, there can be little reason to run him at York, for his value as a stallion is already established. With the exception of Troy and possibly Roberto, the other colts who won the Benson and Hedges Gold Cup in its first ten years – Moulton,

Relkino, Hawaiian Sound, Master Willie, Beldale Flutter, and even the 2000 Guineas winner Wollow – would never have commanded the same value when syndicated as a dual winner of the Derby and Ascot's great race. A Derby winner who runs at York adds little if anything to his value as a stallion, but perforce must forego a reasonable period of rest before running in the Prix de l'Arc de Triomphe. If he wins in Paris his value as a stallion will certainly be enhanced. The fashion of running Epsom Derby winners in the Irish Sweeps Derby is also questionable. The victor at Epsom has already proved himself the best of his generation, and if he has had a hard race at the beginning of June he is on a hiding to nothing, but should he win he is only confirming what we already know. The case of Assert, who had won the Prix du Jockey Club in 1982 but had not met any English classic colts, is of course another matter, as he did have something to prove. Mill Reef, the last colt to win the Derby, the King George VI and the Queen Elizabeth Stakes and the Prix de l'Arc de Triomphe, was rested after Ascot until his victory in Paris. Although he won the Eclipse Stakes at the beginning of July, he only had three races between the end of May and the Prix de l'Arc de Triomphe. Sea Bird only ran in the Grand Prix de Saint-Cloud after the Derby before the Prix de l'Arc de Triomphe, and Vaguely Noble had only one preparatory race between his defeat at Saint-Cloud and his victory at Longchamp. Although there have been several vintage crops of fillies in France in recent years, they do not have such a packed and strenuous programme as the colts in June, July and August. Therefore they arrive at Longchamp in October fresher, having been rested in midsummer and brought back to their peak for the Prix de l'Arc de Triomphe. This less hectic programme may partly account for the number of their victories in the past decade.

There were few excuses for the others who finished behind Three Troikas. Alain Lequeux said that Pevero's loss of ground at the start made little difference as he would have dropped him out anyway. Ile de Bourbon clearly had not got over his illness and ran a long way behind his best form. Yves Saint-Martin reported that Top Ville had only raced for seven furlongs; the only two possible explanations were that he was a much better horse on firm going or that he had gone over the top. Philippe Paquet said that Le Marmot 'ran a great race but came up against a filly who was just that bit better', whilst Lester Piggott reported that Trillion ran her usual honest race but never gave him any hopes of winning.

François Boutin had achieved the remarkable feat of training the second, the fourth and the sixth horse to finish behind Three Troikas, and gained some compensation when his charge Vagaries won the final race of the afternoon, the Prix de la Concorde. Maurice Zilber and Lester Piggott also did not come away empty-handed, for Producer won the Prix de l'Opéra.

The Rest of the Season

Three Troikas did not run again in 1979 but remained in training as a four-year-old. Northern Baby's victory in the Champion Stakes over Town and Country and Haul Knight more than franked the merit of Three Troika's performance. Salpinx beat Brenneville three-quarters of a length in the Prix du Conseil de Paris with Jeune Loup 1½ lengths away fourth, and thus provided further proof of the excellence of Mme Alec Head's filly. Trillion again campaigned in the United States in the autumn; she was a neck runner-up, first to Golden Act in the Canadian International Stakes on 21 October, and then to Bowl Game, who had won the Man o'War Stakes, in the Turf Classic a week later. In the Washington D.C. International Stakes at Laurel Park on 10 November Bowl Game beat Trillion and Le Marmot three-quarters of a length and a nose; the latter was not particularly well suited by the course but neither could be said to have let down the Prix de l'Arc de Triomphe form. In the Premio Roma on 18 November Noble Saint beat Telescopio 1½ lengths, the pair finishing 4 lengths clear of Quadrupler, the St Leger Italiano winner.

Pedigree and Stud Record

Three Troikas was bred by A. Pfaff but raised by the Comte Roland de Chambure at the Haras d'Etreham. She was sent to the Newmarket Houghton Yearling Sales and the British Bloodstock Association Ireland bought her for 41,000 guineas, acting for Mme Alec Head. One of Robert Sangster's trainers, Dermot Weld, had wanted to buy her but was persuaded against her by his veterinary surgeon. Three Troikas was by Lyphard who won six of his twelve races. As a two-year-old he was successful in the Prix Herod and the Prix de l'Hudson River and was fourth in the Prix de Saint-Firmin. As a three-year-old he was reputedly the best colt in Alec Head's yard and won the Prix Lagrange and Prix Daru before finishing an unlucky fourth in the Prix Lupin behind Hard to Beat. He lost all chance in the Derby when running very wide at Tattenham Corner; his jockey, Freddie Head, was severely criticised but it has since been revealed that Lyphard was a rig and he suffered great pain when trying to turn on a left-handed course. He almost certainly did not stay a mile and a half; although he was by Northern Dancer, he was out of a Court Martial mare and his subsequent victories were in the Prix Jacques le Marois and the Prix de la Forêt. He was also second in the Prix du Moulin de Longchamp. He was clearly an extremely good colt but his record as a stallion is even more impressive. He headed the sires list in France in 1978 and 1979, was second in 1977 and third in 1980. His progeny in France up to September 1982 had won 218 races and 26,441,275 francs. In France his best winners besides Three Troikas were Dancing Maid, Reine de Saba, Pharly, Al Nasr (Prix d'Ispahan, Prix de la Côte Normande, Prix La Force, Prix Dollar and Prix Exbury), Calderina (Prix de Malleret), and Bellypha (Prix de la Jouchère, Prix Daphnis, Prix Quincey and Prix Thomas Bryon). In England and Ireland his stock won forty-four races and £348,529 in the same period and his best winners were Durtal, Monteverdi (William Hill Dewhurst Stakes), Lyphard's Wish (Ladbroke Craven Stakes and Mecca Dante Stakes), and Lyphard's Special (Champagne Stakes Salisbury, 2nd Royal Lodge Stakes, 3rd William Hill Futurity Stakes). Lyphard was exported to the U.S.A. at the end of 1977 where, since 1978, he has sired winners of sixty races worth $1,720,356, including Sangue. Lyphard is also the sire of Lydian, the winner of the Gran Premio di Milano and the Grosser Preis von Berlin.

Three Troikas's dam, Three Roses, was by Dual who had been trained by Jack Gosden for Sir Harold Wernher. Dual won the Solario Stakes and was runner-up in the Royal Lodge Stakes as a juvenile. He won the 2000 Guineas Trial Stakes at Kempton, beating the subsequent Derby winner Psidium into third place, and the Doonside Cup at Ayr, and was a short head second to Just Great in the Brighton Derby Trial Stakes. He ran unplaced in the Derby. As a four-year-old his best performances were to finish third in the John Porter Stakes and in the Coronation Stakes at Sandown. Dual stood at the Russelltown Stud in West Meath in Ireland but was by no means a conspicuous success as a stallion. He died in 1976 and his progeny only won ninety-one races worth £77,972 in England and Ireland; the best was probably the handicapper Alloway Lad third in the 1969 Ebor Handicap. Three Roses won twice as a two-year-old, when trained by M. Connolley in Ireland, a maiden race at Navan over 6 furlongs and the Park Stakes at Phoenix Park. She was sold at the Newmarket December Sales for 8,200 guineas and raced in France at three and four but never won again in nine outings, although she was placed three times. Her first foal, Taifun by Linden Tree, won over 1800 metres at Longchamp in 1977 but was nothing exceptional. Three Troikas was her second foal, and the fact she was so inexpensive for a Lyphard filly was almost certainly due to her dam and maternal sire. Her third foal, Three Tops by Kashmir II, only ran once – at three years at Evry.

Three Roses's dam, Always Loyal by King's Troop, was very moderate on the racecourse. She was not even able to win in maiden company at such unpretentious tracks as Galway, Limerick and Tuam, though she was placed. Three Roses was her best produce. She got a minor winner of eight races between the age of three and seven in the U.S.A. named Solemn, and a moderate stayer, Melody Music by Tudor Music, who won three minor events as a four- and five-year-old.

Always Loyal's dam, Constant Worry by Torbido, was trained by Jack Jarvis for H.D. H. Wills as a two-year-old and three-year-old, until she won the Whitby Selling Handicap over a mile. She was then sold for 600 guineas and trained by Harry Blackshaw. She won the Leyburn Handicap Stakes (11f) at Catterick landing quite a gamble in the process. Constant Worry never won again however, either as a four-year-old or as a five-year-old, and only managed three places in selling handicaps in the north of England. She proved a little more successful in the paddocks getting three winners. Ghyll Manor by Giles de Retz won over five furlongs at Kempton and six furlongs at Windsor as a juvenile and as a four-year-old won three handicaps, the Bradwell Handicap at Yarmouth, the Donnetta Handicap at Kempton and the Turn of the Lands Handicap at Newmarket. The latter was also won by his half-brother, Prince Jamie by The Phoenix, who also won the Exning Handicap at Newmarket as a three-year-old and the Abbey Nursery Handicap at Bath as a juvenile. Her two-year-old Knight Alarm won a selling handicap at Lincoln before being exported to the West Indies. He had been sold for 220 guineas after his victory. He was successful seven times in the West Indies.

Constant Worry's dam, Idle Curiosity by Bois Roussel, was only placed and went to the paddocks a maiden. She did produce one other English-trained winner in addition to Constant Worry, Idle Rocks by Rockefella. As a two-year-old he won the Woodcote Stakes at Epsom, the Somerville Stakes (5f) at Newmarket and the Gimcrack Stakes in 1955. He was also third in the Coventry Stakes at Ascot. However he was extremely temperamental, detested the starting gate and nearly always dwelt at the start. As a three-year-old he totally refused to show his ability and was unplaced in the 2000 Guineas, the Derby and the St Leger. He was exported to India, where he must have been miraculously reformed to gain seven victories, including the Jammu Stud Gold Cup. Idle Curiosity also produced Uncle Blue who won thirteen races and $14,039 in the U.S.A. and ran third in the Niagara Handicap.

Idle Curiosity's dam, Fairetta by Fairway, never saw a racecourse but was the dam of two winners, the useful handicap stayer Persetta and the moderate filly Sarala. Fairetta's dam, Caretta, was, however, a very good two-year-old filly who won the Queen Mary Stakes at Ascot and was runner-up, giving a stone to winner Lady Gabriel, in the Cheveley Park Stakes. She was rated the best two-year-old filly in the 1934 Free Handicap, 12 lbs behind Bahram. As a three-year-old Caretta ran third to Mesa in the 1000 Guineas for which she started favourite. She only ran once afterwards, when unplaced in the Coronation Stakes at Ascot. Caretta was also a good broodmare, producing Annetta by Fairway, the winner of the Irish 1000 Guineas in 1944. Caretta's grandam Tillywhim was also an ancestress of Panastrid and Levmoss. This family traced back to Rouge Rose by Thormanby, the dam of Bend d'Or.

Three Troikas was sold in 1981 privately by Mme Head to a syndicate of Americans of which the Overbrook Stud in Lexington had a 50 per cent share. It later bought out the other members for $1,250,000. In 1982 Three Troikas produced her first foal, a filly by Exclusive Native. It will be interesting to see if she can prove as brilliant a broodmare as she showed herself to be on the racecourse.

PRIX DE L'ARC DE TRIOMPHE 2400 metres
at Longchamp on Sunday 7 October 1979. For three-year-olds and upwards. Going: Good to soft.
1,200,000 francs to the winner; 480,000 fr. to the 2nd; 240,000 fr. to the 3rd; 120,000 fr. to the 4th. 300,000 fr. to the breeder of the winner; 120,000 fr. to that of the 2nd; 60,000 fr. to that of the 3rd; 30,000 to that of the 4th. Value to the winner: 1,200,000 fr.

THREE TROIKAS	b.f.	by Lyphard – Three Roses by Dual	3.54½k	Mme Alec Head	F. Head	1
LE MARMOT	b.c.	by Amarko – Molinka by Molvedo	3.56k	Rodolph Schafer	P. Paquet	2
TROY	b.c.	by Petingo – La Milo by Hornbeam	3.56k	Sir Michael Sobell	W. Carson	3
PEVERO	b.c.	by Busted – Caprera by Abernant	4.59k	Gerald Oldham	A. Lequeux	4
TRILLION	b.f.	by Hail to Reason – Margarethen by Tulyar	5.57½k	Edward L. Stephenson	L. Piggott	5
NORTHERN BABY	b.c.	by Northern Dancer – Two Rings by Round Table	3.56k	Mme Anne-Marie d'Estainville	P. Eddery	6
JEUNE LOUP	gr.c.	by Mill Reef – Skelda by La Varende	4.59k	Mme Jacques Herold	A. Badel	7
NOBLE SAINT	b.c.	by Vaguely Noble – Santa Paula by Santa Claus	3.56k	Raymond Guest	J. Mercer	8
PAWIMENT	b.c.	by Mehari – Pytia by De Corte	5.59k	Haras Moritzberg	G. Starkey	9
CRIMSON BEAU	ch.c.	by High Line – Crimson Belle by Red God	4.59k	H.T. Spearing	G. Baxter	10
NOBLE DANCER		by Prince de Galles – Helen Traubel by Sing Sing	7.59k	Haakon Frietheim	J. Vasquez	
ILE DE BOURBON		by Nijinsky – Roselière by Misti	4.59k	Sir Philip Oppenheimer	J. Reid	
VALOUR		by Vaguely Noble – Louisador by Indian Hemp	4.59k	George Ward	P. Cook	
TELESCOPIO		by Table Play – Filipina by Fomento	4.58½k	Mahmoud Fustok	S. Hawley	
KAMARIDAAN		by Djakao – Diamond Drop by Charlottesville	3.56k	H.H. Aga Khan	J. Héloury	
TOP VILLE		by High Top – Sega Ville by Charlottesville	3.56k	H.H. Aga Khan	Y. Saint-Martin	
FABULOUS DANCER		by Northern Dancer – Last of the Line by Northern Dancer	3.56k	Mme Alec Head	J. Taillard	
TWO OF DIAMONDS		by Blakeney – Santa Maria by Tropique	3.56k	A.D. Shead	S. Cauthen	
PLAYER		by Luthier – Illyria by Reform	3.56k	Sir Michael Sobell	E. Hide	
RIVADON		by Riva Bridge – Sarah Bernhardt by Buckpasser	3.56k	Sir Michael Sobell	A. Bond	
PITASIA		by Pitskelly – Asian Princess by Native Prince	3.54½k	Sir Douglas Clague	A. Gibert	
PRINCESS REDOWA		by Prince Regent – Ballydowa by Ballymoss	3.54½k	Sir Douglas Clague	M. Ouzouir	

22 runners.
THREE TROIKAS trained by Mme C. Head; bred by A. Pfaff.
Distances: 3 lengths; 1 length; neck; 4 lengths; ½ length; 2 lengths; 8 lengths; 2 lengths; 5 lengths. Time: 2 mins 28.90 seconds.
Pari-Mutuel: Win: 9fr. 80. Places: 2fr. 40; 2fr. 00; 1fr. 40.
S.P.: 8 to 10 Troy, Player and Rivadon (couplé); 23 to 4 Top Ville and Kamaridaan (couplé); 29 to 4 Le Marmot; 88 to 10 THREE TROIKAS and Fabulous Dancer (couplé); 9 to 1 Ile de Bourbon; 16 to 1 Trillion; 17 to 1 Telescopio; 35 to 1 Pitasia and Princess Redowa (couplé); 38 to 1 Pevero; 82 to 1 Crimson Beau; 83 to 1 Two of Diamonds; 90 to 1 Northern Baby; 99 to 1 Jeune Loup, Noble Saint, Pawiment and Noble Dancer.

THREE TROIKAS bay filly 1976	Lyphard b. 1969	Northern Dancer b. 1961	Nearctic b. 1954	Nearco br. 1935	Pharos
					Nogara
				Lady Angela ch. 1944	Hyperion
					Sister Sarah
			Natalma b. 1957	Native Dancer gr. 1950	Polynesian
					Geisha
				Almahmoud ch. 1947	Mahmoud
					Arbitrator
		Goofed ch. 1960	Court Martial ch. 1942	Fair Trial ch. 1925	Fairway
					Lady Juror
				Instantaneous b. 1931	Hurry On
					Picture
			Barra ch. 1950	Formor ch. 1934	Ksar
					Formose
				La Favorite b. 1934	Biribi
					La Pompadour
	Three Roses b. 1968	Dual b. 1958	Chanteur b. 1942	Château Bouscaut b. 1927	Kircubbin
					Ramondie
				La Diva br. 1937	Blue Skies
					La Traviata
			Duplicity b. 1945	Nearco br. 1937	Pharos
					Nogara
				Doubleton b. 1938	Bahram
					Double Life
		Always Loyal ch. 1963	King's Troop b. 1957	Pricely Gift b. 1951	Nasrullah
					Blue Gem
				Equiria b. 1946	Atout Maître
					Epona
			Constant Worry ch. 1949	Torbido br. 1941	Ortello
					Tempesta
				Idle Curiosity b. 1945	Bois Roussel
					Fairetta

Detroit 1980

Detroit

The Field

Three Troikas and Le Marmot, the two principals in the finish of the previous year's Prix de l'Arc de Triomphe, met again in the 1980 renewal of Longchamp's most prestigious race. They had both made their seasonal début on Easter Monday in the Prix d'Harcourt. Although Three Troikas was reported to be in need of the race, she looked superb in the parade ring and easily beat Gain 1½ lengths with Le Marmot a short head away third. Mme Alec Head's filly therefore started at 3 to 10 in the Prix Ganay a month later, but when Le Marmot took the lead 300 metres out, she did not stride out with her customary zest and was still 1½ lengths behind at the post, with Northern Baby and Gain 3 lengths and 2 lengths away third and fourth. Either Le Marmot had derived greater benefit from his initial outing or more likely Three Troikas for some unknown reason was not at her best that afternoon.

Mme Alec Head's filly next ran in the Prix Dollar on 1 June. She moved up on the leaders 400 metres out but then quickly weakened and finished a length and 2½ lengths behind Northern Baby and Strong Gale. The cause of this defeat was quickly diagnosed; she had severely damaged a metacarpal bone in her off-fore foot. She was confined to her box for the following five weeks and she did not reappear until 21 September in the Prix du Prince d'Orange, when she ran Dunette to half a length with Northern Baby 1½ lengths away third and Providential and Moulouki only fifth and sixth. Christiane Head felt she had run well after such a long lay-off but Freddie Head reported she had not come down the hill well and had changed her legs several times. She did, however, make up considerable ground in the straight and if the fortnight that remained before the Prix de l'Arc de Triomphe was sufficient to bring her to peak fitness it was more than possible that she could emulate Corrida's second consecutive victory in 1937. On Sunday 5 October Three Troikas started at odds of 4 to 1.

Le Marmot's victory in the Prix Ganay over a distance really too short for him had been some compensation for François Boutin's big disappointment at the disqualification of Nureyev in the 2000 Guineas the previous day. Le Marmot was clearly the form horse for his next race, the King George VI and the Queen Elizabeth Diamond Stakes, for which he was made 7 to 4 favourite. What was not generally known at the time was that a minor setback had interrupted his preparation for the race and that he had hit the point of his off-hind hock while travelling over from France. In the parade ring François Boutin instructed his jockey Philippe Paquet to request permission to withdraw the colt if he was not satisfied with the way Le Marmot went down to the start. Le Marmot strode out well, however, and Paquet was happy with him. Rodolph Schafer's colt clearly was not right at Ascot as he finished a well beaten fifth 7 lengths behind Ela-Mana-Mou. Afterwards Paquet said Le Marmot hung very badly when he asked him to go after Ela-Mana-Mou and that it was almost impossible to ride him. He was still hanging in the straight and his action looked all wrong. It was reported from Chantilly that after Ascot Le Marmot was not showing his customary enthusiasm on the gallops. He reappeared in the Prix Foy on 14 September when the firm going did not entirely suit him. Dunette was withdrawn owing to an attack of the colic; Niniski was also an absentee on account of the going. Le Marmot made very heavy weather of beating Anifa and Gain by a short neck and half a length and there seemed a distinct possibility that he had become a little sour, for on his best form he should have had no difficulty in disposing of these two. During the season the Irish owner breeder Tim Rogers had acquired a 50 per cent share in Le Marmot. Since he also had a third share in Ela-Mana-Mou, the two started 2 to 1 favourite couplé.

Ela-Mana-Mou had been trained by Guy Harwood as a three-year-old. He had easily won the King Edward VII Stakes at Ascot, and the Heath Stakes at Newmarket; he had been runner-up 4 lengths behind Gay Mecene in the Grand Prix de Saint-Cloud and third to Troy and Gay Mecene in the King George VI and the Queen Elizabeth Diamond Stakes at Ascot. He had started favourite for the Derby but had only finished fourth 11 lengths behind Troy. It had seemed that he was a few pounds behind the best of his generation. In the winter he had been bought for £500,000 by the Weinstock family and Tim Rogers, and as a four-year-old was trained by Dick Hern. On his first run in 1980 he only beat Haul Knight, the third in the 1979 Champion Stakes, a neck in the Earl of Sefton Stakes. He put up a good performance when giving Moomba Masquerade 8 lbs more than weight for age and beating him 2 lengths in the Prince of Wales's Stakes at Royal Ascot, but the runner-up had only managed eleventh place in the Derby behind Henbit. When Ela-Mana-Mou beat Hello Gorgeous and Gregorian three-quarters of a length and 1½ lengths in the Joe Coral Eclipse Stakes on 5 July at Sandown, he demonstrated he had improved considerably as a four-year-old. Hello Gorgeous had won the Mecca Dante Stakes from Master Willie before finishing only sixth in the Derby through lack of stamina, whilst Gregorian had won both the Westbury Stakes and the Brigadier Gerard Stakes over ten furlongs at Sandown. In the King George VI and the Queen Elizabeth Diamond Stakes Ela-Mana-Mou led four furlongs out and very courageously held off the strong challenge of Mrs Penny throughout the final two furlongs by three-quarters of a length, with Gregorian and Dunette 5 lengths and half a length back third and fourth. He did not race again before the Prix de l'Arc de Triomphe but arrived in Paris unbeaten as a four-year-old. Ela-Mana-Mou had outgalloped his adversaries at Ascot; he did not have a decisive burst of speed and although he was the logical favourite for the Prix de l'Arc de Triomphe, it was more than possible that Three Troikas at her best or any other of his rivals with a sharp turn of foot would catch him in the last 200 metres.

Dick Hern had two other four-year-olds in the race on 7 October. Lindoro, who had won three handicaps in 1979, but had deteriorated in

1980, was to be used as a pacemaker. Lady Beaverbrook's Niniski had won the Irish St Leger by 10 lengths and had beaten Anifa and Bolsa in the Prix Royal Oak the previous October. He had also defeated M-Lolshan in the Geoffrey Freer Stakes and had run third to Son of Love and Soleil Noir in the St Leger. On his début in the John Porter Stakes at Newbury in April he made up at least 8 lengths in the last two furlongs and beat Morse Code, going away by 3 lengths. He then just got the better of Two of Diamonds in a very close finish for the Ormonde Stakes at Chester. Niniski could only manage to finish 2½ lengths runner-up to Sea Chimes in the Coronation Cup at Epsom and never looked like winning. Although Sea Chimes was recording his eighth consecutive win, he was not one of the best winners of the race in the last twenty years, having graduated from handicaps. Niniski had not run since Epsom as he really required an easy surface, and his odds of 21 to 1, considering how long he had been absent from the racecourse, seemed excessively cramped.

Mrs Penny appeared much the best of the other three English challengers. As a two-year-old she had won the Lowther Stakes, the Cherry Hinton Stakes and the William Hill Cheveley Park Stakes, and had been a most unlucky runner-up to The Dancer on her début at Newbury. She had been without question the best two-year-old filly in England. In the spring of 1980 she was beaten in the Fred Darling Stakes at Newbury but ran a close third to Quick as Lightning in the 1000 Guineas and was a creditable third to Cairn Rouge in the Goffs Irish 1000 Guineas. It was clear even at that time that a mile was too short for her. When she appeared in the parade ring for the Prix de Diane de Revlon she looked superb. Lester Piggott always had her up with the leaders and 300 metres out he shot her into the lead. Her principal rival, Aryenne, who had won the Poule d'Essai des Pouliches, detached herself from the remainder and gave furious pursuit. Piggott had chosen his moment to set sail for home exactly right and Mrs Penny held Aryenne by a short head with the Prix Saint-Alary winner Parente 3 lengths away third. As much credit must go to Lester Piggott as to Mrs Penny for this victory. The Prix de Diane heroine had looked the winner two furlongs out at Ascot, but Ela-Mana-Mou had battled on so gamely that Mrs Penny had to be content with second place. In the Benson and Hedges Gold Cup, Mrs Penny was trapped on the rails and when eventually she was extricated from the pocket Master Willie and Cairn Rouge were gone beyond recall and she finished fourth but still full of running. For once Lester Piggott had not been at his most brilliant. As a result of Piggott being suspended, John Matthias, Ian Balding's stable jockey, took the mount on Mrs Penny in the Prix Vermeille. With 300 metres to go she took the lead from Little Bonny but had to fight to hold her advantage over the Irish filly and the fast-finishing and most unlucky Detroit. She beat them half a length and three-quarters of a length with Gold River fourth, Cairn Rouge fifth and Aryenne only seventh. It was far from certain that Mrs Penny would confirm the form with Detroit, but at 12 to 1 she could not be left out of the reckoning on 5 October.

Raymond Guest's Noble Saint again ran in the Prix de l'Arc de Triomphe. In May he had won the Yorkshire Cup by 1½ lengths from Buttress, but was then a close but disappointing fourth in the Henry II Stakes at Sandown and a well-beaten sixth to Le Moss in the Ascot Gold Cup. He was only fifth to the dead-heaters Dunette and Shakapour in the Grand Prix de Saint-Cloud. He made little show under top weight in the Ebor Handicap but it was reported he had pulled a muscle in the race. On balance, however, it seemed unlikely that his odds of 99 to 1 were anything but a fair reflection of his chance on 5 October.

The five-year-old Nicholas Bill, trained by Henry Candy, had won three of his six races in 1979, including the Jockey Club Cup at Newmarket in October when he beat Vincent three-quarters of a length with Le Moss, below par that day, only fifth. He threw a splint in the spring of 1980 and did not appear on the racecourse until the Princess of Wales's Stakes at Newmarket on 8 July. Nicholas Bill put up an excellent performance to beat Son Fils and More Light a neck and 1½ lengths. He next won the Geoffrey Freer Stakes at Newbury by 1½ lengths and half a length from Valour and Son Fils; Main Reef finished a length second but was disqualified. Nicholas Bill was not at all well suited by the slow early pace of the Grosser Preis von Baden and in the circumstances did well to finish 2¼ lengths fourth to the German champion Nebos. It was likely, however, that Nicholas Bill might find a mile and a half a little too sharp for him in Paris, but his chance of finishing in the frame was better than the 70 to 1 offered about him.

Comtesse Margit Batthyany's German-trained Nebos had won ten of his thirteen races as a three-year-old and four-year-old and had been runner-up on the other three occasions. In 1979 his victories included the Grosser Preis von Berlin when he beat the useful English colt Obraztsovy and he had been a very close second to Königsstuhl in the Deutsches Derby. He had also won the Preis von Europa when M-Lolshan had been only third. In 1980, after winning the Grosser Preis von Dortmund on 26 May, Nebos had 2½ lengths to spare over Königsstuhl in the Grosser Preis von Düsseldorf in June. He then beat Marriacci, the winner of the 1979 Derby Italiano and 1980 Gran Premio di Milano, by 3½ lengths in the Grosser Preis von Berlin on 27 July. He was far from disgraced when failing to concede 7 kilos to Wauthi in Aral-Pokal on 17 August and had then won the Grosser Preis von Baden. His chance seemed much more obvious than that of Star Appeal in 1975. His victories over Obtraztsovy, M-Lolshan and Marriacci put him not far behind the very best in England and France and his starting price of 7 to 1 on 5 October reflected the confidence behind him.

The sole challenger from Ireland was the three-year-old filly Little Bonny who had won a maiden race and a three runner event both over a mile at Leopardstown in the spring. She had been second to Shoot a Line in both the Cheshire Oaks and the Irish Guinness Oaks and on 25 August had won the Virginia Stakes (10f) at Newcastle. Her best performance had been when runner-up in the Prix Vermeille, but she was not certain to confirm the form with the third, Detroit, on 5 October and, consistent as she had been, it seemed probable that she was taking on too formidable opposition in the Prix de l'Arc de Triomphe; she was permitted to start at 41 to 1.

Robert Sangster's filly Detroit had not run as a two-year-old. She was not trained in England or Ireland but in France by Olivier Douieb. On her racecourse début Detroit won the Prix Timandra, a débutantes' race at Saint-Cloud, by an effortless 5 lengths. She next beat Gold River 2½ lengths in the Prix Fille de l'Air on the same course on 21 June, and then had 1½ lengths to spare over Acamas's half-sister by Caro, Licara, in the Prix Chloë at Evry on 19 July. In the Prix de la Nonette run at Deauville on 31 August Detroit beat India Song and Luth de Saron, the winner of the Prix Vanteaux and the Prix de Malleret and fourth in the Prix de Diane de Revlon, by 2 lengths and a short neck. Since Douieb's stable jockey was under suspension, Pat Eddery rode Detroit for the first time in the Prix Vermeille. It was his first mount for Robert Sangster since the announcement that he would be riding for Vincent O'Brien and Sangster in 1981 in place of Lester Piggott. Eddery did not shine on Detroit at Longchamp on 14 September; he got hopelessly boxed on the rails in the straight and by the time he extricated himself from the pocket and got Detroit going he had no chance of catching Mrs Penny and Little Bonny. Detroit finished third 1¼ lengths behind Mrs Penny, but all at Longchamp that afternoon regarded her as a most unlucky loser and fully expected her to reverse the form in the Prix de l'Arc de Triomphe. For this reason Detroit started at 67 to 10, almost half the odds of Mrs Penny, on 5 October and she was considered the liveliest threat to Ela-Mana-Mou and Three Troikas.

The other French three-year-old filly in the Prix de l'Arc de Triomphe, André Chollou's Satilla, was one of the total outsiders at 99 to 1. She won a minor race at Rouen on 13 April and after winning a selling race at Longchamp on 11 May she was claimed out of Christiane Head's stable for 83,360 francs. Satilla then ran fourth twice, to Perrault in the Prix Sica Boy and to Noalchoholic in the Prix Colonel Gillois at Evry. She was a close third to Ferragu in the Prix du Perray at Maisons-Laffitte on 24 July and then won the Prix de Thiberville at Deauville in August. Her best performance, however, was when 1½ lengths second to Prince Bee in the Prix Niel, a neck and a short head in front or Ruscelli and Garrido. Although she was certainly improving, Satilla seemed certain to be flying too high in the Prix de l'Arc de Triomphe.

Frederick Tinsley's three-year-old colt Policeman made his racecourse début in the spring at Cagnes-sur-Mer when he won two of his four races. Returning to Paris, he ran third 2 lengths and a neck behind Shakapour and Corvaro in the Prix de Guiche on 20 April. In the Prix Matchem (1800 metres) at Evry on 10 May, Policeman was beaten 2 lengths into third place by the improving Axius who was struck into and injured severely on his subsequent outing in the Prix la Force. Willie Carson, who had won the Derby on Henbit and the Oaks on Bireme in the preceding four days, was engaged by Policeman's trainer Charles Millbank to ride him in the Prix du Jockey Club. Carson rode him very enterprisingly at Chantilly, taking him into the lead before the field passed the château, after which Policeman was never in danger. He had no difficulty in holding Shakapour and Providential by 1½ lengths and 3 lengths with the Vincent O'Brien-trained favourite Gonzales, a half brother to Youth and full brother to Mississipian, only fifth and Argument in rear. It was a remarkable treble for Carson, achieved in only five days. Policeman had started the total outsider at Chantilly at 54 to 1 but in the Grand Prix de Saint-Cloud he showed it was no fluke. Carson again had the mount and led before the field reached the straight. It was only 150 metres from the line that Dunette and Shakapour, who dead-heated for first place, got the better of him and beat him 3 lengths. Carson was claimed for Prince Bee in the Prix Niel and Yves Saint-Martin rode Frederick Tinsley's colt. To general surprise Saint-Martin employed waiting tactics on Policeman who was never better than fifth and finally finished last. These tactics were obviously a mistake and Policeman's performance clearly had to be ignored. Although no Prix du Jockey Club winner had won the Prix de l'Arc de Triomphe since Sassafras ten years earlier, it would have been unwise to discard Policeman entirely and it was quite possible he would be concerned in the finish on 5 October. He was accompanied to the post by the five-year-old Iseo, who had won three races in 1980 including the Prix de Louveciennes at Longchamp, as his pacemaker. The pair started at 12 to 1 couplé.

Serge Fradkoff's Providential, a stable-companion of Detroit, had been trained by François Boutin as a juvenile and in the spring of 1980. As a two-year-old he had beaten Glenorum 3 lengths in a maiden race at Saint-Cloud and won the Critérium de Saint-Cloud in November. On his reappearance he easily defeated Blast Off in the Prix Greffulhe on 13 April. In the Prix Hocquart on 11 May Providential was unable to resist the late challenge of Mot D'Or by a short neck with Belgio and Shakapour third and fourth. After his third in the Prix du Jockey Club, Providential was sold to Fradkoff for a reported million dollars. In the Prix du Prince d'Orange Providential was a mediocre fifth to Dunette and Three Troikas and it was difficult to give him much chance on 5 October.

The mount of Lester Piggott in the 1980 Prix de l'Arc de Triomphe, Glenorum, had improved steadily all season. He won a mile maiden race at Saint-Cloud in March and ran Argument to 2 lengths over the same course and distance on 3 April, and ten days later won the Prix du Bel Air Handicap at Longchamp. He was a close third to Nemr in the Prix la Force on 25 May but then was rested till August. In the Prix de la Côte Normande he was in receipt of 4 kilos from Argument but only beat him a head. A fortnight later, ridden by Piggott, Glenorum beat Perrault 3 lengths in the Grand Prix de Deauville. This victory and the presence of Lester Piggott on his back meant that he started at 12 to 1 on 5 October.

After his success at Saint-Cloud over Glenorum Argument finished fourth about 2½ lengths behind In Fijar, Moorestyle and Ruscelli in the Poule d'Essai des Poulains on 27 April. Ruscelli was placed last after a stewards' enquiry and Argument was promoted third. Argument was then a very creditable half a length runner-up to Belgio in the Prix Lupin, a short head in front of In Fijar, but ran unaccountably badly in the Prix du Jockey Club. On 21 July Argument beat Dhausli and Strong Gale 2 lengths and a nose in the Grand Prix Prince Rose at Ostende. In the Prix de la Côte Normande, Argument's remarkable burst of speed in the final 200 metres almost carried him to victory and it seemed sure that he would reverse the form with Glenorum at level weights in the Prix de l'Arc de Triomphe. The early funereal pace of the Grosser Preis von Baden was a grave disadvantage to Argument and in a truly-run race he certainly would have been much closer to Nebos than his actual fifth place. Argument surprisingly was permitted to start at 74 to 1 in the Prix de l'Arc de Triomphe and as a possible outsider seemed much better value than Glenorum at 12 to 1.

After his disqualification from third place in the Poule d'Essai des Poulains, Ruscelli won the Prix de la Jonchère (1600 metres) at Longchamp on 15 May and then was a most creditable neck runner-up to the 2000 Guineas third, Night Alert, in the Prix Jean Prat (1800 metres) at Chantilly. Ruscelli was a well-beaten and disappointing sixth to Corvaro in the Prix Eugène Adam at Saint-Cloud in July. When third to Prince Bee in the Prix Niel it was his first attempt at a mile and a half. On balance, it appeared likely that his odds of 80 to 1 represented a fair reflection of his chance.

Mahmoud Fustok's Moulouki easily won a newcomers' race at Maisons-Laffitte over 1700 metres on 21 March and the Prix Jean le Gonidec at Saint-Cloud on 15 April. He was a disappointing seventh in the Prix Hocquart but receiving 10 kilos from Hard to Sing, the Prix de Barbeville and Prix Jean Prat (3100 metres) winner, Moulouki beat him a neck in the Prix Jean de Chaudenay on 28 May. In his only other race before the Prix de l'Arc de Triomphe, Moulouki was only sixth to Dunette in the Prix du Prince d'Orange and it was difficult to envisage him taking any part in the finish on 5 October.

The four-year-old mare Dunette was the only horse to defeat Three Troikas as a three-year-old. Mme Alec Head's filly had revenged her Prix de Diane de Revlon defeat in the 1979 Prix Vermeille when Dunette was fourth. Dunette reappeared in the Grand Prix d'Evry on 10 May; looking in need of the race in the paddock she was not disgraced when a short neck and 2 lengths third to Scorpio and Gain. In La Coupe at Chantilly on 15 June Dunette started odds-on favourite but was a very disappointing fifth to Prove it Baby. In the Grand Prix de Saint-Cloud she looked very well in the parade ring; she challenged Policeman in the straight and got the better of him 150 metres from the line. Meanwhile Shakapour had made up a lot of ground and launched a strong challenge. Dunette's jockey Georges Doleuze threw his hand in the air five metres before the winning post as he had done the previous June in the Prix de Diane de Revlon. In doing so, he permitted Shakapour to snatch a dead-heat on the line. Yves Saint-Martin had dropped his whip 200 metres out and possibly, equipped with it, would have won. Dunette was not disgraced behind Ela-Mana-Mou at Ascot, but may have been fortunate to beat a half-fit Three Troikas in the Prix du Prince d'Orange. She was a very difficult ride and did not always consent to show the ability she certainly possessed. Were she to do so on 5 October Dunette could spring a surprise, but she had a reputation of disliking Longchamp and her odds of 15 to 1 did not appear good value.

Aryenne was the intended mount of Yves Saint-Martin in the 1980 Prix de l'Arc de Triomphe, but was found to have an abscess in her mouth on the morning of the race and had to be withdrawn. For the first time for many years Saint-Martin was left without a mount in Longchamp's most prestigious race. Henbit and Master Willie, the first two in the Derby at Epsom, were also absentees. Henbit had cracked his off-fore canon bone at Epsom and Master Willie had the Champion Stakes as his autumn objective. Tyrnavos, the surprise winner of the Irish Sweeps Derby, had run so disappointingly at Ascot behind Ela-Mana-Mou that the Prix de l'Arc de Triomphe never featured in his programme.

The Final Preparations

Ela-Mana-Mou's final gallop was on Saturday 27 September at West Ilsley where Dick Hern trains. He easily beat Prince Bee and Niniski and afterwards Carson stated he was very confident that the favourite would run very well. Nicholas Bill, who did his final work with Master Willie, also satisfied his trainer Henry Candy. Three Troikas, accompanied by Doux Lord and Sealy, did her final gallop on Tuesday 30 September. Although she went well and was clearly getting fitter every day, Christiane Head regretted that she did not have an extra week to complete her preparation. The same morning Argument worked on Chantilly racecourse and his trainer John Cunnington Jr predicted that if the pace were strong at Longchamp, Argument would be able to make use of his strongest asset, his finishing speed, and that he would be concerned in the finish. The most sensational gallop, however, took place on the racecourse at Chantilly on Sunday 28 September. Policeman and Le Marmot worked together with Iseo and Garrido, and the Prix du Jockey Club winner finished 5 lengths clear of Le Marmot. This result naturally raised the hopes of Charles Millbank, but François Boutin was not surprised as his charge had frequently not excelled in the mornings, only to run very well in a race in the afternoon. Detroit more than pleased her young trainer Olivier Douieb when Alain Lequeux worked her on the gallops of Les Aigles the following day and afterwards he affirmed that she had never been better. Mrs Penny covered nine furlongs two days earlier at Newbury and her trainer Ian Balding was glad to see she had suffered no ill-effects from her victory in the Prix Vermeille.

On the Friday evening before the Prix de l'Arc de Triomphe a car bomb exploded outside a synagogue in Rue Copernic in the fashionable sixteenth arrondissement of Paris. On the morning of the Prix de l'Arc de Triomphe another terrorist bomb exploded in Paris, maiming a Dutch tourist. There was talk that some of the new neo-nazis were in fact members of the police and on Saturday 4 October 15,000 Jews marched down the Avenue de Champs-Elysées in protest.

None of these events deterred the normal huge influx of overseas visitors to Longchamp. Not even a General Election in Germany, which resulted in victory for Chancellor Schmidt, kept the German supporters of their champion Nebos away from Paris.

The Race

The going at Longchamp on Sunday 5 October was firm which would suit neither Le Marmot or Niniski. It was so firm that felt pads were fitted under Ela-Mana-Mou's racing plates. Alain Laqueux, who had given up the mount on Detroit to Pat Eddery, easily won the Prix Marcel Boussac on Tropicaro for Maurice Zilber's stable. In the next race, the Prix de l'Abbaye de Longchamp, the French were very hopeful of breaking the English stranglehold on this sprint with the Aga Khan's Adraan, who had broken the 1000-metre record at Longchamp when winning the Prix de Saint-Georges in May in 55.5 seconds. Ardraan, however, could only finish fourth to three British invaders; Lester Piggott took Moorestyle into the lead 200 metres out and he strode away from Sharpo and Valeriga to win in scintillating style by 1½ lengths and a neck.

In the parade ring both Ela-Mana-Mou and Nebos looked superb, especially the latter. It was also difficult to fault Detroit and Nicholas Bill. Three Troikas was also turned out in immaculate condition but she had looked even better the previous year and in the Prix d'Harcourt.

At the start Nicholas Bill lost a little ground and Lindoro set a good pace to Iseo tracked by Little Bonny, Policeman, Ela-Mana-Mou and Providential. Then in a group together came Moulouki, Niniski, Mrs Penny, Detroit, Glenorum, Dunette and Le Marmot. The backmarkers were Argument, Three Troikas and Nebos. There was little change in the order when Lindoro reached the top of the hill except that Nicholas Bill had been pushed up through the field to be on the heels of the leaders. Coming down the hill the pacemakers Lindoro and Iseo quickly weakened and Policeman took a clear lead, followed by Ela-Mana-Mou, Mrs Penny, Providential, Nicholas Bill and Dunette. Little Bonny was already beginning to weaken but Three Troikas had made progress to be about eighth, level with Niniski and Glenorum, with Detroit lying about tenth. At this stage Nebos began to make rapid headway on the outside from the back, but Argument still had not made a move and was last. About 800 metres out, Nebos collided with the weakening Niniski and lost impetus. Le Marmot who was lying about twelfth did not appear to be going very well. Policeman still led Ela-Mana-Mou as he turned into the straight but was clearly coming under pressure. They were closely followed by Providential, Nicholas Bill, Dunette, Three Troikas, the weakening Mrs Penny, Glenorum and Detroit. Nebos was running on again after his collision with Niniski but still had an awful lot to do, as did Argument who was switched off the rails to begin his challenge. A little over 400 metres out, Ela-Mana-Mou went past the fading Policeman into the lead and set sail for home chased by Providential, Nicholas Bill, the rapidly improving Three Troikas, Dunette and Glenorum. At this moment Eddery switched Detroit, who was only ninth, to the outside to make his challenge, while Nebos and Argument were both making rapid progress through beaten horses. Three Troikas had made up so much ground that 200 metres out she was only half a length down on Ela-Mana-Mou but looming about a length behind them was Detroit, who had flown past her other adversaries as soon as she began her run. Nicholas Bill and Dunette were running on at one pace; Providential was going rapidly backwards but both Nebos and Argument were by now finishing fastest of all. For a hundred metres Ela-Mana-Mou held Three Troikas, but Detroit on the outside was now travelling like the winner and she was only half a length down, while Argument was coming like an express train through the centre and Nebos on the very wide outside was finishing fast but clearly too late to take a hand in the finish. With less than fifty metres to go, Detroit went past Ela-Mana-Mou who ran on gamely but was deprived of second place by Argument on the line. Detroit had half a length and a short head to spare over Argument and Ela-Mana-Mou with Three Troikas only a neck away fourth, a length ahead of Nebos. Nicholas Bill was 2 lengths away sixth, a short neck in front of Dunette with

Detroit (19) about to take the lead from Ela-Mana-Mou (8) and Three Troikas (black cap) followed by Argument (7) with Nicholas Bill (1) and Dunette (10) just behind

Le Marmot, who had run on again in the straight, eighth, Ruscelli ninth and Satilla tenth. In winning the 1980 Prix de l'Arc de Triomphe Detroit had beaten the course record by 0·3 seconds, having recorded a time of 2 minutes 28·0 seconds. She thus gained a third success in four years for Robert Sangster in the Prix de l'Arc de Triomphe, a feat not even achieved by Marcel Boussac in his heyday.

Nevertheless the real heroine of the race was Three Troikas who had failed by less than a length to win her second consecutive Prix de l'Arc de Triomphe after Christiane Head had said she would have liked ten days to get her in peak condition. Ela-Mana-Mou had run his usual game and courageous race but his lack of a sharp burst of acceleration had been his downfall. Argument had not enjoyed the best of runs through the centre, but possibly the most unlucky horse was Nebos. His German jockey Lutz Maeder blamed the barging match with Niniski but for which 'Nebos would have been in the first two'. His lack of experience of Longchamp had been evident both before and after that incident and many wondered how the German champion would have fared if Saint-Martin, or another jockey more aware of the difficulties of riding on the track, had been in the saddle. Both Glenorum and Le Marmot finished lame, whilst it seemed Mrs Penny had gone over the top.

Surprisingly, not merely was Detroit's

success the first time Pat Eddery had won the Prix de l'Arc de Triomphe but it was also the first Group One race he had ever won in France. In 1974 he had become the youngest champion jockey in England since Charlie Elliott and Gordon Richards in the 1920s. He was the son of Jimmy Eddery who had ridden Panaslipper to victory in the Irish Derby. On the retirement of Duncan Keith, Pat Eddery became Peter Walwyn's stable jockey. He had been champion jockey in England in 1975, 1976 and 1977. He had been associated with all Grundy's classic victories and had won the Oaks on Polygamy in 1974 and on Scintillate in 1979.

Detroit's trainer, Olivier Douieb, was the son-in-law of the Maisons-Laffitte trainer Pierre Pelat. He had been assistant to Miguel Clément, but took out a licence on his own before the latter's death in a car accident in 1978. In the spring of 1977 he had come to prominence when Sanedtki won the 1000 Guineas Trial at Ascot and landed a gamble. Douieb's success continued in Paris and on Miguel Clément's death Robert Sangster transferred some of his horses in training in Chantilly to the young trainer. He attributed his success to his father-in-law whom he said had taught him enormously well, whilst Sangster called him 'a true professional, he never stops working'.

Immediately after all the congratulations and presentations, Olivier Douieb and Robert Sangster were back in the parade to saddle Hortensia for the Prix de l'Opéra. Ridden by the stable jockey Alain Lequeux, Hortensia beat Moon Ingraver, also trained by Douieb, by half a length, thus adding icing to the cake for both owner and trainer.

The Rest of the Season

Detroit did not run again in 1980 but remained in training as a four-year-old. Three Troikas went to the United States to run in the Turf Classic on 25 October. The going was extremely heavy that afternoon and she was unable to handle it, finishing eighth and last to the virtually unconsidered French filly Anifa. Satilla and Providential ran third and fourth to En Calcat and Lancastrian in the Prix du Conseil de Paris. Niniski was never in the hunt when Gold River beat Monsieur Marcel and Ardross in the Prix Royal Oak. It was left to Argument to uphold the Prix de l'Arc de Triomphe form. Partnered by Lester Piggott he beat The Very One a length in the Washington D.C. International Stakes at Laurel Park.

Pedigree and Stud Record

Detroit had been bred by the Comte Roland de Chambure and Alec Head. She had been bought as a foal by Robert Sangster for a sum reported to be about £100,000.

Detroit's sire, Riverman, was bought by Alec Head as a foal for $41,000 at the Guggenheim dispersal sale in 1969. As a three-year-old he won the Poule d'Essai des Poulains, the Prix Jean Prat and the Prix d'Ispahan. He was third to Brigadier Gerard and Parnell in the King George VI and the Queen Elizabeth Diamond Stakes and runner-up to Brigadier Gerard in the Champion Stakes. Besides Detroit, Riverman has sired Policeman, Gold River, Irish River (Prix Morny, Grand Critérium, Prix de la Salamandre, Poule d'Essai des Poulains, Prix Jacques le Marois, Prix du Moulin de Longchamp and Prix d'Ispahan) and River Lady (Poule d'Essai des Pouliches), all classic winners in France. He topped the French sires list in 1980 and was second in 1978 and 1979. From 1976 to September 1982 his progeny won 271 races worth 28,086,769 francs. He was syndicated for $18 million and was exported to the U.S.A. in 1980. He stands at the Gainsway Stud in Lexington. In American his progeny have won twenty-one races worth $287,881, and in England and Ireland twenty-four races and £76,915.

Detroit's dam, Derna by Sunny Boy, had been bred by Nelson Bunker Hunt but was sold in the mid-1960s to A. Louis-Dreyfus who passed her on to the Comte de Chambure for 100,000 francs in 1972. Detroit was her eighth foal. On the racecourse Derna never won in five outings as a three-year-old but was second at Saint-Cloud and twice third at distances varying from ten to thirteen furlongs. Her first foal, Haschich by Net, won three small races from twenty-nine outings; her second, Darcounette by Dapper Dan, won the Prix des Tuileries. Her third foal, Mégalomane by Snob, never won but was placed thirteen times in France; her next produce, Valderna by Val de Loir, won once and was runner-up in the Grand Prix de Compiègne. Her grey Caro filly, Dyonysies, was unplaced in seven outings at two and three. Derna then produced an exceptional Lyphard filly, Durtal, who was owned by Robert Sangster and trained by Barry Hills. As a two-year-old she won three of her five starts, including the William Hill Cheveley Park Stakes, and was runner-up to J. O. Tobin in the Laurent Perrier Champagne Stakes and second in the Lowther Stakes. As a three-year-old she won the Fred Darling Stakes and was second to Madelia in the Poule d'Essai des Pouliches. After Durtal, Derna's Targowice filly Delagoa won once as a three-year-old in France. In 1978 Derna produced Detroit, and her next foal, Derly by Lyphard, won the Prix Sanguinolette at Maisons-Laffitte in November 1980 and the Prix Perdita at Evry in 1981 when she was also second to Bernicia in the Prix Vanteaux. Derna's colt by Green Dancer, Dillingham, was bought at the Newmarket Houghton Sales in 1980 for 530,000 guineas by Serge Fradkoff. Derna has been, therefore, a prolific and very successful broodmare.

Derna's dam, Miss Barberie by Norseman, was bred by Prince Saud Toussoun. She won three times including the Prix Joubert at Le Tremblay and was placed seven times, including a third in the Prix Messidor and a fourth in the Prix de Royallieu. She produced only two minor winners, Pink Miss by Pink Horse, and Miss Molydal by Piqu' Arrière.

Miss Barberie's dam, Vaneuse by Vatellor, won two races as a three-year-old worth 310,800 francs. She was the dam of four winners, the best of which were Gombar by Goyama, whose three victories included the Grosser Preis von Nordrhein-Westfalen, and Vanette by Tanagrello, who herself became a successful broodmare in the U.S.A.

Vaneuse's dam, Diseuse by Diomedes, was a maiden but produced five winners. She was a half-sister to Fidgette by Firdaussi who won three races in 1942 and 1943, and was the dam of La Cadette by Le Pacha who won the Oaks d'Italia in 1950. She was also the grandam of Herbager (Prix du Jockey Club 1959) and La Canea (Oaks d'Italia 1956) and the third dam of Lauso (Derby Italiano 1961). Diseuse's breeder Henri Coulon also owned her dam Boxeuse by Teddy who won two races as a three-year-old and bred a total of four winners. Her dam, Spicebox by Spion Kop, was bought from Lord Rosebery for 1,800 guineas at the Newmarket December Sales in 1929. Spicebox was in foal to Diomedes at the time and the produce, a filly called Spirituelle, won five races in France including the Prix d'Arenberg as a juvenile in 1932. Spicebox was also the grandam of Espace Vital by Oleander who won the Prix Kergorlay, the Prix Jean Prat and Prix du Conseil Municipal, and the fifth dam of Final Straw, who won the Clerical Medical Greenham Stakes and was runner-up to Posse in the Sussex Stakes and St James's Palace Stakes in 1980 and third to Nikoli in the Airlie/Coolmore Irish 2000 Guineas. Spicebox traced back to Macaroni, the 1883 Oaks winner.

Detroit ran eight times in 1981 and won three races including the Prix Foy but was not as good as she had been as a three-year-old. She has yet to prove herself as a broodmare and it will be interesting to see how she performs in the paddocks. When she retired to stud, she had won eight races and 2,108,000 francs from ten starts.

PRIX DE L'ARC DE TRIOMPHE 2400 metres
at Longchamp on Sunday 5 October 1980. For three-year-olds and upwards. Going: Firm.
1,200,000 francs to the winner; 480,000 fr. to the 2nd; 240,000 fr. to the 3rd; 120,000 fr. to the 4th. 300,000 fr. to the breeder of the winner; 120,000 fr. to that of the second; 60,000 fr. to that of the 3rd; 30,000 fr. to that of the 4th. Value to the winner: 1,200,000 fr.

DETROIT	br.f.	by Riverman – Derna by Sunny Boy	3.54½k	Robert Sangster	P. Eddery	1
ARGUMENT	b.c.	by Kautokeino – Arantelle by Tapioca	3.56k	Mme Pierre Ribes	J.C. Desaint	2
ELA-MANA-MOU	b.c.	by Pitcairn – Rose Bertin by High Hat	4.59k	Simon Weinstock	W. Carson	3
THREE TROIKAS	b.f.	by Lyphard – Three Roses by Dual	4.57½k	Mme Alec Head	F. Head	4
NEBOS	b.c.	by Caro – Nostrana by Botticelli	4.59k	Comtesse Margit Batthyany	L. Maeder	5
NICHOLAS BILL	ch.c.	by High Line – Centro by Vienna	5.59k	William Barnett	P. Waldron	6
DUNETTE	b.f.	by Hard to Beat – Pram by Fine Top	4.57½k	Mme Harry A. Love	G. Doleuze	7
LE MARMOT	b.c.	by Amarko – Molinka by Molvedo	4.59k	Rodolph Schafer	P. Paquet	8
RUSCELLI	b.c.	by Val de l'Orne – Coy Maid by Habitat	3.56k	Zenya Yoshida	M. Philipperon	9
SATILLA	br.f.	by Targowice – Saratoga by Snob	3.54½k	André Chollou	A. Perrotta	10
LINDORO		by Sun Prince – Farce by Bon Mot	4.59k	Lady Beaverbrook	M.L. Thomas	
NINISKI		by Nijinsky – Virginia Hills by Tom Rolfe	4.59k	Lady Beaverbrook	J. Mercer	
NOBLE SAINT		by Vaguely Noble – Santa Paula by Santa Claus	4.59k	Raymond Guest	G. Starkey	
ISEO		by Sir Gaylord – If by Kashmir II	5.59k	Frederick E. Tinsley	J. Héloury	
PROVIDENTIAL		by Run the Gantlet – Prudent Girl by Primera	3.56k	Serge Fradkoff	A. Lequeux	
MOULOUKI		by Sassafras – Senama by Sanctus	3.56k	Mahmoud Fustok	A. Gibert	
GLENORUM		by Prove Out – Cailey Jane by Right Combination	3.56k	Mme Arthur W. Stollery	L. Piggott	
POLICEMAN		by Riverman – Indianapolis by Barbare	3.56k	Frederick E. Tinsley	G. Dubroeucq	
MRS PENNY		by Great Nephew – Tananarive by Le Fabuleux	3.54½k	E.N. Kronfeld	J. Matthias	
LITTLE BONNY		by Bonne Noel – Little Fuss by Sovereign Path	3.54½k	Kevin Prendergast	C. Roche	

20 runners.
DETROIT trained by Oliver Douieb; bred by Comte Rolande de Chambure and Alec Head.
Distances: ½ length; short head; neck; 1 length; 2 lengths; short neck; ½ length; ¾ length; 2 lengths. Time: 2mins 28.00 seconds.
Pari-Mutuel: Win: 7fr. 70. Places: 3fr. 20; 10fr. 90; 2fr. 30.
S.P.: 2 to 1 Ela-Mana-Mou and Le Marmot (couplé); 4 to 1 Three Troikas; 67 to 10 DETROIT; 7 to 1 Nebos; 12 to 1 Policeman and Iseo (couplé), Glenorum and Mrs Penny; 15 to 1 Dunette; 21 to 1 Niniski and Lindoro (couplé); 32 to 1 Providential; 41 to 1 Little Bonny; 70 to 1 Nicholas Bill; 74 to 1 Argument; 76 to 1 Moulouki; 80 to 1 Ruscelli; 99 to 1 Noble Saint and Satilla.

DETROIT bay filly 1977	Riverman b. 1969	Never Bend b. 1960	Nasrullah b. 1940	Nearco br. 1935	Pharos
					Nogara
				Mumtaz Begum b. 1932	Blenheim
					Mumtaz Mahal
			Lalun b. 1952	Djeddah ch. 1945	Djebel
					Djezima
				Be Faithful br. 1942	Bimelech
					Bloodroot
		River Lady b. 1963	Prince John ch. 1953	Princequillo b. 1940	Prince Rose
					Cosquilla
				Not Afraid b. 1948	Count Fleet
					Banish Fear
			Nile Lily br. 1954	Roman b. 1937	Sir Gallahad
					Buckup
				Azalea ch. 1944	Sun Teddy
					Coquelicot
	Derna b. 1961	Sunny Boy b. 1944	Jock br. 1936	Astérus b. 1923	Teddy
					Astrella
				Naic b. 1928	Gainsborough
					Only One
			Fille de Soleil ch. 1935	Solario b. 1922	Gainsborough
					Sun Worship
				Fille de Salut br. 1928	Sansovino
					Friar's Daughter
		Miss Barberie b. 1950	Norseman b. 1940	Umidwar b. 1931	Blandford
					Uganda
				Tara ch. 1932	Teddy
					Jean Gow
			Vaneuse br. 1945	Vatellor br. 1932	Vatout
					Lady Elinor
				Diseuse b. 1935	Diomedes
					Boxeuse

Gold River 1981

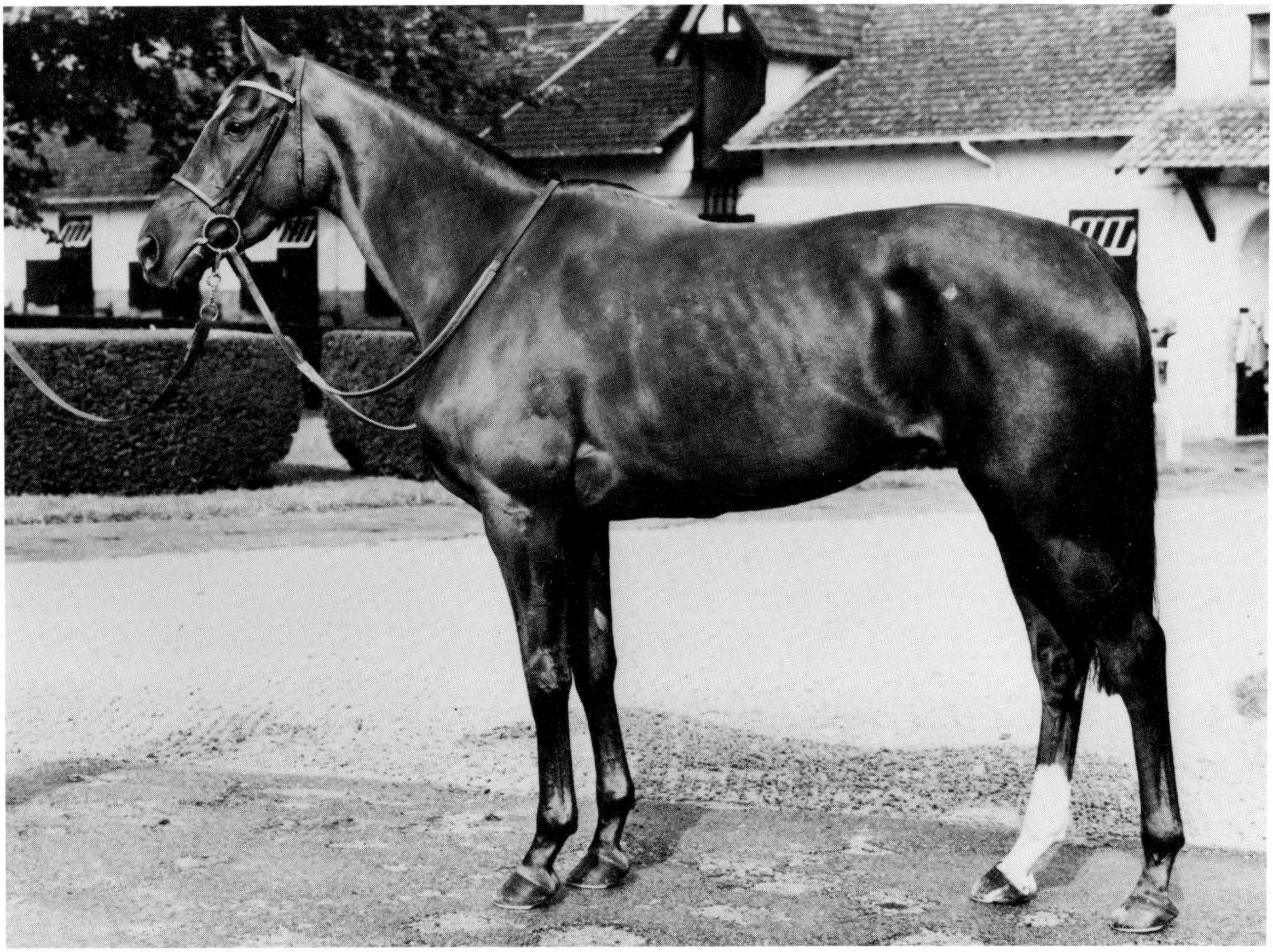

Gold River

The Field

Robert Sangster's 1980 Prix de l'Arc de Triomphe heroine Detroit made her seasonal appearance in the Prix d'Harcourt on 5 April but could only manage fourth place a little over 3 lengths behind Argument. In the Prix Ganay, Detroit fared even worse, finishing fifth 4½ lengths behind Argument. She was off the course till August when she had little difficulty in beating very moderate opposition in the Prix d'Arromanches over 2200 metres at Clairefontaine and then four days later she defeated Ruscelli 1½ lengths in the Prix Ridgway at Deauville. She was receiving weight from both Lancastrian and Gold River when she beat them in the Prix Foy at Longchamp in September. The early slow pace did not suit either Lancastrian or Gold River and it was by no means certain that she would confirm the form with either of them in the Prix de l'Arc de Triomphe on 4 October. Nevertheless she started second favourite at 7 to 2 coupled with King's Lake and Snow Day.

King's Lake, who ran in the colours of Jean-Pierre Binet but was partly owned by Sangster, was trained in Ireland by Vincent O'Brien. He had passed the post first, a neck in front of To-Agori-Mou in the Airlie/Coolmore Irish 2000 Guineas, but was disqualified by the local stewards and placed second. He was controversially reinstated when his connections appealed to the Stewards of the Irish Turf Club. In the St James's Palace Stakes To-Agori-Mou got his revenge by a neck, but on their third meeting in the Sussex Stakes King's Lake beat his rival by a head. In the Prix Jacques Le Marois at Deauville on 16 August King's Lake was again a nose behind To-Agori-Mou, but both their limitations were exposed as Northjet slammed them by 5 lengths. On 19 September King's Lake ran for the first time over ten furlongs in the Joe McGrath Memorial Stakes at the Curragh. He beat Erins Isle and Kind of Hush a length and 3 lengths with a below-par Blue Wind fourth. It seemed improbable that King's Lake would stay the mile and a half of the Prix de l'Arc de Triomphe. The rivalry between him and To-Agori-Mou had provoked much comment in the summer but in retrospect both were probably overrated.

Robert Sangster's three-year-old filly Snow Day won the Prix de la Celle Saint-Cloud at Longchamp in the spring. She beat Léandra a head in the Prix de Royaumont at Chantilly on 7 June and a fortnight later won the Prix Fille de l'Air at Saint-Cloud. She was, however, only fourth to Léandra in the Prix de Malleret on 4 July and fifth to April Run in the Prix Vermeille, though in the latter her saddle slipped and she might have been closer. It was difficult to envisage Snow Day, however, playing a prominent part in the finish on 4 October.

Including King's Lake, there were eleven challengers from abroad in the 1981 Prix de l'Arc de Triomphe, of which Charles St George's five-year-old Ardross appeared to hold the best chance. He was originally trained by Paddy Prendergast in Ireland and then on his death by Kevin Prendergast. The battles between Ardross and Le Moss in the Ascot Gold Cup, the Goodwood Cup and the Doncaster Cup, had been one of the highlights of the previous season. He won the Jockey Club Cup at Newmarket in the autumn and was past his best when third to Gold River in the Prix Royal Oak. In 1981 prior to his arrival in Paris Ardross was unbeaten in four outings. He had been sold early in the year to Charles St George and was trained in 1981 by Henry Cecil at Newmarket. On 14 May Ardross easily defeated Nicholas Bill in the Yorkshire Cup. The ease of his victory over Shoot A Line in the Ascot Gold Cup was in no way conveyed by the official margin of one length. Ardross next won the Goodwood Cup easily. In the Geoffrey Freer Stakes at Newbury Ardross ran out an immensely impressive 5-lengths winner from Castle Keep, who was receiving 8 lbs with the future St Leger winner Cut Above another 1½ lengths further back. Despite a very slow early pace, this victory dispelled all doubts that Ardross would not have the speed to come back to shorter distances. Ardross did not run in the Doncaster Cup as Henry Cecil considered the going too firm. After Castle Keep ran Perrault to a short head in the Grand Prix de Deauville, the chance of Charles St George's fine-stayer in the Prix de l'Arc de Triomphe was there for all to see and he started at odds of 11 to 2 on 4 October.

Sir John Astor's Cut Above was one of two runners trained by Dick Hern, the other being Sir Michael Sobell's Prince Bee. After a promising second to Kalaglow in the Horris Hill Stakes at Newbury as a juvenile, Cut Above easily won the White Rose Stakes on his seasonal reappearance at Ascot in April. In the Irish Sweeps Derby he was flattered by his proximity to Shergar when only beaten 4 lengths into second place. In the Geoffrey Freer Stakes he was well-beaten, 6½ lengths behind Ardross, and it seemed the slow early pace was even more disadvantageous to him than to Charles St George's stayer. Cut Above then caused the biggest upset of the season when he beat Glint of Gold and his better fancied stable-companion Bustomi 2½ lengths and 4 lengths, with the best English three-year-old Shergar 5 lengths away fourth. In the week prior to the race there had been rumours that all was not well with the brilliant Shergar; but these had been quickly denied. Nonetheless, in the pre-parade ring the Aga Khan's colt did not look anywhere near as well as he had been in the Derby and the King George VI and the Queen Elizabeth Stakes. Indeed, the pick of the paddock at Doncaster was unquestionably Glint of Gold, whilst the light-framed Cut Above also looked very well. At the three-furlong marker Shergar was showing signs of distress and the race seemed to rest between the leader Bustomi and Glint of Gold, who was going very easily in second place, for Mercer was already at work on Cut Above to keep him in touch. Glint of Gold went ahead over two furlongs out and looked the assured winner. Mercer on Cut Above then swooped inside the final furlong and strode away for his sensational victory. In addition to his probably not being at his best, Shergar also appeared not to stay the distance. Glint of Gold did not possess a sharp burst of acceleration and was unable to respond to Cut Above's challenge. To most it seemed better to regard the St Leger as a slightly freak result and Cut Above's chance at Longchamp was judged on his other form, which resulted in him starting at 38 to 1.

Sir Michael Sobell's Prince Bee won the Predominate Stakes, the Gordon Stakes, the Great Voltigeur Stakes and the Prix Niel, as a three-year-old and was runner-up to Tyrnavos in the Irish Sweeps Derby. Prince Bee reappeared in the Coronation Cup and was not disgraced when 2 lengths runner-up to Master Willie. Although he was reported to have met interference, he was only sixth to Akarad in the Grand Prix de Saint-Cloud, but he did return to the paddock minus his front shoes. He did beat Noalto in the Valdoe Stakes at Goodwood in September, but on balance it was difficult to be enthusiastic about his chance at Longchamp.

Sir Michael Sobell's other four-year-old, Lancastrian, was trained by D. Smaga in France. In 1981 he comfortably beat En Calcat three-quarters of a length in the Prix d'Hedouville on 20 April and defeated Perrault in the Grand Prix d'Evry by a short head on 23 May. He was 3 lengths and 5 lengths behind Pèlerin and Light Cavalry in the Hardwicke Stakes at Royal Ascot but was not suited by the firm ground. In the Grand Prix de Saint-Cloud he ran Akarad and Bikala to 2 lengths and half a length. He was not disgraced when runner-up to Detroit in the Prix Foy as she was receiving 8 lbs. Nevertheless, whenever Lancastrian took on the best, he normally found one or two of his rivals too speedy for him and the best he could hope for was one of the minor places. He and Prince Bee started at the somewhat cramped odds of 12 to 1 couplé on 4 October.

The best performance of Sir Philip Oppenheimer's four-year-old colt Pèlerin in 1980 had been to run fourth to Henbit in the Derby, but he had been most disappointing after that at the Curragh, at Ascot and in the Great Voltigeur Stakes. In 1981 he regained his best form. On his reappearance in the John Porter Stakes at Newbury he beat Cracaval a head and then won the Ormonde Stakes at Chester. His 3-length victory over Light Cavalry in the Hardwicke Stakes stamped him as a very useful four-year-old. He was unlucky in the King George VI and the Queen Elizabeth Stakes at Ascot when 10½ lengths fifth to Shergar as one of his adversaries pushed him on to the rails four furlongs out. He might well have finished runner-up had that incident not happened. Pèlerin's last race before the Prix de l'Arc de Triomphe was in the Grosser Preis von Baden when he beat Hohritt an easy 2 lengths in record time. He started at odds of 33 to 1 on 4 October and seemed a possible outsider.

A.J. Kelly's Beldale Flutter beat an immature Shergar 2½ lengths in the William Hill Futurity Stakes as a two-year-old when he also won the Grand Critérium International at Ostende and ran fourth to Robellino and Recitation in the Royal Lodge Stakes. He disappointed in the Clerical Medical Greenham Stakes and behind To-Agori-Mou in the 2000 Guineas but then won the Mecca Dante Stakes by three-quarters of a length from Shotgun. A week before the Derby he got loose on Newmarket Heath and collided with Moorestyle. The injuries he suffered meant that he was not seen out again till the Benson and Hedges Gold Cup when he beat Kirtling and Master Willie three-quarters of a length and the same. This was a very encouraging performance but on balance not quite good enough for a potential winner of the Prix de l'Arc de Triomphe. Thus although Beldale Flutter might run into a place, his odds of 12 to 1 seemed rather ungenerous.

Craig B. Singer's Irish trained three-year-old filly, Condessa, caused a major upset in the Musidora Stakes at York in May when she beat Madam Gay 4 lengths with Fairy Footsteps, the 1000 Guineas winner, only third. Earlier in the spring she had won a handicap at Clonmel and had run third to Leap Lively in the Johnny Walker Oaks Trial at Lingfield. Condessa was a dismal eighth to Strigida in the Ribblesdale Stakes at Royal Ascot but then was an excellent 2½-length runner-up to Blue Wind in the Irish Guinness Oaks. She beat Leap Lively a neck in the Yorkshire Oaks, but was a disappointing sixth to Alma Ata in the Park Hill Stakes at Doncaster. Condessa was clearly useful but inconsistent and was almost certainly taking on too stiff a task at Longchamp on 4 October when she started at 99 to 1.

Mrs Bertram Firestone's Blue Wind won the Silken Glider Stakes at two and was sixth in the Critérium des Pouliches. Mrs Firestone acquired Blue Wind for 180,000 guineas at Goffs November Sales at the end of her two-year-old season. On her three-year-old début, Blue Wind won the Edenderry Stakes (8f) at the Curragh in May and then ran Arctique Royale to a short head in Goffs 1000 Guineas on 23 May. In the Oaks, partnered by Lester Piggott, Blue Wind trounced Madam Gay and Leap Lively by 7 lengths and 10 lengths. Blue Wind decisively beat Condessa 2½ lengths in the Irish Guinness Oaks, but she was a disappointing fourth to King's Lake in the Joe McGrath Memorial Stakes on 19 September. It seemed likely that she was a better filly in the spring and summer and had possibly gone over the top by October.

Mrs Bertram Firestone fortunately had another very good three-year-old filly in the Prix de l'Arc de Triomphe, April Run, who was trained in France by François Boutin. Unraced at two, April Run won a maiden race at Saint-Cloud over a mile and a half and then beat Landresse and Léandra 3 lengths and the same in the Prix Cléopâtre at Saint-Cloud on 16 May. April Run was not disgraced when third in the Prix de Diane de Revlon 4 lengths and a short neck behind Madam Gay and Val d'Erica, the winner of the Premio Regina Elena and the Oaks d'Italia. April Run was a creditable fourth to Akarad in the Grand Prix de Saint-Cloud and in August at Deauville she slammed her moderate adversaries by 5 lengths in the Prix de Pomone. She was most impressive in the paddock before the Prix Vermeille and was clearly improving physically. After taking the lead nearly 500 metres from the line, April Run had no difficulty in holding Léandra and Madam Gay by 1½ lengths and a head with Snow Day fifth and Tootens seventh. Many good judges at Chantilly felt she would go very close in the Prix de l'Arc de Triomphe and she and Blue Wind started at odds of 13 to 2 couplé.

Baron de Zuylen de Nyevelt also had two runners in the 1981 Prix de l'Arc de Triomphe, the three-year-old filly Léandra and the four-year-old colt Perrault. Prior to finishing runner-up to April Run in the Prix Vermeille, Léandra had won the Prix de la Nonette at Deauville narrowly from Snow Day, who was later disqualified for carrying the wrong weight. On 4 July she beat Landresse half a length in the Prix de Malleret with Snow Day 4 lengths away fourth, but in the Prix de Royaumont on 7 June the latter had defeated Léandra a head. In the spring she had been runner-up to Last Love in the Prix Pénélope at Saint-Cloud and third to April Run in the Prix Cléopâtre. Although consistent it seemed likely that Léandra would find several of her adversaries too good for her in the Prix de l'Arc de Triomphe.

The four-year-old Perrault won the Prix Sica Boy and a maiden race at Longchamp and was a runner-up in the Grand Prix de Vichy and the Grand Prix de Deauville in 1980; he was some way behind the best of his generation as a three-year-old. In the spring of 1981 he ran second to Armistice Day in the Prix Exbury and was third to Lancastrian and En Calcat in the Prix d'Hédouville. On 10 May Perrault beat Son of Love a head in La Coupe at Longchamp. A little less than a fortnight later he had a terrific duel with Lancastrian in the Grand Prix d'Evry, going down by only a short head. Baron de Zuylen de Nyevelt's colt then easily disposed of Roi Guillaume in the Prix Maurice de Nieuil at Saint-Cloud on 14 July. In the Grand Prix de Deauville he took the lead 150 metres out and held Castle Keep and Glenorum by a short head and three-quarters of a length with such good horses as Nicholas Bill and First Prayer further behind. It was a measure of Perrault's improvement as a four-year-old that he should have defeated such a strong field. He certainly could be considered a possible outsider and he and Léandra started at 27 to 1 couplé.

Bruce McNall's four-year-old Argument was trained in 1981 by Maurice Zilber. On his reappearance he beat Katowice and Armistice Day a head and 3 lengths in the Prix d'Harcourt, with Detroit only fourth. He then was a most impressive 3-length winner over Armistice Day in the Prix Ganay on 3 May with Detroit fifth. Argument was third to Vayrann in the Prix Jean de Chaudenay, but was demoted to fifth place. He sweated up before the Grand Prix de Saint-Cloud when only seventh to Akarad. He was not disgraced when sixth, 5 lengths behind John Henry, in the first running of the Arlington Million at the end of August. It seemed, however, that like Detroit, Argument was not as good at four as he had been at three. The field for the Prix Ganay was one of the least competitive for several years and Argument was permitted to start at odds of 32 to 1 on 4 October.

The two best three-year-old colts in France in 1981 were probably the Aga Khan's Akarad and Jules Ouaki's Bikala. When Akarad and Shergar were both yearlings at stud in Ireland, François Mathet chose Akarad in preference to the Epsom Derby winner, who then went to the yard of Michael Stoute at Newmarket. Akarad was unraced at two and on his racecourse début won the Prix des Gobelins, a newcomers' race over 2100 metres, by 6 lengths at Longchamp on

3 May. Three weeks later he beat his stable companion Vayrann 1½ lengths in the Prix de l'Avre (2400 metres) at Longchamp. He was, therefore, very lightly raced when he met Bikala for the first time in the Prix du Jockey Club. Bikala on the other hand had two outings as a juvenile. On 24 March he was runner-up to No Lute in the Prix Santoi (2000 metres) at Saint-Cloud and then won the Prix du Bel Air (1850 metres) at Longchamp on 12 April. In the Prix Lupin on 17 May Bikala was held up and finished fourth to the subsequently disqualified No Lute, The Wonder and Dunphy. At Chantilly he was the mount of the 18-year-old apprentice Serge Gorli, who sent him into the lead some way before the straight. Bikala was 2 lengths clear 500 metres out, and galloped on resolutely to beat Akarad 4 lengths, with Gap of Dunloe and Rahotep 2½ lengths and three-quarters of a length away third and fourth. Akarad had had a lot to do entering the straight and did not enjoy the best of runs. When he did get clear he made rapid progress and took second place 200 metres from the line, but by then Bikala was beyond catching. In the Grand Prix de Saint-Cloud Yves Saint-Martin kept Akarad in closer touch with the leaders. Bikala took the lead over 600 metres from home, but had no answer to Akarad's challenge and went down by 2 lengths, with Lancastrian and April Run half a length and a neck away third and fourth. Gold River was 2½ lengths further back ahead of Prince Bee and Argument. The Aga Khan's colt had only one more race before the Prix de l'Arc de Triomphe in the Prix Niel, in which he gave Rahotep a kilo and a half a length beating. He looked as if he would benefit from the race and undoubtedly possessed a very live chance in the Prix de l'Arc de Triomphe. His starting price of 2 to 1 was exceptionally ungenerous and must in part have been due to the occasional blind faith certain *turfistes* place in Yves Saint-Martin and François Machet.

Bikala also only had one outing after the Grand Prix de Saint-Cloud, in the Prix du Prince d'Orange. He gave Vayrann a kilo but failed by half a length to hold the Aga Khan's colt, whose autumn objective was the Champion Stakes. Surprisingly Bikala was permitted to start at odds of 12 to 1 on 4 October, which looked much better value than those of Akarad.

Jean-Claude Weill's Rahotep won on his début as a juvenile at Deauville and ran fourth to Dunphy in the Prix des Chênes and fifth to The Wonder in the Prix de Condé on his other two outings. On 10 May he beat Mbaiki 1½ lengths in the Prix Hocquart after he had earlier won the Prix Barneveldt (2100 metres) at Saint-Cloud on 14 April. Nevertheless he had been behind Akarad both at Chantilly and in the Prix Niel and, as there seemed no valid reason why he should reverse the form, he started at 25 to 1 on 4 October.

Alec Head's stable had only one runner in the 1981 Prix de l'Arc de Triomphe, the four-year-old mare Gold River who was owned by Jacques Wertheimer. Unraced as a juvenile on account of a split pastern, she won the Prix de Neuilly and the Prix de Saint-James, both over 2000 metres at Longchamp in May and June 1980. She then ran second to Detroit in the Prix Fille de l'Air and third to Great Verdict in the Prix de Minerve at Evry. At Deauville she had 3 lengths to spare over Mariella in the Prix de Pomone. Gold River was fourth in the Prix Vermeille and then a close second to Mariella, whom she was meeting on 1½ kilos worse terms than at Deauville, in the Prix de Royallieu. She then impressively defeated Monsieur Marcel 3 lengths in the Prix Royal Oak with Ardross a close third, although Charles St George's stayer was probably feeling the effects of his long season. Nevertheless Gold River was an exciting prospect for 1981.

In her first race as a four-year-old, the Prix Jean Prat, Gold River was almost 10 lengths behind the leaders turning into the straight at Longchamp, but she produced such a decisive burst of finishing speed that she got up to beat Proustille a neck. It was the same story in the Prix du Cadran; Gold River was even further behind entering the straight but again showed such electrifying acceleration that she had 3 lengths to spare over Hereas. It was decided not to run her against Ardross in the Gold Cup at Royal Ascot but to put her back to middle distances. Gold River was beaten into fifth place by Akarad in the Grand Prix de Saint-Cloud, but it was clear she did not act on left-handed courses. In the Prix Foy Gold River had to concede Detroit 3 kilos and, considering the funereal early pace of the race, she did very well to finish only 3½ lengths behind Robert Sangster's 1980 Prix de l'Arc de Triomphe heroine. Indeed she was finishing better in the final hundred metres than either Detroit or Lancastrian. Given a strong pace on 4 October, Gold River's chance could not have been much, if at all, inferior to that of Ardross or Detroit. Certainly she represented extremely good each-way value at the amazingly generous odds of 53 to 1.

None of the remaining five runners could be given a serious chance on 4 October. Tootens, a half-sister to Nocturnal Spree, the 1000 Guineas winner, had beaten Tropicaro in a substandard Prix Saint-Alary but was a well-beaten fifth in the Prix de Diane de Revlon. She required heavy going to show her best and it would be a surprise if she reached the frame. Gap of Dunloe was trained in England by John Dunlop as a two-year-old but was unplaced at Sandown and Newmarket. He beat Roi Guillaume 1½ lengths in the Prix des Marronniers at Longchamp but was last in the Prix Hocquart. Although a surprise third in the Prix du Jockey Club, he was a well-beaten sixth in the Irish Sweeps Derby and was trounced 6 lengths when runner-up to Strong Gale in the Grand Prix Prince Rose at Ostende on 21 July.

The four-year-old mare Gilded Vanity had beaten Saint Jonathon over 2000 metres at Cagnes-sur-Mer and landed a gamble in the Brownstown Stakes at the Curragh in August but was little better than a handicapper.

Ring the Bell had won eleven of his twenty-four starts in New Zealand and Australia. In New Zealand he won the Derby, the Avondale Guineas, the Avondale Champion Stakes and the Alison Stakes. In Australia he won the Canterbury Guineas and was runner-up to Our Paddy Boy in the A. J. C. Derby at Randwick. He was easily beaten into third place by Boathouse on his début in England in September at Goodwood and was only sixth, nearly 6 lengths behind Vayrann and Bikala, in the Prix du Prince d'Orange. It was clear he had had insufficient time to acclimatise. It was a sporting gesture to send Ring the Bell to France as it was for the connections of the Danish challenger Action Man to do the same. He had been successful in three of his six outings in 1981. On 6 September he had narrowly won over nine furlongs at Klampenborg in Copenhagen. It was completely impossible to judge his ability compared to his rivals but it was virtually certain that he would face a formidable task at Longchamp.

The absences from the field of the 1981 Prix de l'Arc de Triomphe made almost more news than the participants in the weeks preceding it. Shergar's 10-length victory over Kirtling, later to win the Gran Premio d'Italia, in the Guardian Classic Trial at Sandown and his 12-length success in the Chester Vase, had stamped him as an exceptional colt. His astonishing and superb 10-length victory over Glint of Gold, who would have won the Derby in a normal year, at Epsom was a privilege for all who watched it. When he then facilely beat Madam Gay in the King George VI and the Queen Elizabeth Stakes at Ascot after his equally impressive victory in the Irish Sweeps Derby, it was more than reasonable to compare him with Ribot and Sea Bird. Neither of them raced over the distance of the St Leger and in view of the fact that he was by Great Nephew it was always possible that he would not stay over a mile and a half, more especially as his grandam Ginetta was best at a mile having won the Poule d'Essai des Pouliches in 1959. A victory in the St Leger would have added nothing to his value as a stallion and, seeing that his autumn objective was the Prix de l'Arc de Triomphe, it would have seemed considerably more logical to have run Shergar either in the Prix Niel or the Prix du Prince d'Orange. They would not have taxed his stamina and it would have allowed him to familiarise himself with the Longchamp track. It is a matter of some astonishment that more British trainers do not make use of these preparation races for their runners in the Prix de l'Arc de Triomphe. After his defeat in the St Leger, it was quite understandable that Shergar did not run in Paris for what was supposed to be an easy racecourse gallop had turned into a hard race.

The other absentee who occupied the headlines for several days was Robert Sangster's champion two-year-old of 1980, Storm Bird. After the incident involving his tail and mane being cut in the winter, Storm Bird met with several setbacks in the spring. In late August and early September he was reported to be working in excellent fashion on the gallops with King's Lake and was heavily supported in the antepost market for the Prix de l'Arc de Triomphe. Vincent O'Brien sent him to contest the Prix du Prince d'Orange but clearly he had not recovered from the coughing that had afflicted him, for he finished a dismal seventh of nine to Vayrann, his only outing as a three-year-old. He was, therefore, retired to stud.

It was announced early in September that Glint of Gold, whose victories in 1981 included the Grand Prix de Paris, the Derby Italiano and the Great Voltigeur Stakes, would have the Preis von Europa at Cologne as his autumn objective. He had been the first English-trained winner of the Grand Prix de Paris since Lemonora in 1921 so it was regrettable, after Shergar's withdrawal, that this courageous and consistent colt was not in the field on 4 October. Both Vayrann, who was to be aimed at the Champion Stakes, and Master Willie, who was unsuited by soft ground, were withdrawn from the Prix de l'Arc de Triomphe two days before the race.

The Final Preparations

On Friday 26 September the CGT, the massive communist trade union, gave a warning to the French employers' associations that unless a new collective bargaining agreement was reached by Saturday 3 October industrial action would be taken to stop the Prix de l'Arc de Triomphe. The unions, who represented 80 per cent of those employed in racing attacked this politically motivated stand of the CGT. A meeting took place at the Ministry of Labour two days before the race and when it ended after twelve hours, it was announced that an agreement had been reached which lifted the threat to Longchamp's most prestigious race.

On Thursday 1 October Henry Cecil said at Newmarket that Ardross was very well and 'I am more than hopeful'. It was also reported that Beldale Flutter, who had a reputation as a mudlark, had worked very well the following morning. Pat Eddery, who had been going to ride Beldale Flutter before he was claimed for King's Lake, said that he was delighted to be on Robert Sangster's colt who, he thought, had 'a great chance of lasting the trip at Longchamp'. The mount on Detroit went to Freddie Head who obtained permission from his retaining owner Jacques Wertheimer ten days before the race to ride the 1980 Prix de l'Arc de Triomphe heroine rather than Gold River. Alec Head was not very happy about his son's decision, but contacted Gary Moore, the son of George Moore, who was riding in Hong Kong, and he agreed to take the mount on Jacques Wertheimer's fine stayer. Olivier Douieb felt that Detroit 'could be even better than last year' and was full of confidence, though he thought Bikala would be a danger. The Aga Khan's Akarad worked extremely well with Vayrann in his final gallop at Chantilly and there was a lot of optimism in François Mathet's stable afterwards.

Président Mitterrand did not come to Longchamp in this his first autumn after his election. Neither did his Finance Minister Jacques Delors, who was attending a meeting of all the EEC finance ministers on the European Monetary System, in Brussels. Since the change of tenant in the Elysée Palace, the Banque de France had lost an estimated 35,000 million francs from its reserves attempting to defend the franc. In the week preceding the Prix de l'Arc de Triomphe the Banque de France had spent about 8,500 million francs supporting the currency. It was therefore agreed in Brussels that there should be a realignment of the currencies in the EMS: the Deutschmark and Dutch guilder were revalued by 5·5 per cent and there was a devaluation of 3 per cent of the French franc and the lira. In less than six months of the new Socialist government in France the French franc had been devalued by 8·5 per cent thus decreasing the real value of the Prix de l'Arc de Triomphe for foreign owners. It came too late for the huge international crowd at Longchamp on Sunday 4 October who had had to exchange their dollars and pounds at the old rate. Amongst the well-known English racegoers at the Prix de l'Arc de Triomphe were the financier Sir Gordon White accompanied by the film actress Mary Tyler Moore whom he had escorted across five continents in the previous six months. Antony Gilbey of Gilbey Racing, whose assessment system for the Champion Racehorse of Europe had pointed to Star Appeal's chance in 1975, admitted before the race to a small investment on Gold River who was indicated as the best value at the then current position of the ratings.

It was a cold, cloudy and windy afternoon, conditions more normally associated with Newmarket Heath than Longchamp. Alan Clore's Kirchner won the first race, the Prix de Beauvais. In the next, the Prix Marcel Boussac, Lester Piggott led throughout on Play It Safe and held off his better fancied stable-companion River Lady, ridden by Philippe Paquet, by a head. The Prix de l'Abbaye de Longchamp again turned out to be an English benefit with young Wally Swinburn on Marwell having a neck and a length to spare over Sharpo and Rabdan. The latter's jockey Piggott lodged an objection but it was quickly overruled.

The Race

As the runners started coming into the paddock for the Prix de l'Arc de Triomphe, a few spots of rain began to fall but fortunately not in earnest. Akarad, the favourite, who had looked a little burly before the Prix Niel, stripped much fitter but he did not canter down to the post very well. In fact, the pick of the field was April Run. Both Ardross and Bikala also looked exceptionally fit. Blue Wind, however, appeared a shadow of her former self in midsummer; Detroit was not as impressive as the previous year and Cut Above was broken in his coat. Gold River, who was on her toes in the parade, had retained her summer coat and looked very well. Pèlerin, as usual with horses trained by Harry Wragg, was turned out in magnificent condition but as Sir Philip Oppenheimer's son, Antony, said before the race, he would have preferred a sounder surface.

The crowd was so huge that Ardross's owner Charles St George, who earlier had commented that 'one needs to have at least ten pounds in hand to win this race', decided to watch the Prix de l'Arc de Triomphe on the television in the salon next to that of the stewards where the famous picture of Gladiateur hangs. Since Ardross was drawn the worst at twenty-four, Piggott had the problem of deciding whether to drop him out in the rear or to go quickly out of the stalls and try to get a prominent position early on, an unenviable choice in such a large field.

Condessa dwelt at the start and lost some ground. Bikala and Ardross were both quickly away and on settling down Serge Gorli on Bikala set a reasonable pace to Ardross on the wide outside. Gold River was one of the last out of the stalls and Gary Moore tacked across to the rails behind the majority of the field. After the Petit Bois Bikala held a clear lead from Ardross who was a little in front of Detroit, Beldale Flutter, Akarad and Ring the Bell, ahead of Perrault and April Run, who was tracking the favourite. At the top of the hill, Bikala still led Ardross, Beldale Flutter, Detroit, Akarad and Perrault, whilst Gold River was improving steadily on the rails to be about 8 lengths behind the leader. Coming down the hill Akarad began to pull, and Beldale Flutter and Ring the Bell dropped back; Gold River was making very significant progress to be just behind April Run. On the turn into the straight Bikala led Ardross and Perrault, tracked by Detroit, Akarad, April Run, Gold River and Argument who had also improved on the descent. Early in the straight Detroit quickly weakened and Akarad lost his action and was clearly not going to win. With 400 metres to go, Ardross began to struggle; the effort to take up a prominent position early on had taken its toll. Bikala still led from Perrault who launched a very dangerous-looking challenge 300 metres out, but Bikala gamely fought him off. Gold River meanwhile had come smoothly through, having been switched off the rails to avoid being hampered by April Run, to be poised to challenge Bikala a furlong out. The finishing speed of Jacques Wertheimer's filly was too great for the Prix du Jockey Club winner and she swept past him to win easily by three-quarters of a length. Only a nose behind came the fast-finishing April Run who had been hampered by the tiring Akarad and had not enjoyed the luckiest of runs. She finished 2 lengths and the same in front of Perrault and Ardross. They were followed by Argument and Akarad in sixth and seventh places. Tootens was 3 lengths away eighth, Léandra ninth, Snow Day tenth and King's Lake eleventh. Throughout the race, however, only the first seven had ever held out any hopes of winning. Gary Moore had never touched Gold River with the whip and it is more than probable that her margin of victory could have been more than three-quarters of a length. Nonetheless Bikala had put up a fine display to lead from the start until only fifty metres from the line.

Gold River's success was greeted by almost total silence by the crowd when she returned to the unsaddling enclosure. The French *turfistes* must have been influenced by Freddie Head's decision to switch to the very disappointing Detroit. Gold River was, however, a deserved and meritorious winner and jockeys are notorious for being the world's worst tipsters, with a few select exceptions like Lester Piggott. Alec Head commented that Freddie Head had made 'a big mistake' but added that it was 'a sign of old age' that George Moore on Saint Crespin and twenty-two years later his son Gary on Gold River had won the Prix de l'Arc de Triomphe for his stable. He explained Gold River's failure in the Grand Prix de Saint-Cloud by saying that she could only act on right-handed tracks and that she was much better at Longchamp than elsewhere. He had been tempted only to run her in the Prix Royal Oak instead of both the Prix de l'Arc de Triomphe and the later classic, but

Gold River (11) beating Bikala (rails) and April Run (19) with Perrault (7) fourth

Jacques Wertheimer had insisted that she should run in the much sterner test as well.

Twenty-nine-year-old Gary Moore, who had ridden permanently in France for a few years in the 1970s, had gained his first success in the Prix de l'Arc de Triomphe. On Saturday he had ridden the winner of the last race at Happy Valley on Hong Kong Island. He had then rushed across the beautiful harbour to Kai Tak Airport where he had caught a plane to Paris, on which there was a bomb scare. Eventually after a fourteen-hour flight he arrived at Roissy-Charles de Gaulle Airport at six o'clock in the morning. He related that he 'felt sure we were going to win' halfway down the hill when Gold River made rapid headway on the leaders. He added that he had brought his golf clubs with him and that he was going to play Freddie Head and his brother-in-law Philippe Paquet the next day so that they 'can get some of their money back'. He stayed the Sunday and Monday with Philippe and Michelle Paquet and returned to Hong Kong on Tuesday 6 October in order to ride there the following day.

Gold River gave Jacques Wertheimer his second success in the Prix de l'Arc de Triomphe in six years. It was the fourth success for his trainer Alec Head who thus equalled the record of C.H. Semblat who had trained Djebel, Ardan, Caracalla and Coronation.

There were few valid excuses for the beaten with the possible exception of April Run. Paquet had tracked Akarad and when he weakened about 400 metres out she was unable to find a gap on the inside as Argument blocked her way. Paquet was forced to switch her to the outside of the field to make her challenge and she finished fastest of all and with a clear run would certainly have finished second. Freddie Head reported Detroit was finished early in the straight and Douieb felt she might have begun to go over the top two weeks earlier. Piggott said he had had to make too much use of Ardross early on to obtain a prominent position and thought it might have been a different story if he had not had the worst draw. Tony Murray said that Beldale Flutter was never going well. There were no excuses either for Akarad, who had every chance in the straight but failed to quicken, or King's Lake, who did not stay, or Prince Bee.

The Rest of the Season

Ardross and Gold River met again in the Prix Royal Oak on 25 October in very heavy going. There were only seven runners and the first half of the race, with no one wanting to make the pace, was almost a canter. Piggott was able to settle Ardross in sixth place until about five furlongs from home when he quickly took Charles St George's stayer into the lead and was never thereafter in danger, having Proustille 4 lengths behind him on the line with a disappointing Gold River 2½ lengths away third. Head reported that the Prix de l'Arc de Triomphe winner pulled a lot in the first half of the race and that she produced nothing when she came under pressure. She certainly ran some way behind her best form but she probably would not have beaten Ardross even at her best that afternoon in the very heavy going.

Prince Bee was not disgraced when he ran 5½ lengths fourth to Vayrann and Cairn Rouge in the Champion Stakes on 17 October with Snow Day unplaced. Rahotep, who was never seen with a chance in the Prix de l'Arc de Triomphe, trounced Two Steps 8 lengths in the Prix du Conseil de Paris with Lancastrian and Gap of Dunloe unplaced. April Run and Detroit both ran in the Turf Classic at Aqueduct. Robert Sangster's filly met some interference and could only manage fifth place. April Run emphatically upheld the Prix de l'Arc de Triomphe form, however, by beating the Man o'War Stakes winner Galaxy Libra and The Very One 1½ lengths and 2½ lengths. In the Washington D.C. International Stakes the useful ex-French colt Providential beat her a length with his much more fancied stable-companion Galaxy Libra 3½ lengths away third, Cairn Rouge fourth and Beldale Flutter only seventh.

Pedigree

Gold River's sire was Riverman who had also got the previous year's heroine Detroit. It was the first time in the history of the race that a stallion had been responsible for two different winners of the Prix de l'Arc de Triomphe in consecutive years. Ribot had been the sire of Molvedo in 1961 and Prince Royal II in 1964. Tourbillon had sired two winners, Djebel in 1942 and Caracalla in 1948. Brûleur had been responsible for three individual winners, Ksar in 1921 and 1922, Priori in 1925 and Samos in 1935, but even he had not sired two separate winners in consecutive years as had Riverman.

Gold River's dam, Glaneuse by Snob, was an extremely useful filly. As a juvenile she ran third in the Critérium de Maisons-Laffitte and was not disgraced when sixth to Yelpana in the Grand Critérium. At three she won the Prix Chloë in April and then ran third only three-quarters of a length and 1½ lengths behind Crepellana and Saraca in the Prix de Diane. After she won the Prix de Malleret, she only failed by half a length to give the 1000 Guineas runner-up, Hecuba, 3 kilos in the Prix de la Nonette. Glaneuse then ran fourth, beaten a little over 1½ lengths to Saraca and Crepellana, in the Prix Vermeille. She later beat Bacuco, the St Leger Italiano winner in 1969 and runner-up in the Gran Premio di Milano in 1970, three-quarters of a length in the Gran Premio del Jockey Club, with Remand only fourth. The year after her victory Bacuco was to beat Ortis in the same race, which gives some idea of Glaneuse's merit. In total she earned her owner Mme Pierre Wertheimer 719,635 francs in prize money.

Glaneuse's first foal Gracious by Habitat won three times and is the dam of the useful French sprinter Greenaway who won the Prix d'Arenberg. Her Riverman filly Glena won two races in 1977 but her full brother Sir Godfrey was nothing exceptional and although he achieved three victories was claimed out of a seller at Saint-Cloud for 83,333 francs in April 1979. Glaneuse's next winning produce was Gold

River. In 1978 she was barren to Green Dancer.

Glaneuse's dam, Glamour by Djebé, was bought as a foal for 2100 guineas in December 1960 and imported to France where she won once and was placed five times as a three-year-old, winning 8,225 francs. In all, she produced eight winners, the best of which apart from Glaneuse was Gleam by Spy Well. She was second three times as a juvenile – on her final outing to Vela in the Critérium des Pouliches. Although she only won once as a three-year-old – when she took the Prix de Minerve at Chantilly – she was a high-class filly finishing second to Humble Duty in the 1000 Guineas, second in the Prix de Malleret and third to the subsequent Irish Guinness Oaks winner Santa Tina in the Prix de Royaumont.

Glamour's dam, Tudor Gleam by Owen Tudor, ran five times as a juvenile in the colours of Lady Portal. She was trained by Geoffrey Brooke and won the Cotton Maiden Plate over five furlongs at Manchester. She also was a very good broodmare; she got six winners, two including Glamour in France and four in England. The best was probably Sovereign Gleam by Sovereign Path, whose victories as a two-year-old included the Grand Critérium d'Ostende and the Tankerville Nursery Handicap at Ascot. He passed the post first in the Prix Thebais at Longchamp but was demoted to second place. As a three-year-old, Sovereign Gleam was runner-up to Sparkler in the Thirsk Classic Trial and in the Grand Prix de Bruxelles. He also won the Prix Chevalier de Stuers at Ostende and was a good third to the useful French colt Sharapour in the Prix Caligula at Saint-Cloud. Tudor Gleam was also the dam of Regal Light by Sovereign Path whose four victories included the Royal Hunt Cup and the Rous Nursery Handicap at Doncaster.

Tudor Gleam's dam, Riding Rays by Nearco, was also useful on the racecourse. As a two-year-old she won the Girton Stakes at Worcester and was third to Neolight and Rivaz in the Cheveley Park Stakes from four outings. She ran eight times in the highest class at three but was unplaced in the 1000 Guineas and Coronation Stakes at Ascot. She won the Red Oaks Handicap (7f) at Ascot and was runner-up in the Nassau Stakes at Goodwood. She produced five winners at stud, the best of which was Byng by Vimy who won the Great Metropolitan Handicap amongst his four victories.

Riding Ray's dam, Infra Red by Ethnarch, we have already encountered, as she was the fourth dam of Mill Reef. Gold River's success in the Prix de l'Arc de Triomphe, ten years after Mill Reef, only goes to underline the significance of Black Ray as a broodmare. If Gold River is only half as successful at stud as was her sixth dam, she will become a very significant broodmare.

PRIX DE L'ARC DE TRIOMPHE 2400 metres
at Longchamp on Sunday 4 October 1981. For three-year-olds and upwards. Going: Good to soft.
2,000,000 francs to the winner; 800,000 fr. to the 2nd; 400,000 fr. to the 3rd; 200,000 fr. to the 4th. 500,000 fr. to the breeder of the winner; 200,000 fr. to that of the 2nd; 100,000 fr. to that of the 3rd; 50,000 fr. to that of the 4th. Value to the winner: 2,000,000 fr.

GOLD RIVER	ch.f.	by Riverman – Glaneuse by Snob	4.57½k	Jacques Wertheimer	G.W. Moore	1
BIKALA	b.c.	by Kalamoun – Irish Bird by Sea Bird	3.56k	Jules Ouaki	S. Gorli	2
APRIL RUN	b.f.	by Run the Gantlet – April Fancy by No Argument	3.54½k	Mme Bertram R. Firestone	P. Paquet	3
PERRAULT	ch.c.	by Djakao – Innocent Air by Court Martial	4.59k	Baron Th. de Zuylen de Nyevelt	H. Samani	4
ARDROSS	b.c.	by Run the Gantlet – La Melody by Levmoss	5.59k	Charles St George	L. Piggott	5
ARGUMENT	b.c.	by Kautokeino – Arantelle by Tapioca	4.59k	Bruce McNall	A. Lequeux	6
AKARAD	br.c.	by Labus – Licata by Abdos	3.56k	H.H. Aga Khan	Y. Saint-Martin	7
TOOTENS	ch.f.	by Northfields – Night Attire by Shantung	3.54½k	Mlle Cora Lynch	G. Doleuze	8
LEANDRA	b.f.	by Luthier – Ady Endre by Reliance	3.54½k	Baron Th. de Zuylen de Nyevelt	G. Dubroeucq	9
SNOW DAY	b.f.	by Reliance – Vindaria by Roi Dagobert	3.54½k	Robert Sangster	A. Gibert	10
ACTION MAN		by Belmont – Eminent by Umidwar or Pearl Diver	6.59k	Jorgen Mathiesen	O. Larsen	
PELERIN		by Sir Gaylord – Padrona by St Paddy	4.59k	Sir Philip Oppenheimer	E. Hide	
LANCASTRIAN		by Reform – Rosalie by Molvedo	4.59k	Sir Michael Sobell	M. Philipperon	
PRINCE BEE		by Sun Prince – Honerko by Tanerko	4.59k	Sir Michael Sobell	J. Mercer	
RING THE BELL		by Rangong – Witchcraft by Mystery	4.58½k	Mme E.M. Mackley	N. Tiley	
GILDED VANITY		by Run the Gantlet – Sunset Temple by Golden Horus	4.57½k	C.J. Fergus	G. Curran	
DETROIT		by Riverman – Derna by Sunny Boy	4.57½k	Robert Sangster	F. Head	
CUT ABOVE		by High Top – Cutle by Saint Crespin	3.56k	Sir John Astor	B. Taylor	
KING'S LAKE		by Nijinsky – Fish Bar by Baldric II	3.56k	Jean-Pierre Binet	P. Eddery	
GAP OF DUNLOE		by Sassafras – Absaretch by Dancer's Image	3.56k	Bob Demuyser	W. Swinburn Jr	
BELDALE FLUTTER		by Accipiter – Flitter Flutter by Cohoes	3.56k	A.J. Kelly	A. Murray	
RAHOTEP		by Mattahawk – La Masure by Net	3.56k	Jean-Claude Weill	J.L. Kessas	
BLUE WIND		by Lord Gayle – Azurine by Chamossaire	3.54½k	Mme Bertram R. Firestone	W. Swinburn Sr	
CONDESSA		by Condorcet – Varinessa by Varano	3.54½k	Craig B. Singer	D. Gillespie	

24 runners.
GOLD RIVER trained by A. Head; bred by J. Wertheimer.
Distances: ¾ length; neck; 2 lengths; 2 lengths; 1½ lengths; ¾ length; 2½ lengths; head; short neck. Time: 2mins 35.20 seconds.
Pari-Mutuel: Win: 54fr. 00. Places: 11fr. 90; 4fr. 50; 3fr. 80.
S.P.: 2 to 1 Akarad; 7 to 2 Snow Day, Detroit and King's Lake (couplé); 11 to 2 Ardross; 13 to 2 April Run and Blue Wind (couplé); 12 to 1 Beldale Flutter; 14 to 1 Prince Bee and Lancastrian (couplé) and Bikala; 25 to 1 Rahotep; 27 to 1 Léandra and Perrault (couplé); 32 to 1 Argument; 33 to 1 Pèlerin; 38 to 1 Cut Above; 53 to 1 GOLD RIVER; 99 to 1 Tootens, Action Man, Ring the Bell, Gilded Vanity, Gap of Dunloe and Condessa.

GOLD RIVER chestnut filly 1977	Riverman b. 1969	Never Bend b. 1960	Nasrullah b. 1940	Nearco br. 1935	Pharos
					Nogara
				Mumtaz Begum b. 1932	Blenheim
					Mumtaz Mahal
			Lalun b. 1952	Djeddah ch. 1945	Djebel
					Djezima
				Be Faithful br. 1942	Bimelech
					Bloodroot
		River Lady b. 1963	Prince John ch. 1953	Princequillo b. 1940	Prince Rose
					Cosquilla
				Not Afraid b. 1948	Count Fleet
					Banish Fear
			Nile Lily br. 1954	Roman b. 1937	Sir Gallahad
					Buckup
				Azalea ch. 1944	Sun Teddy
					Coquelicot
	Glaneuse b. 1966	Snob b. 1959	Mourne ch. 1954	Vieux Manoir b. 1947	Brantôme
					Vieille Maison
				Ballynash b. 1946	Nasrullah
					Ballywellbroke
			Senones br. 1952	Prince Bio b. 1941	Prince Rose
					Biologie
				Sif br. 1936	Rialto
					Suavita
		Glamour br. 1960	Djebé b. or gr. 1945	Djebel b. 1947	Tourbillon
					Loïka
				Catherine gr. 1940	Tiberius
					Catherinette
			Tudor Gleam bl. or br. 1952	Owen Tudor br. 1938	Hyperion
					Mary Tudor
				Riding Rays bl. 1943	Nearco
					Infra Red

Akiyda 1982

Akiyda

The Race Gains a Sponsor

Since 1977 when Revlon had first sponsored the Prix de Diane, the Société d'Encouragement had been considering the possibility of obtaining a sponsor for the Prix de l'Arc de Triomphe. Unlike those of England and Ireland, the laws of France forbade sponsors to be the manufacturers either of cigarettes or any kind of alcohol. In the previous couple of years Jean Romanet, the Directeur-Général of the Société d'Encouragement, had been approaching various companies in France to sound them out. French industry was not yet accustomed to the idea of sponsorship of racing. The day after the Grand Prix de Saint-Cloud in 1982, won by Paul Mellon's English-trained Glint of Gold, the author discussed the possibility of trying to find a sponsor for the Prix de l'Arc de Triomphe in England or Ireland since it has become the most important international race in the world. Jean Romanet agreed that no harm could come from such a search especially if a sponsor could be found for the race in the coming October.

On my return to England I immediately enlisted the help of Richard Berens and Lord James Crichton-Stuart, whose premature death was to occur but five months later. We quickly realised that there were only a handful of companies who could make a prompt decision, one of which was the international hotel group, Trusthouse Forte, whose hotels in Paris included the Georges V, Plaza Athenée and la Tremoille. On the day of the King George VI and the Queen Elizabeth Stakes at Ascot we had a meeting with Jean Romanet and his son Louis Romanet to check that Trusthouse Forte and the other companies we envisaged approaching met with the approval of the Société d'Encouragement and to decide the length of the sponsorship and other details. The following week we contacted various public relations executives, including Lord Charles Spencer-Churchill of Trusthouse Forte. Two replied that the time-scale was too short but by 6 August Lord Charles Spencer-Churchill came back to us to say that the Board of Trusthouse Forte was very interested in the idea and that he had been appointed to act as the negotiator for the company with the Société d'Encouragement. Two days later the other remaining company informed Lord James Crichton-Stuart that they were also interested in discussing the idea with the Société d'Encouragement. A meeting was therefore arranged at Deauville between Lord Charles Spencer-Churchill and Jean Romanet and Louis Romanet on the Tuesday after the Prix Morny. Another meeting was also set up with the representatives of the other company at Deauville but this had to be cancelled at the last moment owing to the sudden illness of the Chairman's wife.

The meeting with Jean Romanet and Louis Romanet lasted less than an hour and a half. It was agreed in principal that Trusthouse Forte would sponsor the Prix de l'Arc de Triomphe in 1982 for £70,000 since a limited amount of time remained for the company to benefit from the publicity arising from the deal and that for the next three years they would sponsor the race for £100,000 in 1983, £110,000 in 1984 and £121,000 in 1985, with an option for the following three years. It was further agreed that a Press Conference would be held on 12 September, the day of the Prix Vermeille, to announce the sponsorship to the media. Lord Charles Spencer-Churchill obtained the consent of the Board of Trusthouse Forte to these terms and on the day of the Grand Prix de Deauville he met M. Hubert de Chaudenay, Président du Société d'Encouragement, and they shook hands on the deal.

There then followed a period of frenetic activity at Longchamp. A part of the stands had to be transformed into a restaurant and salon for the guests of Trusthouse Forte and a staircase with an awning in the company's distinctive blue colours over it had to be built. On 4 September Lord Charles Spencer-Churchill came to Paris with representatives of the advertising company appointed by Trusthouse Forte to discuss details. They then visited Longchamp where they were shown what the Société d'Encouragement proposed to do. They also walked the course in order to plan where placards could be placed most advantageously. The Press Conference was attended by the two joint-managing directors of Trusthouse Forte, Rocco Forte and Eric Hartwell. M. Hubert de Chaudenay announced the sponsorship to the assembled press and television reporters and Rocco Forte replied on behalf of Trusthouse Forte. The Prix de l'Arc de Triomphe was, in future, to be known as the 'Trusthouse Forte Prix de l'Arc de Triomphe'.

The Field

Although there was only a field of seventeen for the first running of the Trusthouse Forte Prix de l'Arc de Triomphe, it included six foreign challengers: four from England, Charles St George's Ardross and Garo Vanian's Critique, both trained by Henry Cecil, and the three-year-old fillies Awaasif and Last Feather; one from Ireland, Robert Sangster's Assert; and one from the Soviet Union, the three-year-old colt Kastet.

The 5 to 2 favourite for the Trusthouse Forte Prix de l'Arc de Triomphe on Sunday 3 October was Robert Sangster's Be My Guest colt Assert, who was trained by David O'Brien, Vincent O'Brien's son. As a two-year-old he ran Golden Fleece to 2½ lengths at Leopardstown on his début and then impressively beat Longleat 4 lengths in the Beresford Stakes (8f) at the Curragh on 10 October. He was, however, only sixth to Count Pahlen in the William Hill Futurity Stakes at Doncaster. After running Golden Fleece, owned by Robert Sangster but trained by Vincent O'Brien, to 2½ lengths in the Nijinsky Stakes (10f) at Leopardstown on 8 May, Assert beat moderate opposition in a canter by 10 lengths in the Gallinule Stakes at the Curragh on 22 May. In the Prix du Jockey Club at Chantilly on 6 June Assert was by far the most impressive in the paddock. Before the turn into the straight it was clear he was going much better than his rivals. He took the lead from Real Shadai more than 400 metres out and thereafter never looked in danger. He beat Real Shadai and Bois de Grace 3 lengths and 2 lengths with Cadoudal only seventh and Newjdar tenth. Assert contemptuously disposed of the Derby third, Silver Hawk, by 8 lengths in the Irish Sweeps Derby on 26 June. Robert Sangster's colt started odds-on favourite at 10 to 11 for the King George VI and the Queen Elizabeth Diamond Stakes at Ascot on 24 July. He was not so impressive in the paddock as he had been at Chantilly and was very much on his toes. When his full brother and leader, Bikala, began to fade a furlong and a half out, his jockey Christy Roche was forced to take the lead. Assert was caught and beaten a neck in the last fifty yards by the fast-finishing Joe Coral Eclipse Stakes winner Kalaglow, who was winning over twelve furlongs for the first time in his career. Glint of Gold followed the first two 3 lengths in arrears with Critique and Bikala a neck and the same away fourth and fifth. If Roche had not taken on Bikala so early in the straight, Assert might have had more in reserve to fight off Kalaglow's late challenge. In the Benson and Hedges Gold Cup, Assert took the lead after a furlong and trounced Norwick 6 lengths on 17 August. Robert Sangster's colt was eased a furlong out when effortlessly beating Kind of Hush in the Joe McGrath Memorial Stakes (10f) at Leopardstown on 18 September. On the book he certainly had a favourite's chance at Longchamp but all his victories in 1982 had been on good or firm going; he had won on the soft as a two-year-old but it was reported that his trainer David O'Brien was worried about his stamina over a mile and a half in the soft going he would encounter on 3 October. There was also a body of opinion who felt in such conditions Assert's best distance might be ten furlongs.

If there was any doubt about Assert's stamina, it was certain that Charles St George's six-year-old Ardross would exploit the weakness. On his seasonal reappearance Ardross beat Glint of Gold half a length in the Jockey Club Stakes (12f) at Newmarket on 30 April more easily than the bare verdict suggests. Ardross then won the Yorkshire Cup on 13 May – covering the final five furlongs in about 58 seconds – the Henry II Stakes (16f) at Sandown on 1 June, and the Ascot Gold Cup which he won as he pleased by 3 lengths and a neck from Tipperary Fixer and El Badr, the first two in the Prix du Cadran. Ardross met his first defeat of the season when beaten 2 lengths and 4 lengths by Height of Fashion, who broke the course record, and Amyndas in the Princess of Wales's Stakes run on good going over a mile and a half at Newmarket on 6 July. He was however, giving 26 lbs to the winner and 9 lbs to the second, so was meeting them on much worse terms than weight for age, and when Lester Piggott saw he could not win a furlong out, he accepted the situation. In the Geoffrey Freer Stakes at Newbury, Ardross was an exceptionally impressive 4 lengths and a short head winner from Baffin and the Coronation Cup victor Easter Sun. He had an easy task when he won the Doncaster Cup on 9 September. On account of his high draw in the Prix de l'Arc de Triomphe the previous season, Lester Piggott had kept him up with the leaders from the outset and he had faded in the straight. Ardross, however, always preferred to lob along at the rear of the field for the first mile of a race. It was, therefore, decided some considerable time before 3 October to employ these tactics regardless of his draw. In order to make full use of his stamina the softer the going at Longchamp the better and it seemed certain that Charles St George's magnificent stayer, who started at odds of 25 to 4, would give a very good account of himself on 3 October.

It was more difficult to envisage Ardross's stable-companion, Critique, taking a hand in the finish on 3 October. On his only previous visit to Longchamp two years earlier Critique had run Recitation to a head in the Grand Critérium. Since then Critique had run his best races on right-handed courses, seeming to reserve his very best performances for Ascot where he had most impressively won the Cumberland Lodge Stakes in September 1981. One of his two victories in eight outings in 1982 before the Trusthouse Forte Prix de l'Arc de Triomphe was in the Hardwicke Stakes when, receiving 5 lbs, he beat Glint of Gold a length, and he was only a neck behind the same horse when fourth to Kalaglow in July. He only scraped home by a neck and the same from Commodore Blake and Criterion in the September Stakes at Kempton Park on 4 September. Moreover Critique was inclined to tail himself off in the early stages of a race which would not be in his favour at Longchamp. If he did consent to show his undoubted ability, Critique might spring a surprise but his odds of 69 to 1 were really a true reflection of his chance.

There were two English-trained three-year-old fillies in the field for the first Trusthouse

Forte Prix de l'Arc de Triomphe, Gerald Leigh's Last Feather and Sheik Mohammed's Awaasif. Last Feather made all the running when easily winning the Musidora Stakes at York on 11 May. She was caught at the distance and beaten 1½ lengths and three-quarters of a length by Time Charter and Slightly Dangerous in the Oaks, a length and a half in front of Awaasif who was fourth but did not come down the hill at all well and was finishing best of all. All Along was another 2½ lengths behind her in sixth place. Awaasif had effortlessly beaten Dreaming Away 6 lengths in the Kama Stakes at Lingfield on 29 May. She had found the mile of the Masaka Stakes at Kempton when fifth to Time Charter on 12 April too sharp for her. Awaasif and Last Feather met again in the Yorkshire Oaks on 17 August. Not merely did Awaasif gain her revenge but she beat the Irish Guinness Oaks heroine Swiftfoot and Dish Dash, the winner of the Ribblesdale Stakes, a neck and a head with Last Feather 8½ lengths away sixth. In the Prix Vermeille on 12 September Awaasif was a never-dangerous seventh 4½ lengths behind All Along. Last Feather did not run again before 3 October after her defeat in the Yorkshire Oaks. It seemed likely that Awaasif would again prove the better of the two fillies at Longchamp but surprisingly it was she who started at 90 to 1 and Last Feather at 58 to 1. It was difficult to give much chance however to either filly.

On the other hand, the French three-year-old fillies in the field for the Trusthouse Forte Prix de l'Arc de Triomphe, Harbour, Akiyda and All Along, all had their supporters. The Ecurie Aland's Harbour had won once as a juvenile from two starts. On her reappearance she comfortably beat Estere and Akiyda half a length and the same in the Prix Vanteaux on 18 April. In the Prix Saint-Alary Harbour produced a tremendous burst of speed to outclass All Along by 4 lengths. Harbour then started favourite for the Prix de Diane Hermès where her greatest threat appeared to be River Lady, the easy winner of the Poule d'Essai des Pouliches. Tragically the latter broke her pelvis and left femur in the straight and died from an internal haemorrhage. Harbour led early in the straight and was not pressed to hold the late challenge of Akiyda by 2 lengths with All Along 3¾ lengths away fifth. Harbour was rested till the Prix Vermeille but was clearly unsuited by the firm going and only managed fourth place 1½ lengths, a head and three-quarters of a length behind All Along, Akiyda and Grease. Nevertheless she started the shortest priced of the three fillies on 3 October at 25 to 4 as many felt she would improve with the race behind and as the going was soft. On the other hand, she had not seemed as well in the paddock before the Prix Vermeille as she had done at Chantilly in June and there lurked the suspicion in some minds that she might have gone over the top.

Before her authoritative victory in the Prix Vermeille, Daniel Wildenstein's All Along had gamely beaten No Attention by half a length in the Prix Maurice de Nieuil at Saint-Cloud on 14 July. Daniel Wildenstein's filly had won her sole outing as a juvenile in a maiden race at Amiens and beat Zalataia 4 lengths at Saint-Cloud on 27 February. She won the Prix Pénélope by 4 lengths from Paradise at Saint-Cloud a month later with Akiyda 2 lengths away fourth. Harbour had beaten her decisively in both the Prix Saint-Alary and the Prix de Diane, and like Awaasif she had not negotiated the hill at Epsom well. Certainly All Along was one of the best three fillies of her generation in France but, while she had won on soft going, it was reported that she ran best on the fast going she had met in the Prix Vermeille. These conditions she would not find at Longchamp on 3 October and as a result she started at the generous odds of 17 to 1.

The Aga Khan's Akiyda had also won her sole outing as a two-year-old over ten furlongs at Saint-Cloud in heavy going on 7 November. When fourth to All Along on her three-year-old reappearance in the Prix Pénélope, it was quite clear she needed a longer distance and softer going, which was even more evident when she was third to Harbour in the shorter Prix Vanteaux. On 25 May Akiyda recorded her first victory of the season in the Prix des Tuileries (2400 metres) run at Longchamp in soft going when she effortlessly beat First Water 1½ lengths. Over the shorter distance of the Prix de Diane Akiyda did not begin her challenge until halfway up the straight but her strong finish, which carried to within 2 lengths of Harbour, again indicated that she needed a mile and a half. Unfortunately the going at Deauville on the day of the Prix de Pomone was good which did not entirely suit Akiyda's long action. After looking the winner 400 metres out, the Aga Khan's filly was unable to stride out on the fast going and was half a length adrift of Zalataia on the line with April Run 1½ lengths away third. In the Prix Vermeille Akiyda found the fast going prevented her striding out early in the straight. She was catching All Along in the final 200 metres and headed the non-staying Grease by a head almost on the line to take second place. Had the going been soft either in the Prix de Pomone or in the Prix Vermeille the results might well have been different. On 3 October, however, this full-sister to Akarad was running over her best distance and in the soft going she needed so much. Although she had yet to win a group race, she had been extremely consistent and with conditions in her favour, Akiyda looked good each-way value at 11 to 1.

April Run had only four races in 1983 prior to the Trusthouse Forte Prix de l'Arc de Triomphe. On her seasonal reappearance in the Prix d'Hédouville, April Run was about a length fourth to Gap of Dunloe, finishing well but never looking a danger. A fortnight later she was a disappointing eighth to Bikala in the Prix Ganay on 2 May. April Run was rested until the Prix de Pomone at Deauville. In the paddock she looked as if she would benefit from the race, but got into all kinds of trouble during the course of it. She flyjumped in front of the stands shortly after the start and then found herself trapped on the rails in the straight. When she got clear she finished fastest of all but the issue was by then decided. April Run only scraped home by a short neck and a nose from Mariacho and No Attention in the Prix Foy at Longchamp and needed all of Piggott's assistance to last home. Subsequently the stewards reversed the placings of Mariacho and No Attention for interference. It was not a particularly impressive victory but afterwards François Boutin explained that, after the Prix de Pomone, in which April Run had sustained some superficial cuts, he had put her on the easy list for a few days. In the Prix Foy, therefore, she was short of work and he felt the fast going did not suit her. If April Run were back to her three-year-old form on 3 October, her odds of 15 to 1 were more than reasonable value.

Jules Ouaki's Bikala, who had finished a nose in front of April Run when second to Gold River in the 1981 Prix de l'Arc de Triomphe, had also been very lightly raced as a four-year-old. He made all the running on his reappearance in the Prix Ganay at Longchamp on 2 May and held Lancastrian's challenge with no difficulty by half a length with Kalaglow seventh and April Run eighth. Although he did not appear on a racecourse again till the King George VI and the Queen Elizabeth Diamond Stakes, his trainer Patrick-Louis Biancone took him twice to Compiègne, kept him overnight in the racecourse stables and gave him two private gallops on the racecourse on the following days. At Ascot, however, although his connections felt he was in peak condition, he weakened in the straight to finish only fifth to Kalaglow. Had his jockey Serge Gorli set a stronger gallop it was possible Bikala would have been closer. He set a very strong pace to the unbeaten three-year-old Général Holme, who was receiving 5 kilos, in the Prix du Prince d'Orange on 19 September and only narrowly failed to resist the younger colt's challenge by a short neck. It was a very impressive trial for the Trusthouse Forte Prix de l'Arc de Triomphe and Bikala started at odds of 13 to 2 on 3 October.

In spite of their proximity to April Run in the Prix Foy, it was difficult to give much chance to No Attention and Mariacho, the other two French four-year-olds in the field on 3 October. Although No Attention had never been out of the first three in ten runs in 1982, he had only won twice, the Grand Prix de Marseille-Vivaux (2700 metres) on 3 July and the Prix de Reux (2500 metres) at Deauville when he beat Moulouki 2 lengths with Mariacho about a length away fifth. He was also runner-up to Real Shadai in the Grand Prix de Deauville and to All Along in the Prix Maurice de Nieuil but seemed likely to find several of his rivals too fast for him. Mariacho, who was one of only two foals ever sired by Mariacci and had been third to Rahotep in the Prix Hocquart in 1981, won four of his first five races in 1982, three minor events, including one for gentleman riders at Evry, and the Prix Monarque (2400 metres) at Saint-Cloud on 18 July. He was tackling a much higher calibre of opposition in the Trusthouse Forte Prix de l'Arc de Triomphe and was unlikely to make any significant impression.

Mahmoud Fustok's Bon Sang had been a very useful two-year-old winning three of his four starts, including the Prix des Chênes and the Critérium de Saint-Cloud. The whole of the Fustok stable, however, had been under a cloud in the spring and early summer in 1982 as the horses had been given too strong a dose of the anti-worms vaccine. Bon Sang did not run until the Prix Eugène Adam at Saint-Cloud on 18 July when he ran What a Guest to a short head. In the Prix de la Côte Normande at Deauville on 14 August Bon Sang ran Général Holme to a neck;

but he reserved his best performance till 12 September in the Prix Niel at Longchamp. In front from the start he trounced Alfred's Choice, who had been fourth in the Prix du Jockey Club and the Grand Prix de Saint-Cloud, by an impressive 6 lengths with Cadoudal 5½ lengths away sixth. On the strength of that performance some felt he was the best three-year-old colt in France and he started second favourite on 3 October at 11 to 2. Before the Prix Niel he was mounted in the stable area and required two lads to hold him. He could become very fractious which might not be to his advantage in Longchamp's most prestigious race.

The most consistent and probably the best French three-year-old colt was Zenya Yoshida's Real Shadai. He reappeared at Longchamp on 18 April and easily beat moderate opposition by 3 lengths in the Prix des Marronniers. Real Shadai then ran third 3 lengths behind Cadoudal in the Prix Hocquart on 9 May. He gained his revenge in the Prix du Jockey Club when runner-up to Assert with Cadoudal only eighth. He again put up a fine performance when running Glint of Gold and Lancastrian to a head and 3 lengths in the Grand Prix de Saint-Cloud. He was favourite for the Grand Prix de Deauville and eventually won by 1½ lengths from No Attention, but in the back straight seemed to run his race in snatches and Maurice Philipperon had to give him a couple of slaps of the whip to keep him in touch. This victory was not impressive enough to make one feel one had seen a future winner of a Trusthouse Forte Prix de l'Arc de Triomphe and Real Shadai's odds of 23 to 1 reflected this opinion.

Cadoudal had won on his racecourse début over 2000 metres at Saint-Cloud on 27 March. He then was only beaten a head and a nose by Bois de Grace and Nahirpour in the Prix Greffulhe at Longchamp on 12 April but suffered a great deal of interference when trying to challenge and was patently unlucky. He gained his revenge in the Prix Hocquart when he had a length and 2 lengths to spare over Noble Bloom and Real Shadai with Newjdar a nose away fourth and Bois de Grace only eighth. The going was very soft that afternoon and it turned out that Cadoudal needed very soft or heavy ground to show his best form. Conditions were against him when second favourite but a disappointing eighth to Assert in the Prix du Jockey Club and when sixth in the Prix Niel his only other race before 3 October. The going in the Trusthouse Forte Prix de l'Arc de Triomphe would suit him, but it was still difficult to envisage him reaching the frame.

Yurk Skalka's Newjdar won the Prix Hallebardier (2100 metres) at Maisons-Laffitte in heavy going on 12 March. His best performance in five outings after that was to run third half a length and a short neck behind Coquelin and My Midway in the Prix du Lys at Chantilly on 13 June. It seemed certain that he would be outclassed in the Trusthouse Forte Prix de l'Arc de Triomphe and he started the extreme outsider at 99 to 1.

The challenger from the Soviet Union, Kastet, was very difficult to classify in any accurate manner. He had won his last four races including the Grand Prix National de l'Union Soviétique, the Russian equivalent of the Derby, and the Grand Prix de la Ville de Moscou on 29 August when his opposition included older horses. Even the Russians, however, admitted that they had no idea of his chance at Longchamp as they had no collateral lines of form because Kastet had only run in the Soviet Union. Albert Klimscha Jr provided the Russians with two galloping companions, Dumplino and Dalby Mustang, for Kastet. On the Wednesday morning partnered by his Mongolian jockey Youri Chavouev and watched by his trainer Kasbek Dsalaiev and the head of the Soviet studs, Alexandre Martinenkov, Kastet, Dumplino and Dalby Mustang covered seven furlongs. Serge Gorli who also watched the gallop did not think, however, that Kastet would be a danger to his mount Bikala.

The Final Preparations

Assert arrived by special plane at Beauvais on Friday 1 October. He had done his last work the previous Tuesday and had satisfied both David O'Brien and Pat Eddery. David O'Brien nevertheless warned that if he considered the going too heavy on Sunday morning, it was quite possible that Assert would be withdrawn even if the statutory fine of 10 per cent of the prize money to the winner had to be paid.

Akiyda did her last gallop on Tuesday 28 September on the racecourse at Chantilly watched by both François Mathet and the Aga Khan. It was clear that she had recovered completely from the Prix Vermeille. All associated with her were certain that the softer the going the greater her chance in the Trusthouse Forte Prix de l'Arc de Triomphe.

Ardross more than satisfied Henry Cecil in his last work at Newmarket on Tuesday 28 September and he also was delighted that the going in Paris was going to be very soft. The same applied to the connections of Harbour. She also did her last gallop at Les Aigles the same morning. The one that created most impression that morning was Bon Sang, and his trainer Mitri Saliba was most confident about his chance afterwards.

The most notable absentee from the Trusthouse Forte Prix de l'Arc de Triomphe was Glint of Gold who had won the Grand Prix de Saint-Cloud and the John Porter Stakes. When he had finished third to Kalaglow and Assert it was the first time in his career he had been out of the first two. He was injured during his preparation in September and was retired to stud. Robert Sangster's Derby winner, Golden Fleece, was another absentee as he never ran again after Epsom. Touching Wood, the runner-up in the Derby and winner of the St Leger, had the Irish St Leger as his autumn objective, whilst the first two in the Grand Prix de Paris, Le Nain Jaune and Chem, had been well beaten at Doncaster. Kalaglow, whose stamina was suspect over a mile and a half, was withdrawn a few days before the race but ran disappointingly in the Champion Stakes, which was also the objective of the Oaks winner Time Charter, who won the Sun Chariot Stakes at Newmarket on Friday 1 October.

Sunday 3 October was a fine warm day and as the crowd began arriving at Longchamp they found Longchamp transformed. Placards for Trusthouse Forte abounded. The far end of the stands had been altered into a restaurant for the hotel group's guests, access to which was through a staircase guarded by a doorman from one of their hotels in Paris. The previous day, however, the track workers, on the instructions of the communist trade union, the CGT, had gone on strike. It was not a strike against the Société d'Encouragement but part of a general one against the imposition of a wage freeze by the government. David O'Brien walked the course and threatened to pull out Assert if that part of the course, which was dolled off in the straight so that there was fresh ground for Longchamp's most prestigious race, was not opened up. Very fortunately the Société d'Encouragement successfully negotiated the end of the strike at Longchamp a short while before the first race – the newly sponsored Prix Horse Racing Abroad. Horse Racing Abroad, who for several years had organised trips from England to the Prix de l'Arc de Triomphe and other major international races, had refurbished the old stand and provided a marquee for its clients behind it. The race was formerly known as the Prix des Champs-Elysées and was won by Mme Alec Head's Riverhill.

The Prix Marcel Boussac went to the English outsider Goodbye Shelley who held Mysterieuse Etoile by a short head. Although she was returned at 36 to 1 she had been supported from 50 to 1, so that many of the huge crowd of English must have collected. Apart from the guests of Lord and Lady Forte, who included many editors from Fleet Street – amongst them the editor of the *Daily Telegraph*, a former Minister in the Macmillan government William Deedes, and his wife – such regular visitors to the track as the Earl of Suffolk, William Rees-Davis, the Conservative M.P. for Thanet, and the veteran racegoer Lady Osborne, John Aspinall's mother, were seen amongst the crowd, as were the Senior Steward Lord Manton and Louis Freedman. Amongst the many Americans was William T. Young, the owner of the Overbrook Stud, which had purchased Three Troikas; he was paying his first visit to Longchamp.

The Prix de l'Abbaye de Longchamp again provided a benefit for England when Miss Monica Sheriffe's Sharpo, ridden by Pat Eddery, beat Fearless Lad, the winner of the King's Stand Stakes, by a length.

The Race

Lord and Lady Forte with Lady Thorneycroft and Rocco Forte came down from entertaining their guests to the paddock. The Russian colt, Kastet, seemed rather small whilst Bon Sang was not as fractious as usual. Ardross looked superb and so did Bikala. April Run was turned out in very good condition and had never seemed better all season and, as usual with François Mathet trained runners, Akiyda also appeared fit to run for her life. Harbour, however, did not look as well as she had done in mid-summer and she was sweating; Assert also had appeared better at Chantilly in June than he did that afternoon. Rumours had persisted throughout the first

Akiyda (Yves Saint-Martin) on the rails beating Ardross (Lester Piggott, 4) by a head, with Awaasif and April Run half a length and a head back third and fourth, and Real Shadai and No Attention fifth and sixth

three races that Robert Sangster's colt might be withdrawn, so it was a relief to see him enter the paddock.

There were no incidents during the parade or at the start. April Run, Awaasif and Bon Sang were the quickest away, but on settling down Bikala showed the way to Bon Sang, Mariacho, Assert and Akiyda. No Attention was slowly away and amongst the backmarkers were Kastet, Critique, Real Shadai and Newjdar. Harbour, Last Feather, All Along and Ardross were in the mid-division. At the top of the hill, Bon Sang led Bikala by a length followed by Awaasif, Akiyda, Assert and Harbour. Coming down the hill Ardross began to make headway towards the leaders. On the turn into the straight Bon Sang still led Bikala, Akiyda, Assert, Harbour and Awaasif with Ardross and April Run still making progress on the leading group. Soon after the straight was reached, first Bon Sang and then Bikala dropped out and Yves Saint-Martin slipped Akiyda into the lead a little over 400 metres out. Assert and Harbour were both under pressure and clearly beaten. Awaasif set off in pursuit of the Aga Khan's filly and both Ardross and April Run were beginning their challenges. With 200 metres to go Akiyda still had a length advantage over Awaasif, the hard-ridden but closing Ardrois and April Run on the outside. One hundred metres out Ardross had got the better of Awaasif and with each stride to the line the gap between Charles St George's great stayer and Akiyda became narrower and narrower. At the winning post Akiyda still retained an advantage of a head over Ardross with Awaasif half a length away third a head in front of April Run. With another few strides Lester Piggott and Ardross might have prevailed; the crowd at Longchamp had been treated to the dramatic spectacle of two of the greatest jockeys in the world fighting out a pulsating finish on two very game and courageous racehorses. Real Shadai and No Attention ran through beaten horses to finish 4 lengths and a length back fifth and sixth. They were followed by Cadoudal, Critique, Harbour, Newjdar and the disappointing favourite Assert.

A delighted Aga Khan and his wife the Begum led the winner into the unsaddling enclosure. A disconsolate Lester Piggott told Charles St George on dismounting before the result of the photograph was announced that he was beaten. The presentations of the *objet d'art* and the trophy given by Trusthouse Forte were made on a podium in the centre of the paddock to attempt to protect the winning owner from being jostled by the photographers. The Aga Khan, who had only been hopeful before the Prix Marcel Boussac that Akiyda would finish in the first three, had to leave Longchamp shortly after the race as he was travelling to Kenya the following morning on business. After being entertained by the Président du Société d'Encouragement, M. Hubert de Chaudenay, Lord and Lady Forte returned to their guests in the stands.

John Dunlop quickly gained compensation for Awaasif's defeat when Willie Carson rode his charge Dionne to a length victory over Unknown Lady in the Prix de l'Opéra. Dionne was supported from 12 to 1 to 36 to 10 and there were many happy English faces at Longchamp at the end of the day's racing.

Akiyda's success in the Trusthouse Forte Prix de l'Arc de Triomphe was her first in a pattern race, although she had been placed in several. François Mathet had even considered bypassing the race and running her instead the following Sunday in the Prix de Royallieu. It was a first success for the Aga Khan in Longchamp's most prestigious race. His father had won with Saint Crespin in 1959 five months before his death.

Yves Saint-Martin was gaining his third victory in the Prix de l'Arc de Triomphe, the others having been on Sassafras and Allez France. In many respects, much of the credit for Akiyda's triumph must go to him. By sending her into the lead over 400 metres out, he gained an invaluable advantage over Ardross who was unable to accelerate as rapidly as the filly.

For François Mathet it was a fourth victory in the Prix de l'Arc de Triomphe. Tantième had won twice for him and Sassafras had beaten Nijinsky in 1970. He thus equalled the record for the race held by Alec Head and C.H. Semblat. It was, however, to be his last since he died in his sleep at the age of 74 three months later. His death was an immense loss to French racing, for he was one of the greatest trainers in France since the Second World War.

There were no real excuses after the race. Alec Head said: 'Harbour ran out of gas when it was needed and obviously ran below par.' His daughter, Criquette Head felt she may have gone over the top. Pat Eddery reported that once he let Assert down 'he could not produce his action on the soft ground'. Willie Carson was full of praise for Awaasif commenting: 'She ran a terrific race but was just unable to quicken.'

The Rest of the Season

Neither Akiyda nor Ardross ran again in 1982 but both were retired to stud. No Attention ran Ataxerxes to three-quarters of a length in the Preis von Europa on 10 October; a week later he was a length runner-up to Marasali in the Prix du Conseil de Paris but could only manage fourth place to Denel in the Prix Royal Oak. There were no runners from the Trusthouse Forte Prix de l'Arc de Triomphe in the Dubai Champion Stakes won by Time Charter. Cadoudal could only manage fourth place to Friendswood in the Gran Premio del Jockey Club. On 23 October April Run beat Naskra's Breeze, the Man o'War Stakes winner, by 6½ lengths in the Turf Classic at Aqueduct. She then defeated the Rothmans International Stakes winner at Woodbine, Majesty's Prince, by 6½ lengths in the Washington International D.C. Stakes on 6 November with Awaasif only fifth. In the Japan Cup in Tokyo on 28 November All Along ran the surprise winner Half Iced to a short neck with April Run a neck away third.

Pedigree

Akiyda's sire was Labus by Busted, a stallion virtually unknown until the previous year when his son Akarad was the best three-year-old colt in France. Labus did not run at two as he injured a hind leg, but on his racecourse début he ran out an easy winner over a mile at Chantilly. He tried to make all in the Prix du Conseil Municipal when 4 lengths runner-up to the very useful Kamaraan. This was an extremely good performance for such a lightly-raced colt so it was disappointing that he broke down when third to Card King in the Grand Prix de Marseille on his only other outing. When the Aga Khan acquired all the Boussac bloodstock, he presented Labus to the French National Stud. He stood for the 1981 season at Saint-Lo for 600 francs, but after the success of Akarad his fee was increased in 1982 to 6,000 francs, still a paltry sum. Labus was a very handsome black horse with a lot of his dam's sire Pharis about him. Marcel Boussac's 1969 Prix de Diane heroine Crepellana was by Crepello. It was probable that it was for this reason that he sent Labus's dam Cordovilla to Crepello's son Busted. She was a daughter of Cordova by Djebel. Cordova was the best two-year-old filly in France in 1953, winning both the Prix Robert Papin and Prix Morny. As a three-year-old she won the Prix Eugène Adam and was fifth to Sica Boy in the Prix de l'Arc de Triomphe. At four Cordova won the Prix du Prince d'Orange and was seventh to Ribot at Longchamp. Her dam, Caravelle by Abjer, was the best French two-year-old in France in 1942, winning the Grand Critérium; at three Caravelle won the Prix de Diane, the Poule d'Essai des Pouliches and the Prix de la Forêt, which she also won at four. Thus although Labus had little opportunity to show his undoubted ability on the racecourse, both his top and bottom lines were sufficiently good not to cause great surprise at his recent success as a stallion.

Akiyda's dam, Licata by Abdos, was more than useful on the racecourse. She was only narrowly beaten in the Prix Chloë by La Troublerie who went on to win the Prix de la Côte Normande with San San only fourth. Licata easily won the Prix Cléopâtre and amongst her victims was Decigale, subsequently winner of the Prix de Royaumont and third to the deadheaters San San and Paysanne in the Prix Vermeille. Licata also beat a useful field in the Prix de Malleret including the Prix Imprudence and the Prix Messidor winner Arosa. She finished fifth to Rescousse in the Prix de la Nonette but was injured and never ran again. Nevertheless she was a more than reasonable racemare. In the paddocks Licata has been a remarkable success. Her first foal was Acamas by Mill Reef whose career we have already seen. Her second, Eupalinos by Mill Reef, won a small race at Dundalk. Her third foal, Licara by Caro, won twice as a two-year-old; at three she ran second to Detroit in the Prix Chloë and second to Firyal in the Prix Imprudence. In addition she was a close fourth to Hortensia in the Prix Cléopâtre and sixth to Mrs Penny in the Prix de Diane de Revlon. Licata was then mated with Labus, probably on account of Marcel Boussac's declining financial affairs, but the produce was the excellent colt Akarad. Licata's next foal was Akiyda. Her 1980 foal by Mill Reef, Aktari, was unraced at two and in 1981 she got a bay filly by Top Ville. Licata's sire Abdos won the Grand Critérium in 1961 but injury prevented him running as a three-year-old.

Licata's dam, Gaia by Shantung, only ran twice without success but she was placed once. Licata has ensured her place in the Stud Book, however, and her full brother Eclectic was runner-up in the Prix Daphnis and was exported to Brazil.

Gaia's dam, Gloriana by Pharis, also did not win in two outings but was runner-up in the Prix Quincey; her reputation rests on the fact that Licata is descended from her. Gloriana's dam, Tourzima by Tourbillon, however, was an influential broodmare. Tourzima ran only once as a two-year-old when she easily beat Martia who later dead-heated with Nosca in the Grand Critérium. She produced two full sisters to Gloriana, Floriana and Albanilla, who both won the Critérium de Maisons-Laffitte. Her best produce nonetheless was Corejada by Pharis who won the Cheveley Park Stakes, the Poule d'Essai des Pouliches and the Irish Oaks, and was a close second to Algae Grace in the Prix de Diane in 1950. Corejada also became an extremely good broodmare producing the brilliant Apollonia by Djebel, the winner of the Prix de Diane, the Poule d'Essai des Pouliches, the Grand Critérium and the Prix Morny; Macip by Marsyas, who took the Prix Royal Oak and the Ascot Gold Cup; and Arcor by Arbar, second to Larkspur in the Derby.

Tourzima's dam, Djezima by Astérus, won

two races as a two-year-old. Amongst her winners was the brilliant Priam by Pharis, who beat Ardan in the Grand Critérium. Unfortunately he was ill in the spring and did not take part in the classics but he beat the good miler Fanatique in the Prix des Acacias and won the Prix Jacques le Marois. At four he won the Prix Daphnis et Chloë, the Prix d'Harcourt and the Prix d'Ispahan. Priam was runner-up to Ardan in the Grand Prix de Saint-Cloud and was the first French horse to cross the Channel after the war when he was unlucky to be beaten a short head by the 2000 Guineas winner Court Martial in the Champion Stakes. Another of Djezima's sons, Djeddah by Djebel, won the Eclipse Stakes and the Champion Stakes in England and the Prix Daphnis et Chloë and La Bourse in France.

Djezima's dam, Heldifann by Durbar, was a full sister to Durban, the dam of Banstar by Sunstar and the even more important Tourbillon by Ksar. They traced back through Banshee to Frizette, who was by the great American stallion Hamburg out of the English broodmare Ondulee, a daughter of St Simon and the granddaughter of Shotover, the famous mare who won the Derby and 2000 Guineas exactly one hundred years before Akiyda's victory in the Trusthouse Forte Prix de l'Arc de Triomphe.

TRUSTHOUSE FORTE PRIX DE L'ARC DE TRIOMPHE 2400 metres
at Longchamp on Sunday 3 October 1982. For three-year-olds and upwards. Going: Soft.
2,000,000 francs to the winner; 800,000 fr. to the 2nd; 400,000 fr. to the 3rd; 200,000 fr. to the 4th. 500,000 fr. to the breeder of the winner; 200,000 fr. to that of the 2nd; 100,000 fr. to that of the 3rd; 50,000 fr. to that of the 4th. Value to the winner: 2,000,000 fr.

AKIYDA	br.f.	by Labus – Licata by Abdos	3.54½k	H.H. Aga Khan	Y. Saint-Martin	1
ARDROSS	b.c.	by Run the Gantlet – Le Melody by Levmoss	6.59k	Charles St. George	L. Piggott	2
AWAASIF	b.f.	by Snow Knight – Royal Statue by Northern Dancer	3.54½k	Sheikh Mohammed	W. Carson	3
APRIL RUN	b.f.	by Run the Gantlet – April Fancy by No Argument	4.57½k	Mme Bertram R. Firestone	C. Asmussen	4
REAL SHADAI	b.c.	by Roberto – Desert Vixen by In Reality	3.56k	Zenya Yoshida	M. Philipperon	5
NO ATTENTION	b.c.	by Green Dancer – No No Nanette by Sovereign Path	4.59k	Minoru Hieda	A. Lequeux	6
CADOUDAL	br.c.	by Green Dancer – Come to Sea by Sea Hawk	3.56k	André Boutboul	J.L. Kessas	7
CRITIQUE	br.c.	by Roberto – Cambrienne by Sicambre	4.59k	Garo Vanian	J. Mercer	8
HARBOUR	ch.f.	by Arctic Tern – Heres to You by Molvedo	3.54½k	Ecurie Aland	F. Head	9
NEWJDAR	ch.c.	by New Chapter – Ardja by Djakao	3.56k	Yurk Skalka	G. Guignard	10
ASSERT		by Be My Guest – Irish Bird by Sea Bird	3.56k	Robert Sangster	P. Eddery	
BIKALA		by Kalamoun – Irish Bird by Sea Bird	4.59k	Jules Ouaki	S. Gorli	
BON SANG		by Gyr – La Caldera by Mourne	3.56k	Mahmoud Fustok	A. Gibert	
LAST FEATHER		by Vaguely Noble – Quill by Princequillo	3.54½k	Gerald Leigh	S. Cauthen	
ALL ALONG		by Targowice – Agujita by Vieux Manoir	3.54½k	Daniel Wildenstein	G. Starkey	
MARIACHO		by Mariacci – Sea Queen by Le Fabuleux	4.59k	Mme Alexandre Savin	G. Dubroeucq	
KASTET		by Souzdale – Calipso by Le Loup Garou	3.56k	Haras de Dnepropetrovsky	A. Chavouev	

17 runners.
AKIYDA trained by F. Mathet; bred by H.H. Aga Khan.
Distances: head; ½ length; head; 4 lengths; 1 length; 1½ lengths; ¾ length; short head; ¾ length. Time: 2 mins 37.00 seconds.
Pari-Mutuel: Win: 11fr. 70. Places: 3fr. 40; 4fr. 30; 16fr. 90.
S.P.: 5 to 2 Assert; 11 to 2 Bon Sang; 25 to 4 Ardross, Harbour; 13 to 2 Bikala; 43 to 4 AKIYDA; 15 to 1 April Run; 17 to 1 All Along; 22 to 1 Real Shadai; 26 to 1 Kastet; 39 to 1 No Attention; 54 to 1 Cadoudal; 56 to 1 Mariacho; 58 to 1 Last Feather; 69 to 1 Critique; 90 to 1 Awaasif; 99 to 1 Newjdar.

<table>
<tr><td rowspan="32">AKIYDA
brown filly 1979</td><td rowspan="16">Labus
br. 1971</td><td rowspan="8">Busted
b. 1963</td><td rowspan="4">Crepello
ch. 1954</td><td rowspan="2">Donatello II
ch. 1934</td><td>Blenheim</td></tr>
<tr><td>Delleana</td></tr>
<tr><td rowspan="2">Crepuscule
ch. 1948</td><td>Mieuxcé</td></tr>
<tr><td>Red Sunset</td></tr>
<tr><td rowspan="4">Sans le Sou
b. 1957</td><td rowspan="2">Vimy
b. 1952</td><td>Wild Risk</td></tr>
<tr><td>Mimi</td></tr>
<tr><td rowspan="2">Martial Loan
gr. 1950</td><td>Court Martial</td></tr>
<tr><td>Loan</td></tr>
<tr><td rowspan="8">Cordo Villa
b. 1957</td><td rowspan="4">Pharis
br. or bl. 1936</td><td rowspan="2">Pharos
b. 1920</td><td>Phalaris</td></tr>
<tr><td>Scapa Flow</td></tr>
<tr><td rowspan="2">Carissima
b. 1923</td><td>Clarissimus</td></tr>
<tr><td>Casquetts</td></tr>
<tr><td rowspan="4">Cordova
b. 1951</td><td rowspan="2">Djebel
b. 1937</td><td>Tourbillon</td></tr>
<tr><td>Loïka</td></tr>
<tr><td rowspan="2">Caravelle
br. 1940</td><td>Abjer</td></tr>
<tr><td>Albarelle</td></tr>
<tr><td rowspan="16">Licata
b. 1969</td><td rowspan="8">Abdos
br. 1959</td><td rowspan="4">Arbar
b. 1944</td><td rowspan="2">Djebel
b. 1937</td><td>Tourbillon</td></tr>
<tr><td>Loîka</td></tr>
<tr><td rowspan="2">Astronomie
b. 1932</td><td>Astérus</td></tr>
<tr><td>Likka</td></tr>
<tr><td rowspan="4">Pretty Lady
b. 1942</td><td rowspan="2">Umidwar
b. 1931</td><td>Blandford</td></tr>
<tr><td>Uganda</td></tr>
<tr><td rowspan="2">La Moqueuse
b. 1925</td><td>Teddy</td></tr>
<tr><td>Primrose Lane</td></tr>
<tr><td rowspan="8">Gaia
br. 1962</td><td rowspan="4">Shantung
b. 1956</td><td rowspan="2">Sicambre
br. 1948</td><td>Prince Bio</td></tr>
<tr><td>Sif</td></tr>
<tr><td rowspan="2">Barley Corn
b. 1950</td><td>Hyperion</td></tr>
<tr><td>Schiaparelli</td></tr>
<tr><td rowspan="4">Gloriana
b. 1950</td><td rowspan="2">Pharis
br. or bl. 1936</td><td>Pharos</td></tr>
<tr><td>Carissima</td></tr>
<tr><td rowspan="2">Tourzima
b. 1939</td><td>Tourbillon</td></tr>
<tr><td>Djezima</td></tr>
</table>

APPENDIX: Complete List of Results 1920 to 1982

PRIX DE L'ARC DE TRIOMPHE 2400 metres

Year	*Winner*		*Owner*	*Jockey*	*Trainer*	*Second*	*Third*
1920	COMRADE	3.55k	Evremond de Saint-Alary	F. Bullock	P.P. Gilpin	King's Cross	Pleurs
1921	KSAR	3.55k	Madame Edmond Blanc	G. Stern	W.R. Walton	Fléchois	Square Measure
1922	KSAR	4.60k	Madame Edmond Blanc	F. Bullock	W.R. Walton	Fléchois	Relapse
1923	PARTH	3.55k	A.K. Macomber	F. O'Neill	J.H. Crawford	Massine	Filibert de Savoie
1924	MASSINE	4.60k	Henry Ternynck	A. Sharpe	E. Cunnington	Isola Bella	Cadum
1925	PRIORI	3.55k	Comte Gérard de Chavagnac	M. Allemand	P. Carter	Cadum	Tras los Montes
1926	BIRIBI	3.55k	Simon Guthmann	D. Torterolo	J. Torterolo	Dorina	Ptolemy
1927	MON TALISMAN	3.55k	E. Martinez de Hoz	C.H. Semblat	F. Carter	Nino	Felton
1928	KANTAR	3.55½k	Ogden Mills	A. Esling	R. Carver	Rialto	Finglas
1929	ORTELLO	3.55k	Guiseppe de Montel	P. Caprioli	W. Carter	Kantar	Oleander
1930	MOTRICO	5.60k	Vicomte Max de Rivaud	M. Fruhinsholtz	M. d'Okhuysen	Hotweed	Filarete
1931	PEARL CAP	3.54k	Mlle Diana Esmond	C.H. Semblat	F. Carter	Amfortas	Prince Rose
1932	MOTRICO	7.60k	Vicomte Max de Rivaud	C.H. Semblat	M. d'Okhuysen	Goyescas	Macaroni
1933	CRAPOM	3.55½k	Mario Crespi	P. Caprioli	F. Regoli	Casterari	Pantalon
1934	BRANTÔME	3.55½k	Baron Edouard de Rothschild	C. Bouillon	L. Robert	Assuérus	Felicitation
1935	SAMOS	3.54k	Evremond de Saint-Alary	W. Sibbritt	F. Carter	Péniche	Corrida
1936	CORRIDA	4.58½k	Marcel Boussac	C. Elliott	J. Watts	Cousine	Fantastic
1937	CORRIDA	5.58½k	Marcel Boussac	C. Elliott	J. Watts	Tonnelle	Mousson
1938	ECLAIR AU CHOCOLAT	3.55½k	Baron Edouard de Rothschild	C. Bouillon	L. Robert	Antonym	Canot
1939	No Race						
1940	No Race						
1941	LE PACHA	3.55½k	Philippe Gund	P. Francolon	J. Cunnington	Nepenthe	Djebel
1942	DJEBEL	5.60k	Marcel Boussac	J. Doyasbère	C.H. Semblat	Tornado	Breughel
1943	VERSO II	3.55½k	Comte de Chambure	G. Duforez	C. Clout	Esméralda	Norseman
1944	ARDAN	3.55½k	Marcel Boussac	J. Doyasbère	C.H. Semblat	Un Gaillard	Deux Pour Cent
1945	NIKELLORA	3.54k	Madame Robert Patureau	W. Johnstone	R. Pelat	Ardan	Chanteur
1946	CARACALLA	4.60k	Marcel Boussac	C. Elliott	C.H. Semblat	Prince Chevalier	Pirette
1947	LE PAILLON	5.60k	Madame L. Aurousseau	F. Rochetti	W. Head	Goyama	Madelon
1948	MIGOLI	4.60k	H.H. Aga Khan	C. Smirke	F. Butters	Nirgal	Bey
1949	CORONATION	3.54k	Marcel Boussac	R. Poincelet	C.H. Semblat	Double Rose	Amour Drake
1950	TANTIÈME	3.55½k	François Dupré	J. Doyasbère	F. Mathet	Alizier	L'Amiral
1951	TANTIÈME	4.60k	François Dupré	J. Doyasbère	F. Mathet	Nuccio	Le Tyrol
1952	NUCCIO	4.60k	H.H. Aga Khan	R. Poincelet	A. Head	La Mirambule	Dynamiter
1953	LA SORELLINA	3.54k	Paul Duboscq	M. Larraun	E. Pollet	Silnet	Worden
1954	SICA BOY	3.55½k	Madame Jean Cochery	W. Johnstone	P. Pelat	Banassa	Philante
1955	RIBOT	3.55½k	Marchese Incisa della Rocchetta	E. Camici	U. Penco	Beau Prince II	Picounda
1956	RIBOT	4.60k	Marchese Incisa della Rocchetta	E. Camici	U. Penco	Talgo	Tanerko
1957	OROSO	4.60k	Raoul Meyer	S. Boullenger	D. Lescalle	Denisy	Balbo
1958	BALLYMOSS	4.60k	J. McShain	A. Breasley	V. O'Brien	Fric	Cherasco
1959	SAINT CRESPIN	3.55½k	Prince Aly Khan	G. Moore	A. Head	Midnight Sun	Le Loup Garou
1960	PUISSANT CHEF	3.55½k	Henry Aubert	M. Garcia	C. Bartholomew	Hautain	Point d'Amour III
1961	MOLVEDO	3.55½k	Egidio Verga	E. Camici	A. Maggi	Right Royal	Misti
1962	SOLTIKOFF	3.55½k	Madame Cino del Ducca	M. Depalmas	R. Pelat	Monade	Val de Loir
1963	EXBURY	4.60k	Baron Guy de Rothschild	J. Deforge	G. Watson	Le Mesnil	Misti
1964	PRINCE ROYAL II	3.55½k	Rex Ellsworth	R. Poincelet	G. Bridgland	Santa Claus	La Bamba
1965	SEA BIRD	3.55½k	Jean Ternynck	T.P. Glennon	E. Pollet	Reliance	Diatome
1966	BON MOT	3.55½k	W. Burmann	F. Head	W. Head	Sigebert	Lionel
1967	TOPYO	3.55½k	Madame Léon Volterra	W.B. Pyers	C. Bartholomew Jr	Salvo	Ribocco
1968	VAGUELY NOBLE	3.55½k	Madame R.A. Franklyn	W. Williamson	E. Pollet	Sir Ivor	Carmarthen
1969	LEVMOSS	4.60k	Seamus McGrath	W. Williamson	S. McGrath	Park Top	Grandier
1970	SASSAFRAS	3.55½k	Arpad Plesch	Y. Saint-Martin	F. Mathet	Nijinsky	Miss Dan
1971	MILL REEF	3.55½k	Paul Mellon	G. Lewis	I. Balding	Pistol Packer	Cambrizzia
1972	SAN SAN	3.54k	Comtesse Margit Batthyany	F. Head	A. Penna	Rescousse	Homeric
1973	RHEINGOLD	4.60k	Henry Zeisel	L. Piggot	B. Hills	Allez France	Hard to Beat

Year	Winner		Owner	Jockey	Trainer	Second	Third
1974	ALLEZ FRANCE	4.58½k	Daniel Wildenstein	Y. Saint-Martin	A. Penna	Comtesse de Loir	Margouillat
1975	STAR APPEAL	5.60k	Waldemar Zeitelhack	G. Starkey	T. Grieper	On My Way	Comtesse de Loir
1976	IVANJICA	4.57½k	Jacques Wertheimer	F. Head	A. Head	Crow	Youth
1977	ALLEGED	3.56k	Robert Sangster	L. Piggott	M.V. O'Brien	Balmerino	Crystal Palace
1978	ALLEGED	4.59k	Robert Sangster	L. Piggott	M.V. O'Brien	Trillion	Dancing Maid
1979	THREE TROIKAS	3.54½k	Madame Alec Head	F. Head	Madame C. Head	Le Marmot	Troy
1980	DETROIT	3.54½k	Robert Sangster	P. Eddery	O. Douieb	Argument	Ela-Mana-Mou
1981	GOLD RIVER	4.57½k	Jacques Wertheimer	G.W. Moore	A. Head	Bikala	April Run
1982	AKIYDA	3.54½k	H.H. Aga Khan	Y. Saint-Martin	F. Mathet	Ardross	Awaasif

NOTE: In 1943 and 1944 the Prix de l'Arc de Triomphe was run over 2,300 metres at Le Tremblay

INDEX TO RUNNERS

Page numbers in italics refer to illustrations

GENERAL INDEX